TWILIGHT

TWILIGHT

A STUDY IN
ATMOSPHERIC OPTICS

Georgii Vladimirovich Rozenberg
Deputy Director
Institute of Physics of the Atmosphere
Academy of Sciences of the USSR

With a Preface by
J. V. Dave
National Center for Atmospheric Research

Authorized translation from the Russian by
Richard B. Rodman
Harvard College Observatory

SPRINGER SCIENCE+BUSINESS MEDIA, LLC
1966

The Russian language text was published by the State Press for
Physicomathematical Literature in Moscow in 1963.

TWILIGHT

SUMERKI

Сумерки

Георгий Владимирович Розенберг

Library of Congress Catalog Card Number 65-11345

© 1966 Springer Science+Business Media New York
Originally published by Plenum Press in 1966
Softcover reprint of the hardcover 1st edition 1966
ISBN 978-1-4899-6176-1   ISBN 978-1-4899-6353-6 (eBook)
DOI 10.1007/978-1-4899-6353-6

FOREWORD

The first phase of the passive sounding of the planetary atmosphere can be considered as a sort of "pioneering" in which some sweeping claims are made on the basis of rather meager observational material and theoretical understanding. This helps to attract well-qualified persons from other scientific disciplines such as applied mathematics and classical physics. This in turn creates newer and stronger controversies which demand direct sounding of the atmosphere. As a result, more experimental physicists, engineers, and technologists appear on the scene, and the subject comes to the notice of the public and the government. At this stage, one would expect the passive sounding techniques to die a natural death. But this is not what has happened. On the contrary, history has shown that, in general, they have returned in a new role — either complementing the direct observations or exploring further unknowns on a much sounder basis.

The idea of extracting information concerning the atmospheric structure from ground-based observations of the twilight sky is an old one. But little progress has been made because of the difficulty of obtaining reliable and absolute measurements and of understanding the process of radiation transfer in a nonhomogeneous, spherical atmosphere. Now, however, spacecraft provide many opportunities for observing twilight on our own and on other planets from various directions. Measurements can be made from the ground and also from orbiting and fly-by vehicles. In addition, the development of high-speed, electronic digital computers is bound to aid in the solution of the problems of radiative transfer.

This treatise on twilight by Prof. G. V. Rozenberg appears at an opportune moment when young researchers are being attracted to this field. The book provides a good historical survey for these newcomers. Extensive citations by Prof. Rozenberg of twilight papers published in many countries clearly demonstrate the truly international appeal of the subject. Prof. Rozenberg has provided many references to Russian papers previously unfamiliar to most English-speaking scientists. The need has long existed for relating all these materials in one volume. This work by Prof. Rozenberg should prove extremely useful as a textbook for graduate students in atmospheric physics and as a reference book to persons already actively engaged in the field.

J. V. Dave

National Center for Atmospheric Research
Boulder, Colorado
February 1966

PREFACE TO THE RUSSIAN EDITION

In this book the reader will find the twilight process treated as an optical phenomenon and as a tool for upper-atmosphere research. Our understanding of twilight has advanced considerably over the past 30 years. The acquisition mainly of descriptive material through rudimentary observing techniques has now given way to purposeful quantitative research which exploits every facility of modern experimental technology. And the frankly speculative arguments that had widely prevailed have been replaced by theory, developed in detail and tested experimentally, so that various aspects of the twilight phenomenon can now be treated from a rigorous mathematical standpoint to yield much valuable data on the structure of the atmosphere. But although these results have furnished a fairly comprehensive understanding of twilight, and have in fact established that the twilight method can be applied within certain limits for sounding the atmosphere, they remain scattered through papers in many different journals. It is therefore opportune now to collect the results and examine them from a unified viewpoint.

The author has by no means attempted an exhaustive review of all twilight research — of the earlier work in particular. This book aims to outline the current status of the problem. Considerable space is devoted to ideas that the author has developed in recent years at the Atmospheric Optics Laboratory of the Institute of Atmospheric Physics, USSR Academy of Sciences. He has often reported this material verbally and much of it has been incorporated in published papers, but the results are here presented for the first time in complete form, together with certain data from experimental studies of the twilight sky conducted at the Atmospheric Optics Laboratory over the past few years. As is customary, omission of references to the literature indicates that the material is original.

G. V. R.

BIOGRAPHICAL NOTE

Professor Georgii Vladimirovich Rozenberg, Doctor of Physical and Mathematical Sciences, has been Chief of the Section on Atmospheric Optics of the Institute of Physics of the Atmosphere, Academy of Sciences of the USSR, since 1956.

His field of specialization is theoretical and experimental investigations in the field of the optics of nonhomogeneous media. His principal studies relate to the theory of light propagation in a scattering medium (general formulation of the transfer equation with allowance for the effects of polarization of light, the optical properties of thick layers of a scattering medium, spectroscopic study of dispersive media), experimental and theoretical investigations of polarization effects during the scattering of light, and experimental and theoretical investigations in atmospheric optics (especially the development of optical methods for investigating the atmosphere and interference phenomena in thin films).

Born in Smolensk in 1914, he holds degrees in optics and theoretical physics from Moscow State University. Initially, he worked at the Physics Institute, Academy of Sciences, under Professor G. S. Landsberg. From 1946 to 1961, he was senior scientific editor of the journal Uspekhi Fizicheskikh Nauk (Soviet Physics — Uspekhi).

Professor G. V. Rozenberg is the author of two other books: "Optika tonkosloinykh pokrytii" (Optics of Thin Films), published in 1958; and "Prozhektornyi luch v atmosfere" (Searchlights in the Atmosphere), published in collaboration with a group of specialists in 1960.

CONTENTS

INTRODUCTION

The daily alternation between day and night is one process of nature whose regular repetition exerts a striking impact on the structure of organisms and the way of life of every inhabitant of our planet. Although the effects are often indirect ones, their ultimate source is always the remarkable change in illumination — by a billion times twice a day. The character of the energy balance between the earth's surface and the surrounding atmosphere is qualitatively rearranged, the atmosphere moderating the transition because of its ability to scatter light. The change does not occur instantaneously but over the rather protracted period known as twilight, when many transient and surprisingly varied phenomena take place.

One can, in a sense, grasp the biological, geophysical, and social significance of this transition period from its scale. As viewed from space, by an astronaut, the globe would appear encircled by a broad, multicolored, penumbral twilight zone, invariably covering 20-25% of the surface, depending on atmospheric conditions. On one side of the zone, daylight prevails over 42-45% of the earth's surface area, while on the other side 33-35% of the surface is submerged in night. Thus man lives and works under twilight conditions almost one-fourth of the time. In the tropics, where the sun drops more steeply toward the horizon, the fraction of twilight is reduced to some 10-15% of the time. But in high latitudes, where the entire night may be bright, twilight occupies 30-40% of the annual cycle, and during spring and autumn a continuous twilight lasts for weeks in the polar regions. Transportation, the schedule of work in the fields, artificial lighting, in fact the whole routine of daily life must often be adapted to the varying illumination of twilight.

One might think that twilight directly affects our living conditions on so great a scale that science would long since have reached some depth of understanding as to the nature and behavior of twilight processes. If we turn to the scientific literature, however, we at once find the opposite to be true. Despite a persistent interest in twilight, knowledge of the phenomenon has always lagged considerably behind that of the day and night periods. One reason is the complexity and variety of twilight processes and their sharply defined

1

dynamical character — features which seriously impede scientific analysis and call for application of the most advanced technical and theoretical research tools. Another is the fact that twilight culminates largely in the upper atmosphere, in layers which until very recently had remained unfamiliar and inaccessible to scientific study.

It was realized long ago that the colors of sunset and dawn, which for ages have stimulated the imagination and have invariably figured prominently in the religious beliefs and arts of all peoples, arise in the earth's atmosphere and are governed by its optical properties. Alhazen, the 11th-century pioneer of experimental physics, was apparently aware of this fact. He was also the first to attempt a study of the atmosphere itself by interpreting the variations in sky brightness during twilight [1]. Alhazen pointed out that the earth's shadow rises continually higher during evening twilight, and that night falls as soon as the whole atmosphere is enveloped in shadow. By measuring the duration of twilight in terms of the sun's depression below the horizon at nightfall, Alhazen was able to estimate the height of the atmosphere as 52,000 paces — not too far from the true value since we now know that less than 0.1% of the entire air mass lies above this level. (These computations were subsequently corrected by Kepler [2].)

Afterward the twilight problem aroused the interest of such scientists as J. and N. Bernoulli, Maupertuis, d'Alembert, Clausius, and many others, and it has remained a continuing subject of observational and theoretical research. Yet is was not until 1863 that von Bezold [3] arrived at the presently accepted explanation for twilight phenomena as a consequence of the scattering and attenuation of sunlight in the atmosphere. It is interesting that von Bezold's interpretation appeared only 11 years after Brücke's discovery of the scattering effect, three years after Hovey's discovery that light is polarized when scattered (so that the daytime airglow could be attributed to atmospheric scattering of sunlight), and six years before Tyndall's noteworthy work which led to general acceptance of the explanation [4]. In our opinion, this timing indicates an unremitting attention to twilight studies at that period.

So effective was this theoretical progress that it promptly spawned many attempts to relate various twilight properties, such as the neutral points and the purple light, to meteorological conditions. But the difficulty of developing a fully adequate quantitative theory forced a purely empirical "observational" approach, remote from any efforts to get at the heart of the phenomenon. Even Lord Rayleigh's theory for the molecular scattering of light and the theory developed by Lamb and Mie for scattering by large particles, which completely transformed the atmospheric optics of that time, failed to sidetrack the persistent empirical search for all kinds of twilight "indicators" to aid in forecasting the weather.

Not until the 1920's and 1930's did this slow, essentially aimless acquisition of observational data give way to a systematic study by observation and theory of the general laws characterizing the phenomenon itself. This definite turn toward identifying the real connection between twilight processes and atmospheric structure came at the beginning of man's active mastery of the atmosphere as a theater for practical pursuits. Aviation and radio communications, geophysics and astronomy, nuclear physics and many other branches of science and technology — all began insistently to demand reliable information on the earth's atmosphere, particularly the regions located at heights of tens and hundreds of kilometers.

In the early stages of this penetration far up into the atmosphere, the limits of the directly accessible region remained quite modest: 22 km by manned balloon and 38 km by sounding balloon are the record heights reached by man and his instruments through the 1930's. Only by indirect techniques could information be obtained on the conditions prevailing at higher levels. These techniques therefore assumed prime importance; developed successfully during the second quarter of this century, they have stimulated a thorough revision of our entire knowledge about the upper atmosphere. In particular, we owe to them such major discoveries as the existence of the ionosphere, the ozone and sodium layers, the temperature maximum at the 50-60 km level, and the minimum near 80 km, as well as the refutation of the prevailing opinion that the upper atmosphere is a static region. Indirect methods have established that the whole atmosphere, even the highest layers, is an arena for the most varied processes, on too great a scale to reproduce in the laboratory; thus they take on a unique significance, intimately influencing our living conditions.

With the advent of space rockets and artificial earth satellites, which have made all heights accessible to man as well as to instruments, the situation changed radically and for a while indirect methods were relegated to the background. However, one important outcome of rocket and satellite research has been to confirm the general views on the upper atmosphere that had arisen through applying indirect research techniques — observation from the ground of processes taking place at great heights. It has turned out, nevertheless, that many of the physical parameters of the upper atmosphere remain far more accessible to investigation by indirect methods than by direct measurement from rockets or satellites. Indirect methods for studying the high layers therefore have certainly not diminished in value, especially since their relatively low cost permits regular observing services to be maintained.

Among the indirect methods of upper-atmosphere research that have been important in the past and still are today is the twilight-sounding method, with its detailed analysis of how twilight progresses with time.

After Alhazen and Kepler it was not until 1923 that Fesenkov [5] first proposed using twilight phenomena to survey the atmosphere optically. Over the next 40 years many eminent scientists contributed to the development, considering various aspects of the problem. Their steadfast efforts have clarified or removed numerous obstacles that had thrown doubt upon the reliability of the method. Moreover, the opportunity arose to make a direct quantitative comparison between theory and experiment, something still quite rare in atmospheric optics; and it was shown directly that twilight observations do yield realistic data if details are handled properly.

Meanwhile a very important aspect of the problem has emerged that attracted little attention in the early developments. Since the real interest had not been in explaining the twilight phenomenon itself but in using it to learn about the factors responsible, every circumstance materially affecting its course had unavoidably to be considered. Yet even today some of these factors remain little understood or are difficult to treat effectively, either empirically or by theory. Thus a direct appeal to models, with their indeterminate choice of basic parameters, is not effective and may often lead one astray, especially in view of the extreme variability of the atmosphere itself.

One must therefore determine the auxiliary parameters of the model empirically as well as measure the quantities under study. This can only be done if the observing conditions and procedures are properly matched; a certain minimum of information has to be acquired simultaneously, without which no reliable interpretation of the observations can be feasible. Neglect of this obvious requirement has repeatedly invalidated the efforts of observers and has served as the main source of doubts, once especially widespread among scientists abroad, as to the efficacy of twilight sounding of the atmosphere. Today we are capable of defining quite clearly the scope of this minimal information, which depends on the problem at hand, and of estimating how reliable are the resulting data. Evidently such estimates may rest wholly on a thorough analysis of the contribution of various factors toward twilight phenomena. The book is intended to furnish such an analysis.

We shall refer the entire treatment to the atmosphere of the earth, the only planet for which adequate observational material is available. Twilight, however, occurs on every planet having an atmosphere, where its behavior is governed by essentially the same factors as on earth. Thus the terrestrial twilight theory and the methods for using it to study the earth's atmosphere may at the same time be regarded as preparatory to a study of other planetary atmospheres — externally, such as from space vehicles. We may anticipate that before long twilight investigations of planetary atmospheres will form a leading branch of experimental and theoretical astrophysics.

CHAPTER I

SURVEY OF OBSERVATIONAL RESULTS

§1. Some Properties of the Daytime Sky

The optical characteristics of the atmosphere, particularly the scattering of sunlight, control the appearance of the sky under both twilight and daytime conditions. One may therefore gain a better understanding of the behavior of twilight phenomena by comparing them with the phenomena that are observed under the simpler conditions of a bright cloudless day. Moreover, because of its relatively constant and much greater brightness, factors which facilitate measurement and analysis, the daytime sky has been studied far more thoroughly than the twilight sky.

The first attempts to measure quantitatively the brightness of the clear daytime sky were undertaken by the Swiss physicist de Saussure toward the end of the 18th century. Numerous further efforts followed, but it was not until 1898 that Jensen [6] set out to perform a systematic photometry of the celestial sphere. From that time to the present extensive research has been devoted to acquiring and systematizing observational data. Particularly complete programs, occupying many years, have been carried out by Dorno [7, 8] at Davos, and quite recently by Pyaskovskaya-Fesenkova [9] at a number of stations. Taken together, all these investigations have provided us with a fairly clear understanding of the brightness map of the sky, as well as its dependence on such factors as the sun's altitude and meteorological conditions. Developments in the theory of radiative transfer in scattering media [10-16 and elsewhere] have made possible a penetrating analysis of how light is propagated through the earth's atmosphere during daylight hours, and have yielded good qualitative agreement with the situation actually observed. A full quantitative agreement was not to be expected, for idealizations of natural conditions were unavoidable in computing the models.

Both the total sky brightness and its distribution over the celestial sphere depend heavily on the atmospheric transparency and the form of the scattering diagram. In particular, one finds a strong dependence of the aspect of the sky

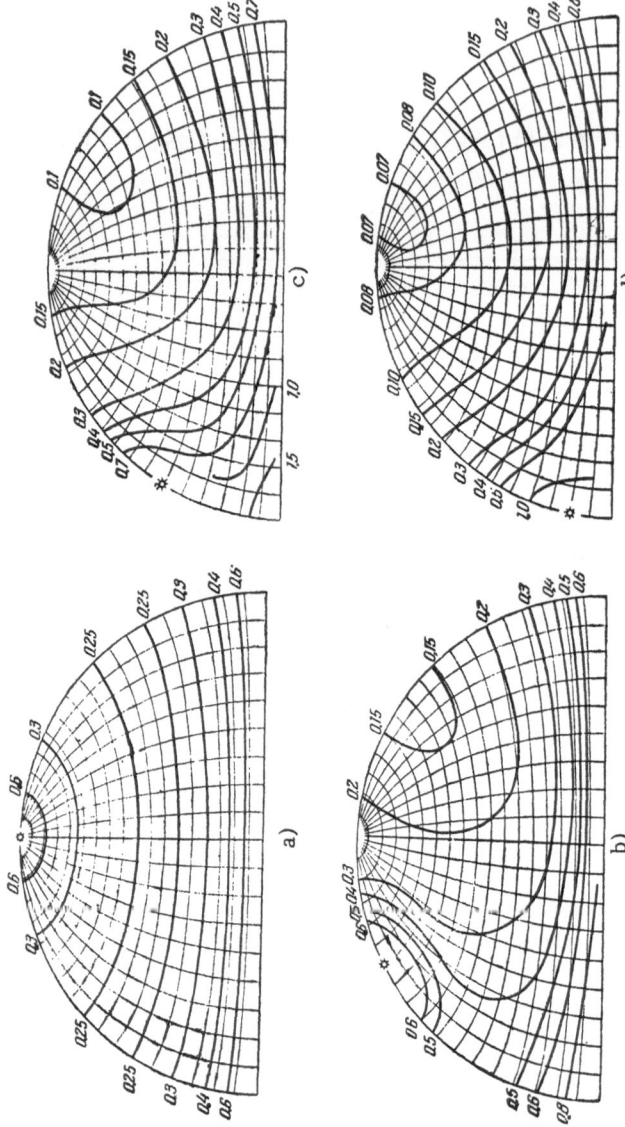

Fig. 1. Smoothed isophotes of the daytime sky for a highly transparent atmosphere (P = 0.87) and various zenith distances ζ of the sun. The surface brightness of the sky is expressed in stilbs. a) ζ = 0°; b) ζ = 30°; c) ζ = 60°; d) ζ = 80°.

on the wavelength observed and on meteorological conditions. As the turbidity of the atmosphere increases, the scattering coefficient usually varies more
weakly with wavelength [17, 18], and there is a marked change in the scattering diagram [9, 19], with its "directivity" correlating well with the value of
the scattering coefficient [20]. But there is also an increase in the role of multiple scattering. Measurements [9] and calculations [13] have shown that if
the atmosphere has the high transparency P ≳ 0.85, the day being particularly
clear at long wavelengths especially, then secondary scattering will contribute
only a few percent of the sky brightness sufficiently far above the horizon. If
the transparency is 0.6-0.7, however, the contribution will reach 50%; and in
the ultraviolet region, where the transparency may fall to 0.2 or even lower
on a perfectly clear day, multiple scattering becomes the dominant factor
governing the brightness distribution over the celestial sphere.

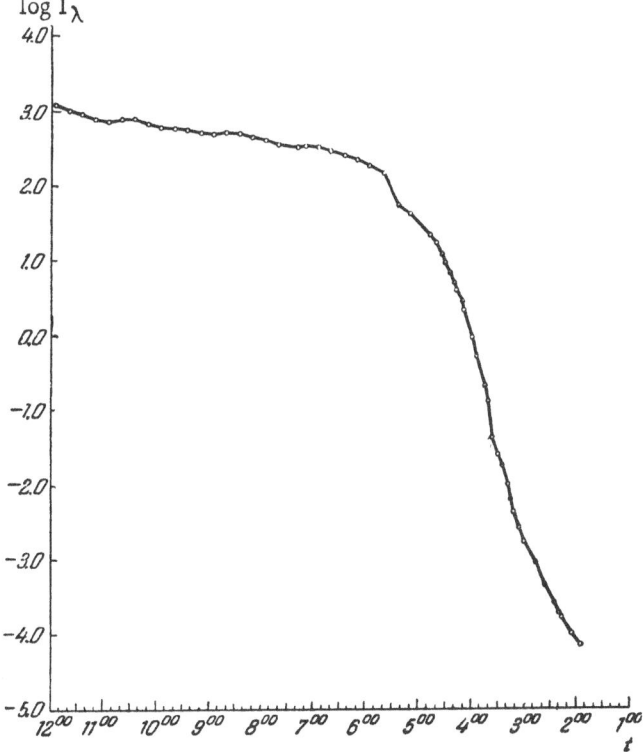

Fig. 2. A sample record of the sky brightness at the zenith as a
function of time from measurements in a narrow spectral band
($\lambda \approx 0.5 \ \mu$, $\Delta\lambda = 6$ A) on a July day that was not very clear.

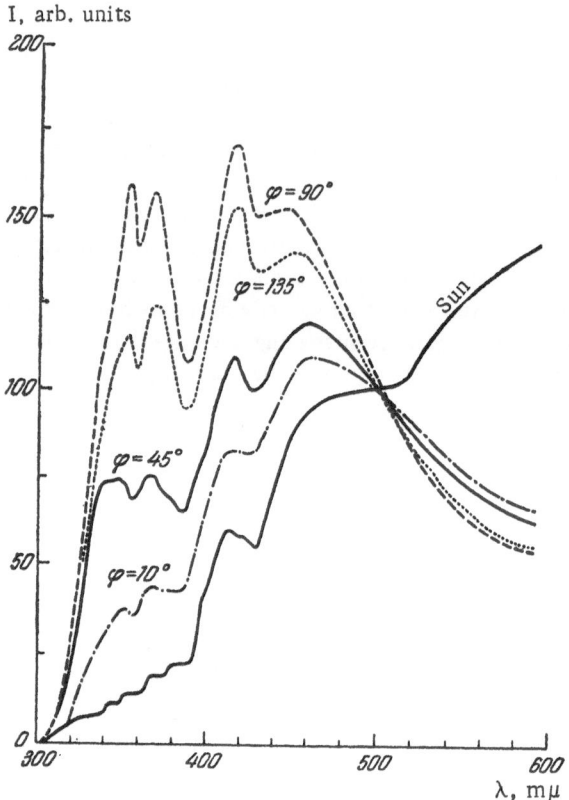

Fig. 3. Spectral energy distribution in the light of the daytime sky at various points along the solar meridian (φ is the scattering angle).

Reflection of light from the earth's surface also plays an important part. Navigators in polar waters are familiar with the fact that the sky appears much darker above open water, which is a comparatively poor reflector of light, than above ice fields, whose albedo is nearly unity (water-sky and ice-sky, or "ice-blink"). This observation is confirmed by direct measurements (for example, [21]) and theoretical calculations [13, 15, 16].

Figure 1 illustrates typical brightness distributions over the sky for a highly transparent atmosphere, according to Pyaskovskaya-Fesenkova's observations [9]. We see that the surface brightness ranges over several tenths of a stilb, or one order of magnitude; the maximum brightness always occurs in the aureole around the sun and at the horizon, while the minimum brightness is not very far from the zenith on the side opposite the sun. As the turbidity of the atmosphere increases, the brightness gradations diminish (except that the aureole intensifies), but the total sky brightness is greater when the transparency is low.

With increasing zenith distance of the sun, the total illuminance produced by scattered light from the sky falls off slowly. The surface brightness at the zenith decreases with time (that is, with ζ) in a similar fashion, as demonstrated by measurements in monochromatic light, $\lambda \approx 0.5\ \mu$, carried out by G. V. Rozenberg and L. Lobova above Moscow during July (Fig. 2). At other points of the sky the relation between brightness and ζ is more complicated, since in many respects, particularly if the atmosphere is very clear, the brightness is determined by the form of the scattering diagram (see [8, 9] and especially [13] for more detail).

The spectral composition of scattered light from the daytime sky is quite unlike the composition of sunlight. It arises largely from the interplay of two factors: the attenuation of light rays along their path through the atmosphere, and the atmospheric scattering event itself. Since the path of a ray through the atmosphere depends on both the altitude of the sun and the direction of the line of sight, a strong dependence of the spectral composition, and hence the color of the sky, on both variables is inevitable. Further changes in the spectrum will be induced by the wavelength dependence of the form of the scattering diagram, and by the component of multiply scattered light, which has an entirely different spectral composition. Moreover, the spectral behavior of the scattering coefficient in air is highly variable, depending on the

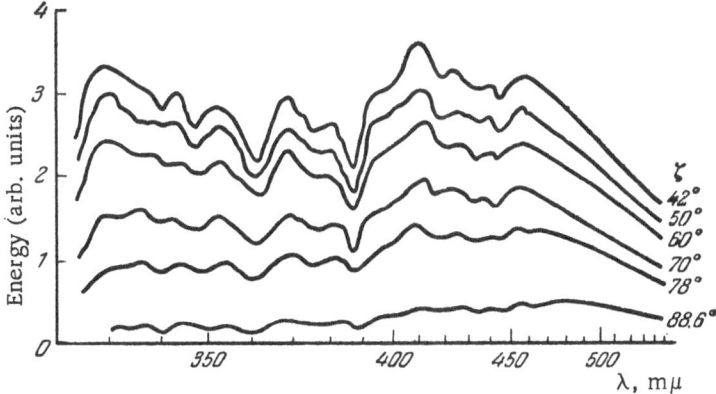

Fig. 4. Spectral energy distribution in the light at the zenith of the daytime sky, with the sun at various zenith distances ζ.

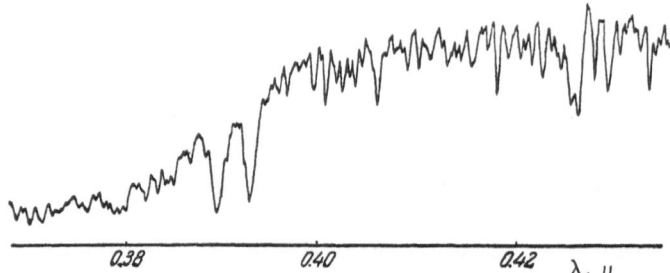

Fig. 5. Portion of a spectrogram recorded at the zenith of the
daytime sky with comparatively high resolution ($\Delta\lambda = 6$ A).

aerosol content of the atmosphere — on meteorological conditions. Finally, at
long wavelengths a considerable role is due to selective absorption of light by
various atmospheric constituents, particularly ozone, and to some extent water
vapor. The color of the sky is therefore far less stable than its brightness, and
is more strongly subject to a variety of contributory effects.

Attempts to measure the color of the sky and its variations were first
made by de Saussure in 1790, and somewhat later by Humboldt. At the end of
the last century, a number of investigations of this kind were carried out as an
experimental check on the scattering explanation for the light of the daytime
sky (Rayleigh, Fogel, Krova, Tsvetukha, Bok, Abney, Fasting, and other authors).
From time to time until the present day, these measurements have continued
to be performed in different variants and with more modern equipment. But
because the spectrum of the daytime sky is so highly variable and so difficult
to interpret, little more than qualitative findings has been extracted from the
data. Of the more recent investigations, we mention those by Krinov [22],
Hess [23], and Götz and Schönmann [24].

In exceptionally clear weather, the spectral brightness distribution of the
daytime sky differs from the solar spectrum in a general way by a factor proportional to λ^{-4} (λ is the wavelength), as would have been expected from the
Rayleigh theory. However, even in the most favorable cases there are considerable departures from this behavior. In a turbid atmosphere, the proportionality corresponds more closely to λ^{-n}, where n varies over an extremely
wide range, depending on the meteorological conditions.

It is important to note that power laws of this type are merely qualitative, and are intended only as a rough guide. The smooth spectral variation
is actually interrupted by a number of well-defined regions of selective extinction or scattering of light in the earth's atmosphere (for example, the

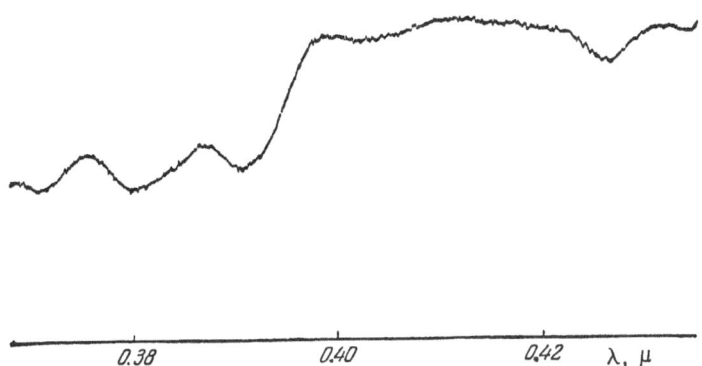

Fig. 6. Portion of a spectrogram recorded at the zenith of the
daytime sky with low resolution ($\Delta\lambda = 70$ A).

Chappuis absorption bands of ozone, water-vapor and oxygen bands, selective
scattering and attenuation of light by aerosols). If we add to these effects the
complicated structure of the solar spectrum itself, we see that the resulting
spectral composition of the scattered light from the daytime sky is very far
from a power law.

Figure 3 shows the spectral energy distribution in the light coming from
different parts of the daytime sky, according to the data of Hess [23]. The
measurements were made in the solar meridian (we use this term to designate
the entire vertical circle through the sun), with the sun at zenith distance
$\zeta = 72°$. The solid curves refer to the half of the meridian containing the sun
(the "solar vertical"), the other curves to the opposite half (the "antisolar
vertical"). Figure 4 demonstrates how the spectral energy distribution in the
light coming from the daytime zenith varies with the zenith distance ζ of the
sun, as determined by Götz and Schönmann [24]. In both cases the spectral
resolution was quite low, and most of the detail in the spectra has been lost.
Nevertheless, the spectra have a fairly complicated structure, and obviously
depend strongly on the geometric parameters, quite apart from the conditions
prevailing in the atmosphere.

Since the structure of the spectrum for the scattered light from the sky
is so complex, its form depends materially on the resolving power of the instrument used. As an example, Figs. 5 and 6 show spectrograms of the daytime sky at the zenith, as obtained by A. Ya. Driving, I. M. Mikhailin, G. V.
Rozenberg, and G. D. Turkin with a self-recording photoelectric spectrophotometer equipped with a polarizing prism (45° angle between the planes of polarization and scattering). Figures 5 and 6 represent approximately the same

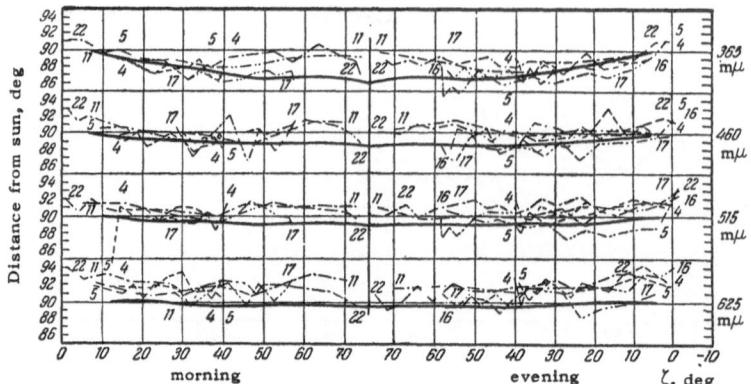

Fig. 7. Short-period variations in the position of maximum polarization of the light from the daytime sky at several wavelengths. Solid curves, computed values; dot-dash curves, observational results.

solar zenith distance. We note that selective extinction in the telluric bands is responsible for the appearance of the Forbes effect (§2), even in relatively narrow spectral intervals.

We have seen that the scattering of light in the real atmosphere is complicated by a number of factors that produce very considerable departures from what molecular scattering of direct sunlight would have led us to expect. These departures are also manifest in the polarization of the light that we receive from the celestial sphere. Like the spectral composition, the polarization of the light from the daytime sky is highly sensitive to geometric factors (the altitude of the sun and the direction of observation) and to meteorological conditions. It is, however, much simpler to determine the polarization than to measure the spectrum, and throughout the century and a half since it was discovered by Arago [25], the polarization has continued to be the subject of many investigations.

If only single molecular scattering of light occurred in the atmosphere, then the degree of polarization would be given independently of wavelength by the simple Rayleigh-Cabannes expression

$$p = \frac{\sin^2 \varphi}{1.06 + \cos^2 \varphi},$$

$$(I.1, 1)$$

which includes a correction for depolarization due to anisotropy of the air molecules. Here φ denotes the scattering angle, the angle between the direc-

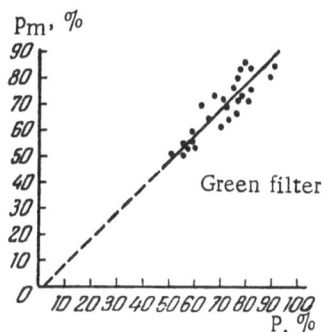

P_m, %

90
80
70
60
50
40   Green filter
30
20
10
0   10 20 30 40 50 60 70 80 90 100
                         P, %

Fig. 8. Maximum degree of polarization of the light from the daytime sky as a function of the vertical transparency of the atmosphere.

tion of the sun's rays and the line of sight. Actually, however, as shown by extensive research which need not be itemized here, this formula holds only in a very crude approximation for extremely high atmospheric transparency, and for regions of the sky far from the sun and from the horizon.

As originally noted by Arago, studied thoroughly by Brewster [26] and Rubenson [27], and observed subsequently by almost all investigators, maximum polarization occurs on the solar meridian at a point about 90° away from the sun. However, depending on the state of the atmosphere and apparently the earth's surface, the position of maximum ordinarily fluctuates by 3-5°. Displacements of the maximum both toward and away from the sun have often been observed — for example, by Rubenson [27], R. Wild, Jensen [6], Sekera [28], Stamov [29], Pyaskovskaya-Fesenkova [30], and the author and his colleagues [19]. Detailed measurements [28] have shown that the position of maximum polarization depends on wavelength, and undergoes random short-period displacements (Fig. 7).

Moreover, the degree of polarization at maximum, P_m, is usually considerably less than 0.94, the value given by Eq. (I.1, 1), and it never exceeds this value. Even under the most favorable conditions (exceptionally pure atmosphere, sun at the horizon), P_m does not exceed 0.84-0.88. In fact, the degree of polarization is usually between 0.5 and 0.7, and changes strongly with the conditions of the atmosphere. Several authors have investigated the relation between P_m and the atmospheric turbidity [31-35]. Figure 8 shows P_m as a function of the vertical transparency of the atmosphere, according to measurements obtained in the Crimea in 1958 by G. V. Rozenberg and N. K. Turikova [19]. The dependence of P_m on the altitude of the sun is considerably weaker [19, 28-30]; it becomes important only with the onset of twilight, as the sun approaches the horizon [6-8, 31-33, 36, 37]. But Sekera [37] found a distinct anomaly in the dependence of polarization on the sun's zenith distance, apparently caused by unstable atmospheric conditions or by horizontal inhomogeneity [19, 29]. The albedo of the earth's surface has also been discovered to have a marked effect on the degree of polarization (J. Correy, Dorno, McConnell, and Stamov [29], Sekera [28], Rozenberg and Turikova

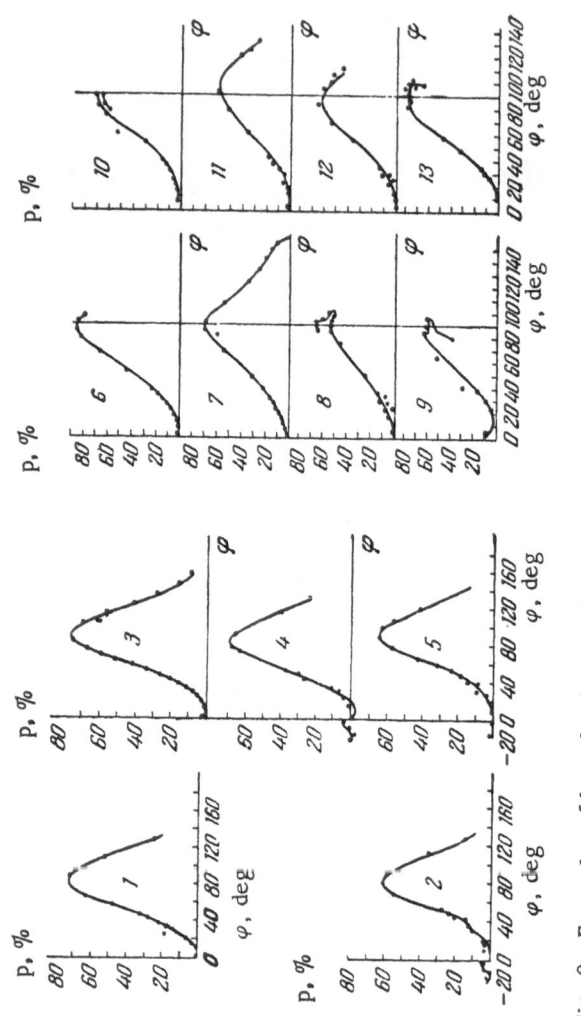

Fig. 9. Examples of how the degree of polarization of the light from the daytime sky depends on the scattering angle along the solar vertical (1-5) and almucantar (6-13). Crimea, 1958. 1) Blue filter, $\zeta = 30°.2$, $P = 80.2\%$; 2) blue, $\zeta = 36°.3$, $P = 75.9\%$; 3) green, $\zeta = 14°.6$, $P = 90\%$; 4) green, $\zeta = 45°.6$, $P = 86.5\%$; 5) green, $\zeta = 45°.5$, $P = 77\%$; 6) blue, $\zeta = 39°$, $P = 80.6\%$; 7) blue, $\zeta = 15°.1$, $P = 71.1\%$; 8) blue, $\zeta = 41°.2$, $P = 70\%$; 9) blue, $\zeta = 43°.6$, $P = 64.8\%$; 10) green, $\zeta = 43°$, $P = 63\%$; 11) green, $\zeta = 27°.9$, $P = 80.7\%$; 12) green, $\zeta = 30°.4$, $P = 76.2\%$; 13) green, $\zeta = 47°.3$, $P = 80\%$.

[19], and others). Moreover, the relation between degree of polarization and direction of observation also fails to satisfy Eq. (I.1, 1). Examples of these effects, as measured by Rozenberg and Turikova [19] along the solar vertical and almucantar, are shown in Fig. 9.

Taking into account the extensive observational material, Stamov [29] suggests that Eq. (I.1, 1) be replaced by the empirical formula

$$p = p_m \frac{\sin^4 \varphi}{1 - p_m \cos^4 \varphi},$$ (I.1, 2)

which appears to satisfy the observed polarization of the light from the day-time sky far enough from the sun and horizon, provided that the sun is not too low, that the weather is clear and stable, and that the atmosphere is in a hori-zontally homogeneous state.

Measurements by Fesenkov [38], Rozenberg [19], and apparently those by Sekera [37] have shown that the light from the daytime sky is linearly polar-ized; any traces of elliptical polarization are very weak. At points far from the sun and the horizon, the preferential direction of polarization coincides, in general, with that predicted by the Rayleigh law. However, one occasion-ally observes small deviations of the plane of polarization from this position; the effect is a comparatively regular one, but depends strongly on the solar azimuth and the weather conditions [19, 30]. It probably derives either from light reflected by the earth's surface, or from horizontal inhomogeneities in the atmosphere [39].

In 1892, Pil'chikov [40] reported the discovery of a wavelength depend-ence, not at all comprehended by the Rayleigh theory, of the degree of polar-ization for the light from the daytime sky. The phenomenon, subsequently in-vestigated and confirmed by a number of authors [28, 36, 41-43], has been called "polarization dispersion" by Schirmann [44]. As a rule, the degree of polarization of the daytime sky shows an extremal spectral variation, with maxi-mum polarization being displaced toward long wavelengths when the trans-parency is high, and in the other direction when it is low. Polarization dis-persion is accompanied by the so-called "polarization pleochroism," a differ-ence in the color of differently polarized components of the scattered light from the daytime sky, first predicted and observed by Rayleigh for the case of light scattered by large particles [45].

The most characteristic departure from the predictions of the Rayleigh theory is the behavior of the polarized light from the daytime sky in the vicin-ity of the sun and the antisolar point. Arago, followed by Babinet and Brewster, long ago found that the solar meridian contains three "neutral points" where

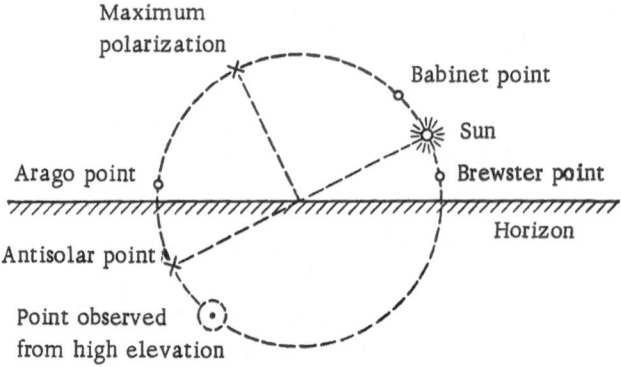

Fig. 10. Diagram of the neutral points.

the degree of polarization vanishes; each point is named for its discoverer (Fig. 10). When the atmosphere is observed from a high elevation, as from a high-flying aircraft, one finds a fourth neutral point approximately symmetric to the Arago point.

Innumerable papers report determinations of the distances of the neutral points from the sun or the antisolar point. There are tens of thousands of separate measurements (see, for example,[6-8, 27, 41, 46-49]). It has been established that the distances fluctuate, ranging from 12 to 30° depending on the wavelength, altitude of the sun, meteorological conditions, and albedo of the earth's surface. In particular, the distances increase with the atmospheric turbidity and with the albedo of the underlying surface. Knopf [50], Jensen and Süring [6], Sekera [28], and Rozenberg and Turikova [19] have observed deviations of the neutral points from the solar meridian, and even displacements toward the solar almucantar.

Between the pairs of neutral points near the sun and the antisolar point, the direction of the plane of polarization is found to be perpendicular to that implied by the Rayleigh theory — a "negative polarization," with the electric vector of the light wave lying in the plane of scattering. For this reason the polarization diagram of the sky is a very complicated one, and a comparison of the polarization maps compiled by different observers reveals much variability.

For a long time there has been a controversy regarding the explanation for the departures from the Rayleigh theory of the polarization of the light from the daytime sky, particularly the neutral-point phenomenon and the regions of negative polarization. One school of thought, represented by Soret [51] and subsequently by Ahlgrimm [52] and Tikhanovskii [53], seeks an explanation in multiple scattering of light. On the other hand, Rubenson [27], Schirmann [44, 54], and Milch [55] have ascribed the effect to scattering of light by the aerosol particles that are always suspended in the air. The development of transfer theory has permitted a detailed study to be made of the implications of the first suggestion. A qualitative calculation performed by the author [39] has shown that if multiple molecular scattering is taken into account, the polarization of the light scattered in a homogeneous layer of the atmosphere will be described by an ellipsoid of polarization (analogous to the Fresnel ellipsoid) whose umbilical points correspond to the neutral points in the daytime sky. The horizontal inhomogeneity of the atmosphere distorts the shape of the ellipsoid, and tends to displace the neutral points from the solar meridian. More recently Chandrasekhar and Elbert [15], Sekera [28], and Coulson, Dave, and Sekera [16] have performed rigorous calculations of the polarization of the light from the daytime sky, allowing for multiple molecular scattering and for the effect of the albedo of the underlying surface; they have shown that qualitatively the aspect of the sky corresponds well to theoretical expectations. Quantitatively, however, both systematic and random departures are observed to be significant, presumably because of the effect of aerosols in the atmosphere. If we recall, now, that the spectral dependence of the sky brightness and the character of the brightness maps of the daytime sky in no way agree, in their basic features, with the assumption of molecular scattering, then it will be clear that both factors are important here, but that their relative role is highly dependent on atmospheric conditions.

§2. The Twilight Phenomenon

The term t w i l i g h t refers to the entire complex of optical phenomena that take place in the atmosphere when the sun is near the horizon. It occupies the interval separating daytime conditions of illumination from night.

The main factor controlling the course of these phenomena is the scattering of sunlight in the earth's atmosphere, and the accompanying attenuation of the direct solar rays. As the sun drops toward the horizon, the path of its rays through the atmosphere lengthens and their brightness weakens, leading to a decrease in the illumination of the earth's surface not only by the direct sunlight, but also by the light scattered in the atmosphere. Under daytime conditions, the combined illuminance shows only a minor dependence on the altitude of the sun. But when the sun has sunk to within 5-10° of the horizon, the progressive drop in the illuminance begins to accelerate sharply. It is with this accelerated darkening that twilight is considered to have set in.

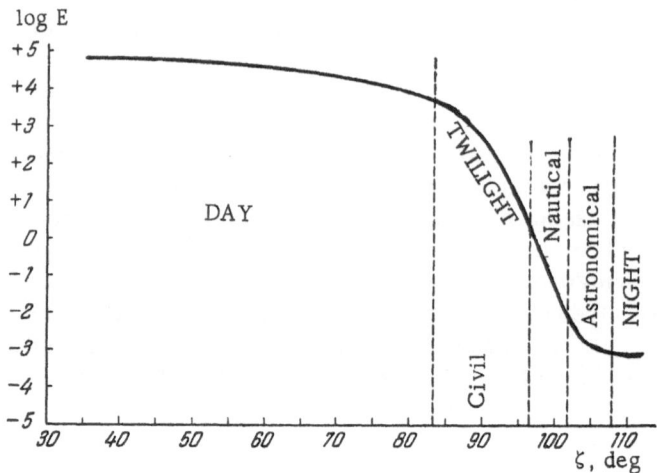

Fig. 11. Smoothed illuminance E(in lux) on a horizontal surface as a function of the zenith distance ζ of the sun.

Simultaneously, the relative contribution of the rapidly weakening direct solar rays in illuminating the earth's surface begins to decrease. The atmosphere penetrated by the rays of the setting sun increasingly serves as the
source of illumination. From the moment of sunset it remains the sole source
of light. But the earth's shadow gradually rises upward and claims an increasingly large portion of the atmosphere. The lower atmospheric layers, now submerged in shadow, no longer contribute to the sky brightness, and the scattered
light comes more and more from the higher layers which are still illuminated
by direct sunlight. Since the air density falls off rapidly with height, the scattering coefficient does also, sunlight is scattered increasingly more weakly, the
sky brightness diminishes, and so the illuminance at the earth's surface falls.
When the sun reaches a depression of 10-15° below the horizon, the intrinsic
glow of the upper atmospheric layers begins to appear together with starlight,
and the illumination conditions gradually approach those of night. The transition to night is usually complete when the sun is depressed 17-19° below the
horizon, although sometimes, if the atmosphere is exceptionally turbid, twilight will persist to a depression of 22-23°. In this event, one usually observes
a weak minimum in the illuminance and sky brightness when the sun is depressed 20-23° below the horizon [56-58 and elsewhere]. Thus the boundary
of the twilight period is very diffuse and is subject to a shift depending on atmospheric conditions. The duration of twilight is controlled by the speed with
which the sun sinks below the horizon, and hence by the geographic latitude
and the time of year. Appropriate tables may be found in several places [59,
60, 234, among others].

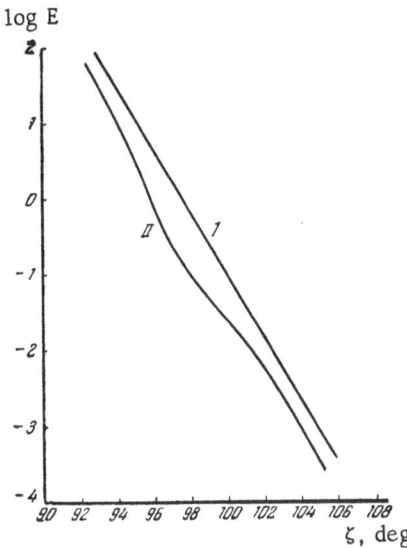

Fig. 12. Twilight illuminance (in lux) on
a vertical surface facing toward the sun (I)
and away from the sun (II), as a function
of the sun's zenith distance.

During the course of twilight the illuminance at the earth's surface varies by a factor of almost a billion, from 10^4-10^5 lux in daytime to 10^{-4} lux at night. This situation greatly impedes measurement of the phenomenon, and so far comparatively few such investigations have been made [56, 59, 61]. There have also been several studies of the twilight illuminance on a vertical surface oriented in various ways [62-67], and of the illuminance on a horizontal surface due to a restricted but fairly extensive region of sky [68-70].

Figure 11 illustrates how the smoothed illuminance E on a horizontal surface varies with the zenith distance ζ of the sun, according to Sharonov's data [59]. The ζ-dependence of the illuminance on variously oriented vertical surfaces is shown in Fig. 12 for visual measurements, and in Fig. 13 for photoelectric measurements [67].

The measurements of Figs. 11-13 refer to a cloudless summer sky. The presence of snow slightly (by a factor of 1.5-2) increases the illuminance for large depressions of the sun below the horizon, but has hardly any effect in the bright portion of twilight [59]. Much more significant is the influence of the type of cloud cover on the illuminance [59, 67]. The presence of clouds normally will reduce the illuminance, sometimes (with overcast skies) by an order

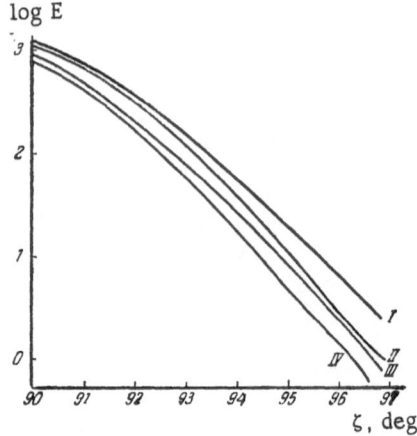

Fig. 13. Photoelectric measurements of
the twilight illuminance (in lux) on variously oriented surfaces, as a function of
the sun's zenith distance. I) Facing toward
the sun; II) horizontal; III) toward the north;
IV) away from the sun.

of magnitude, but cloudiness has almost no effect on the progress of darkness
as the sun declines. Detailed tabulations of the twilight illuminance are given
in [59] and [67].

From the practical need to estimate conditions of visibility at various
times of day, there has historically grown up a tradition of dividing twilight
into three stages, depending on the amount of twilight illumination (Fig. 11).
The brightest portion of twilight, when the natural light in an open place is
enough to allow any task, including reading, to be carried on, has received
the name c i v i l twilight. There is no unanimous agreement regarding its
boundary; depressions of the sun from 6 to 8° below the horizon have been
variously adopted. In recent years, however, most authors have preferred the
first figure. Next comes n a u t i c a l (or navigational) twilight, during which
small details are lost in darkness, but outlines of large objects such as shore-
lines are fairly distinct. The recognized boundaries are solar depressions of
6 and 12° below the horizon, so that by the end of nautical twilight one can
only make out clearly the horizon line. Finally, nautical twilight gives way
to a s t r o n o m i c a l twilight, which continues until the sun is depressed by
18° below the horizon, whereupon night sets in. During this period the illumination conditions can hardly be distinguished from night, but the sky is
still noticeably bright, enough to hinder astronomical observations. The levels

of illuminance corresponding to the boundaries between the several stages of twilight are, on the average, 2.5 lux for a solar depression angle $\xi = \zeta - 90°$ $= 6°$, $6 \cdot 10^{-3}$ lux for $\xi = 12°$, and $6 \cdot 10^{-4}$ lux for $\xi = 18°$ [61].

Lunar twilight, which begins when the lunar disk approaches the horizon, entirely resembles solar twilight but is very much weaker. The brightness of lunar illumination fluctuates, depending on the age of the moon, between 10^{-9} (at new moon) and $2 \cdot 10^{-6}$ (at full moon) of the brightness of solar illumination for corresponding positions of the bodies on the celestial sphere. Thus the level of illumination for a high full moon corresponds approximately to the middle of solar nautical twilight, while the end of lunar astronomical twilight (again at full moon) almost coincides with moonset [59]. Nevertheless, the contribution of the full moon to the sky brightness is best taken into account until the moon has dropped to 5-7° below the horizon [58].

We turn now to the appearance of the sky itself and its variations during the progress of twilight. In §§3 and 4 of this chapter, we review in detail the photometric, spectral, and polarimetric research on the twilight sky. At this point we merely wish to give a qualitative picture of the phenomenon in order to facilitate the more detailed discussion.

With its approach to the horizon, the sun not only loses its brightness but it gradually begins to change its color: the short-wave portion of its spectrum becomes increasingly suppressed. This phenomenon, whose explanation was recognized as early as 1840 by Forbes, is due to the fact that the atmospheric transparency, which generally speaking increases with wavelength, becomes more perceptible as the path of the solar rays through the atmosphere lengthens. Simultaneously the sky itself begins to take on coloration. It acquires yellowish and orange tints in the vicinity of the sun, while above the opposite section of the horizon a pale band with a weakly displayed range of colors appears.

By the time the sun, which has now assumed a dark red color, has set, a bright afterglow extends along the horizon near the sun; the color of this band varies in an upward direction from orange-yellow to greenish-azure. Above the band spreads a round, bright, almost colorless glow. At the same time a dull bluish-gray segment of the earth's shadow slowly begins to rise above the opposite horizon, bordered by a rosy zone, the "belt of Venus."

As the sun continues to sink beneath the horizon, the adjoining part of the sky exhibits a more saturated coloration. The sky becomes deep red at the horizon itself, while at an altitude of 20-25°, at the top of the glow above the sun, a rapidly spreading rose-colored patch begins to form; this is the "purple light," and it reaches its maximum development when the sun is depressed 4-5° below the horizon. Clouds and mountain peaks are flooded with crimson and purple hues, and if there are clouds or high mountains at the very

horizon, their shadows will stretch across the brightly colored sky from horizon to horizon in the form of distinct radial bands, the "rays of Buddha." Meanwhile the earth's shadow rapidly advances into the sky, its outlines become
blurred, its rosy border now barely noticeable.

Toward the end of civil twilight the purple light dies away, the darkening clouds stand prominently silhouetted against the fading sky background,
and only at the horizon where the sun is hidden does a bright, multicolored
glow remain. But this too gradually contracts and fades, until by the beginning of astronomical twilight it has turned into a narrow, pale greenish-white
band. By the end of astronomical twilight even the faint band has vanished —
night has fallen. However, according to some observers [71], even during the
night the atmosphere maintains a faint glow due to scattering (evidently multiple) of sunlight; this effect is called n o c t u r n a l twilight. It is concentrated in a narrow segment at the horizon, and moves in azimuth along with
the sun throughout the night. The altitude of the nocturnal twilight above the
sun is fully 40-55°; it disappears when the sun is depressed still farther below
the horizon. Since this glow is very faint and is accompanied by the zodiacal
light and the intrinsic glow of the earth's atmosphere, the nocturnal twilight
has so far received no study whatever, and its very existence is not reliably
established.

The events as we have described them (and many details have been
omitted here; see, for example,[41, 46, 71-73]) should be regarded as typical
for clear weather only. The progress of twilight phenomena is actually subject to wide variation. If the air is relatively turbid, the twilight colors will
usually be paler, especially at the horizon where instead of red and orange
tints, only a faint brownish-gray coloration is sometimes observed [72]. Occasionally, as Kucherov [72] has noted, phenomena normally occurring at the
same time will develop differently in different parts of the sky. Every twilight is individual, unrepeatable — and this should be regarded as one of its
most characteristic features.

Unquestionably the pattern of twilight phenomena is wholly governed
by the optical structure of the atmosphere, both the ground and the upper air
layers. Yet its involvement with meteorological processes, as we shall see
presently, is considerably more complicated than might seem at first glance.
It is for this reason that lengthy searches, described in [41, 46], for a direct
correlation between twilight phenomena and the weather have been unsuccessful.

The extremely distinctive behavior of individual twilights and the great
variety of optical phenomena accompanying them result from the fundamental fact that the several parameters of the twilight event all depend differently
on the different optical properties of the atmosphere. The spectral, altitude,

horizontal, and angular variation of these optical properties – particularly the extinction and scattering coefficients – vary in different ways with the sun's zenith distance, the direction of observation, and the elevation of the observer. Both the regular progress of the twilight phenomenon and the individual variations are gigantic in scale compared with the analogous changes in the daytime. Along with the relatively static daytime conditions, the swift twilight process offers us, through space and time scans by photometric, spectral, and polarimetric means, an opening into a systematic study of the optical structure of the atmosphere and its characteristic variations under different conditions. It is here that we find the main distinction between atmospheric research under twilight and daytime conditions, for the relatively small scope of the changes during the day sharply limits the resolving power of integrated optical methods, and renders them less fruitful for an understanding of the optical structure of the atmosphere itself [4, 19].

§3. Spectrophotometry of the Twilight Sky

A great many direct measurements have been made of the brightness of the cloudless twilight sky [4, 5, 7, 19, 57, 58, 61, 67-71, 74-124, 279-284]. The individual twilights for which the sky brightness has, to a varying extent, been measured as a function of the sun's zenith distance ζ number in the thousands. One must recognize that the measurements entail considerable difficulties. In particular, the brightness varies over an immense range; the glow becomes extremely faint when the sun is depressed far below the horizon, necessitating receivers of radiation having very high sensitivity; the processes · take place so rapidly as to prevent the use of cumulative-type receiving systems, such as photographic plates, which require long exposures; operations must be carried on under field conditions; and the waiting time for favorable meteorological conditions can be very lengthy. For these reasons, most of the measurements prior to the mid-1940's were obtained visually, without the use of filters. Nevertheless, a number of investigations [84, 86, 87, 90, 93, 94], especially the extensive program of Megrelishvili [97, 101], have established that the ζ-dependence of the sky brightness differs materially for different wavelengths as shown, in particular, by the play of colors that is so characteristic of the twilight phenomenon. But Megrelishvili [101] also was concerned about the possibility that these variations in the spectral composition might produce serious errors in measurements taken over wide wavelength ranges. This situation has impelled most subsequent investigators [19, 67, 103, 109-117, 121-124, 281-284] to use filters in their measurements; and there has been an accompanying progress in the development of high-sensitivity receivers of light, photocells and photomultipliers, so that visual photometry has been rendered almost obsolete.

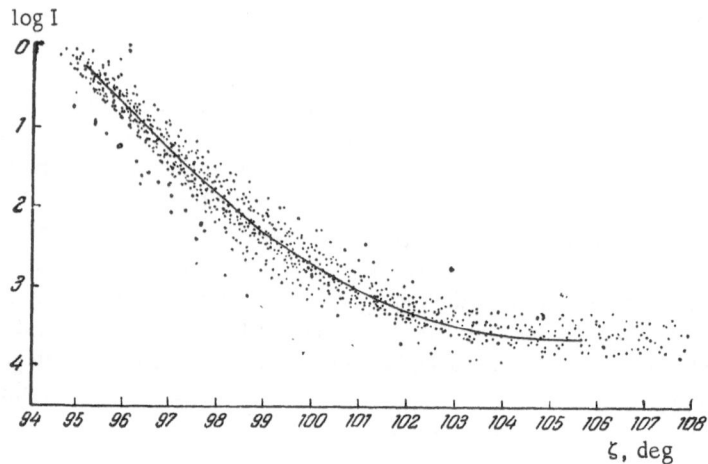

Fig. 14. Brightness of the twilight sky at the zenith as a function of the sun's zenith distance for measurements on 32 evenings, with $\lambda = 0.63\ \mu$.

In recent years, then, a fund of observations has accumulated relating not only to the brightness, but also to the spectral composition of the light from the twilight sky. The overwhelming majority of these measurements refer to the zenith, and only a few to other regions of the sky, such as the neighborhood of the north celestial pole [57, 75], points along the solar vertical [83, 107, 109, 112, 121, 124, 282-284], or the purple light and the twilight glow following it [71, 88, 89, 91, 105, 120]. There have also been series of measurements intended for compiling brightness maps of the twilight sky [58, 67, 81, 104, 105, 118].

Observers who have carried out photometry of the twilight sky have most frequently been interested only in the brightness variation with respect to ζ. They have therefore been satisfied with relative measurements. Absolute determinations have been made by comparatively few observers [6, 7, 58, 67, 74-76, 78, 79, 91, 103, 118, 124, 282-284], and some of this work is not very reliable.

To begin with, we must emphasize the remarkably good agreement among the individual measurements. With the sky brightness varying by a factor of 10^7-10^8 during the twilight process, the values of the brightness as determined by different authors at the same ζ, and for the same direction of observation (but not too close to the horizon), differ by no more than a factor of three to five in the short-wave portion of the visible spectrum (and similarly for measurements without filters). Measurements taken by the same author on

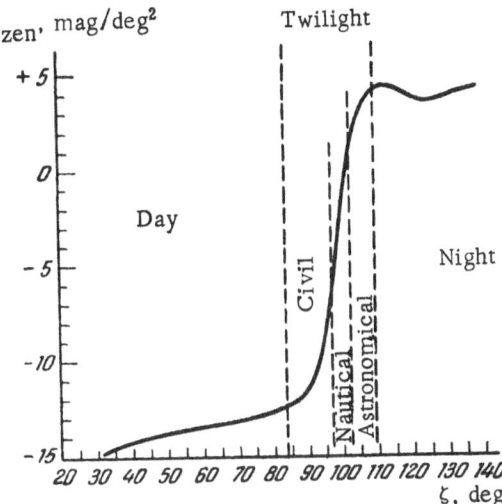

Fig. 15. Sky brightness at the zenith as a function of the sun's zenith distance.

different days tend to show just as great a deviation [101, 117, 121, 282]. This
is illustrated in Fig. 14, after a paper by Link, Neužil, and Zacharov [117],
which collects the data obtained by these authors on the brightness of the twilight sky at the zenith at Ondřejov on 32 evenings in 1955-1956, using an interference filter with $\lambda_{eff} = 6300$ A. The size of each dot represents the error
of an individual measurement. The scatter increases sharply in the long-wave
portion of the spectrum.

According to the many years of measurements by T. G. Megrelishvili,
data which she has kindly communicated to the author, in the region near 1 μ
the brightness sometimes undergoes a 70-fold variation from day to day. Furthermore, all observers have unanimously remarked on the virtual independ-
ence of the brightness of the twilight sky far from the horizon from the weather
at the observing station. In particular, the sky brightness measured through
gaps in the clouds is the same as that in completely clear weather. This is in
marked contrast to daytime conditions, where cloud cover strongly influences
the sky brightness; the theory of twilight phenomena provides the explanation
(Chapter IV).

By combining the measurements secured (without filters) up to 1935,
Brunner [58] established the mean curve shown in Fig. 15 for the sky brightness at the zenith, from $\zeta = 30°$ to $\zeta = 140°$. The results of subsequent studies
follow this curve closely (see, for example, [67]).

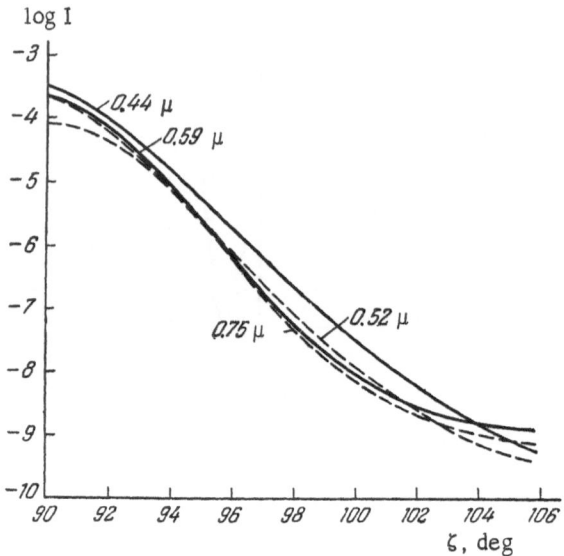

Fig. 16. Sky brightness at the zenith in different spectral regions as a function of the sun's zenith distance.

Figure 16 presents analogous curves obtained by Ashburn [103], but now
with filters. The ordinate is the common logarithm of the sky surface brightness I at the zenith, expressed in erg cm^{-2} sec^{-1} deg^{-2}. The interference filters with effective wavelengths $\lambda_{eff} = 0.44$, 0.52, and 0.59 μ had transmission
bands of 150-A halfwidth; for $\lambda_{eff} = 0.75$ μ the effective halfwidth was 0.23 μ.

Brunner [58] had called attention to the characteristic bend in the brightness curve near the limit of astronomical twilight ($\zeta \approx 102$-$104°$), and he suggested that it might be due to the relatively constant brightness of the intrinsic
airglow of the upper atmospheric layers being superposed on the changing
brightness of scattered sunlight. Indeed, if we subtract the night-sky from the
twilight-sky brightness, we practically remove the bend in the curve [58, 96].
Moreover, Fig. 16 shows that as the wavelength increases — as we pass into a
spectral region where the brightness of scattered light is lower, but the night-
sky brightness is higher — the bend decidedly moves toward smaller ζ [90, 110,
117]. This would seem to indicate that the variable behavior of this part of
the curve arises mainly from variations in the intrinsic glow of the atmosphere
[58, 124-127]. This conclusion does not, however, exclude additional variations in the intensity of the sunlight scattered by the atmosphere, including multiply scattered light.

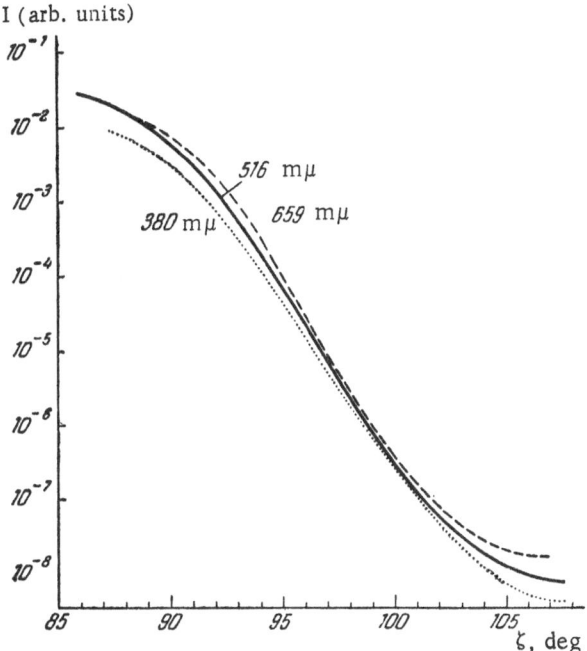

Fig. 17. Sky brightness at the point A = 0°, z = 70° in several regions of the spectrum as a function of the sun's zenith distance.

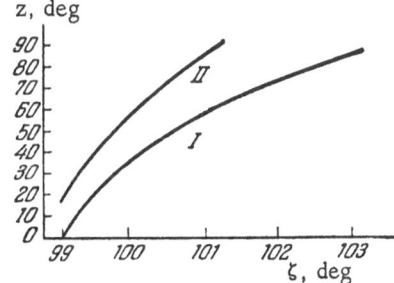

Fig. 18. Displacement of points of given surface brightness along the solar (I) and antisolar (II) verticals with increasing solar zenith distance ζ.

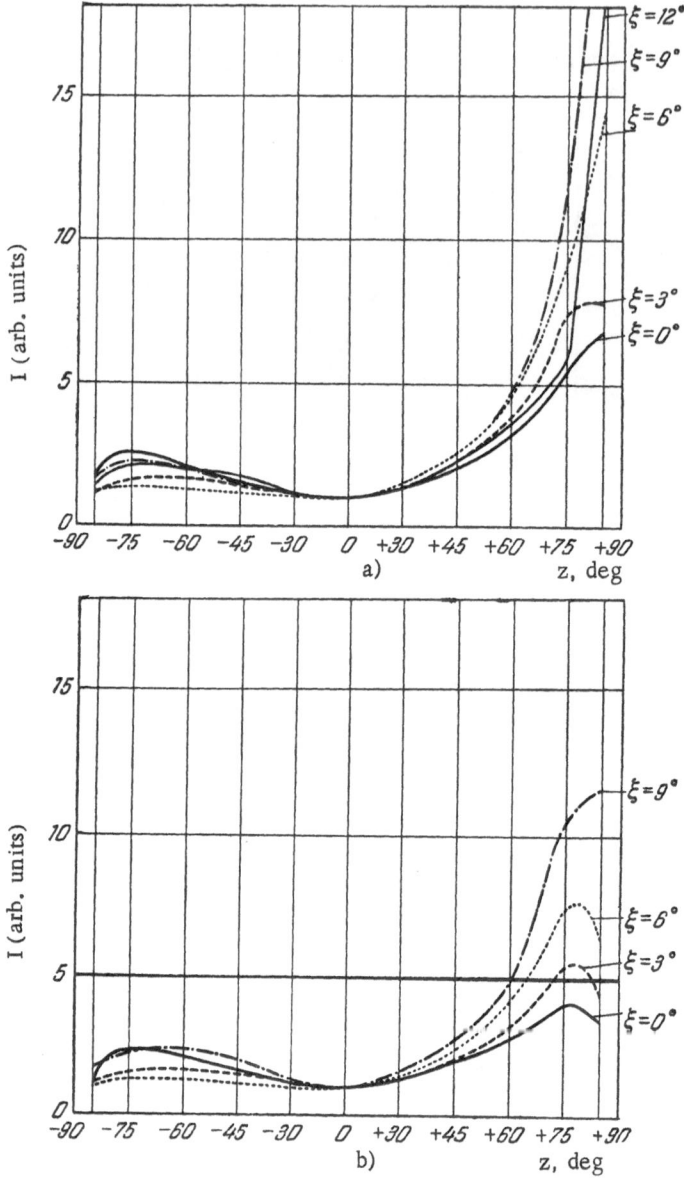

Fig. 19. Sky brightness relative to the zenith along the anti-solar — solar vertical for selected depressions ξ of the sun. a) No filter, $\lambda_{eff} = 4600$ A; b) $\lambda_{eff} = 4350$ A; c) $\lambda_{eff} = 5275$ A; d) $\lambda_{eff} = 6350$ A.

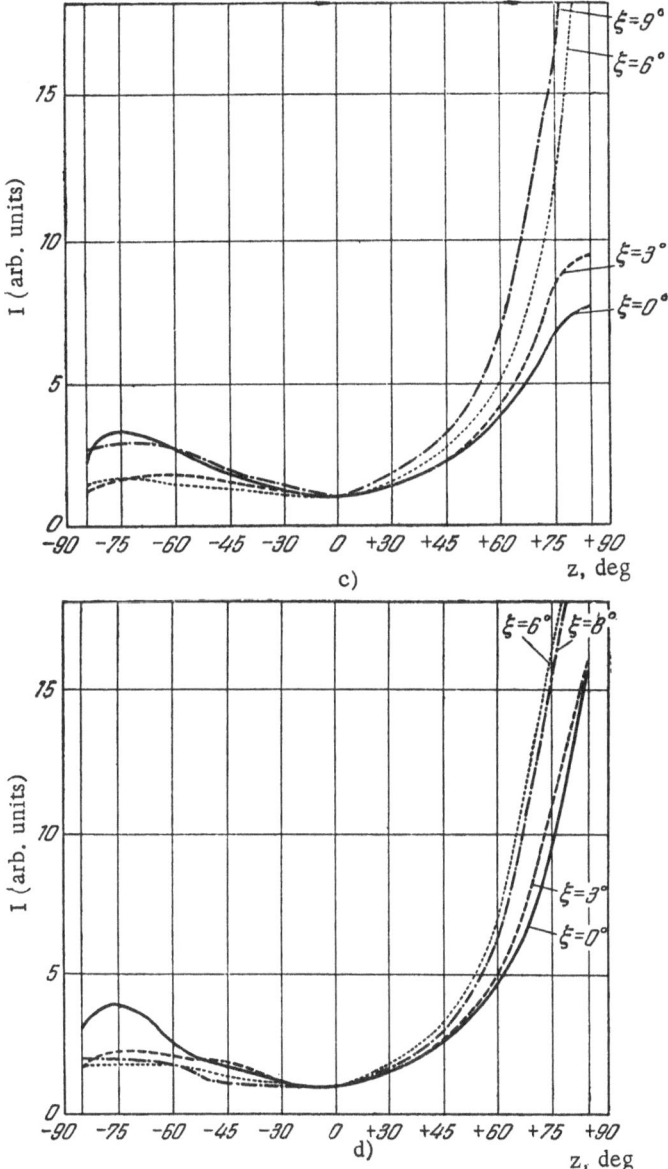

Fig. 19. (Continued)

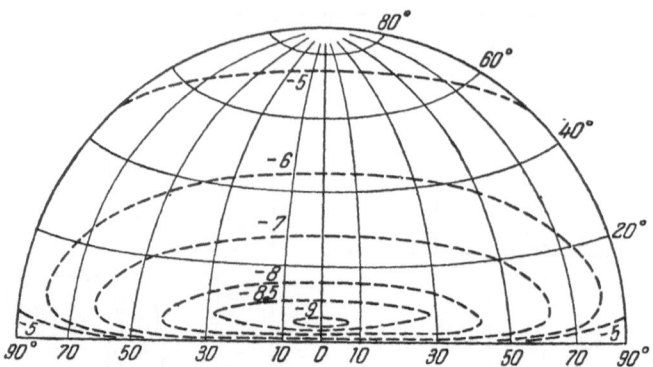

Fig. 20. Isophotes of the solar half of the twilight sky for $\zeta = 90°$, computed with allowance for single molecular scattering only. Each isophote is labeled with the brightness in stellar magnitudes.

Again, Fig. 15 shows a distinct minimum near $\zeta = 123$-$125°$; it coincides approximately with a small extremum in the illuminance at the earth's surface, as noted by Feofilov [56]. Divari [57] has also observed this extremum in the brightness curve of the night sky at the edge of twilight. There is every reason to believe that it is due to some properties of the diurnal variation of the night-sky glow.

Figure 17 demonstrates how the sky brightness measured in the solar vertical (line of sight at zenith distance $z = 70°$) varies with the zenith distance of the sun for three spectral regions [124]. The interference filters of effective wavelengths $\lambda_{eff} = 380$ and 516 mμ had transmission bands of 100- and 140-A halfwidth, respectively, and the absorption filter used at $\lambda_{eff} = 659$ mμ had a 1050-A effective halfwidth. Note that the relative behavior of the curves in Figs. 16 and 17 is entirely different, corresponding to the differing coloration of the sky at the zenith and in the "twilight glow."

We pass now to the brightness distribution on the celestial sphere. Figure 1 (§1) shows that as ζ increases during the day, the minimum in the sky brightness gradually approaches the zenith but at the same time recedes from the sun. This process continues also during the bright portion of civil twilight [48]. The minimum arrives at the zenith only with the onset of total darkness; throughout the twilight period it remains slightly displaced from the zenith along the antisolar vertical. Figure 18 displays the motion of points of constant brightness along the solar and antisolar verticals as the sun's zenith distance varies, according to the data of Graff [81]. Figure 19, which is taken from [67], shows the relative sky-brightness variations along the solar and anti-

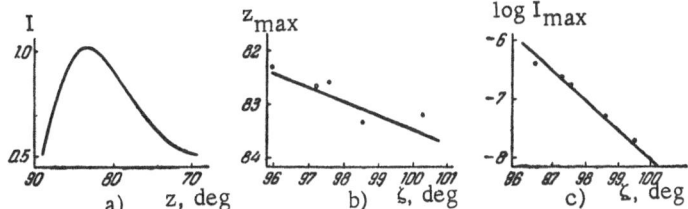

Fig. 21. a) Brightness distribution in the twilight glow along the solar vertical; b) zenith distance z_{max} of maximum brightness; and c) value I_{max} of maximum brightness, as functions of the sun's zenith distance.

solar verticals at selected solar zenith distances. The group of curves illustrated in Fig. 19 essentially represents a quantitative analysis of the course of the twilight phenomenon as described qualitatively in §2.

Note first of all in Fig. 19 that the twilight sky brightness is concentrated in the relatively small twilight-glow segment at the solar horizon, and that this area contracts rapidly in size as ζ increases [91]. For comparison, we present in Fig. 20 a chart of isophotes for the solar half of the twilight sky, as computed by Shtaude [128] with allowance for single molecular scattering only. Returning to Fig. 19, we should further point out that the size and brightness of the twilight glow depend markedly on wavelength, and the dependence itself varies with ζ. Moreover, maximum brightness does not occur at the very horizon, as Hulburt [91] believed, but somewhat higher. Figure 21 shows the sky brightness in the twilight glow (for $\zeta = 96$-$101°$) as a function of the zenith distance z of the line of sight, the mean for a series of measurements by Vasil'ev [129], and also how the position of maximum sky brightness and the brightness there vary with the sun's zenith distance. The measurements were made photographically with a broad-band filter. The position and value of maximum brightness depend, however, on wavelength, as is clear from Fig. 19. This behavior is illustrated by Fig. 22, which was kindly prepared at our request by A. Kh. Darchiya from the results of her spectrographic study of the twilight-glow region [120]. Figure 22 refers to a solar zenith distance $\zeta = 90°$, although a similar behavior is observed at other solar zenith distances as far as $\zeta = 94°$ One often finds that the horizon is obscured by dense haze which conceals the lower portion of the twilight glow. In this event, the position of maximum brightness no longer depends on wavelength, and lies approximately at $z = 83$-$85°$ independently of the varying ζ. In other words, the brightness distribution next to the horizon will then depend only on the structure of the haze layer with respect to height. The considerable scatter of the points in

Fig. 22b, an effect found also in the data of other authors on the region of sky next to the horizon, is naturally due to the variable turbidity of the lower atmospheric layers.

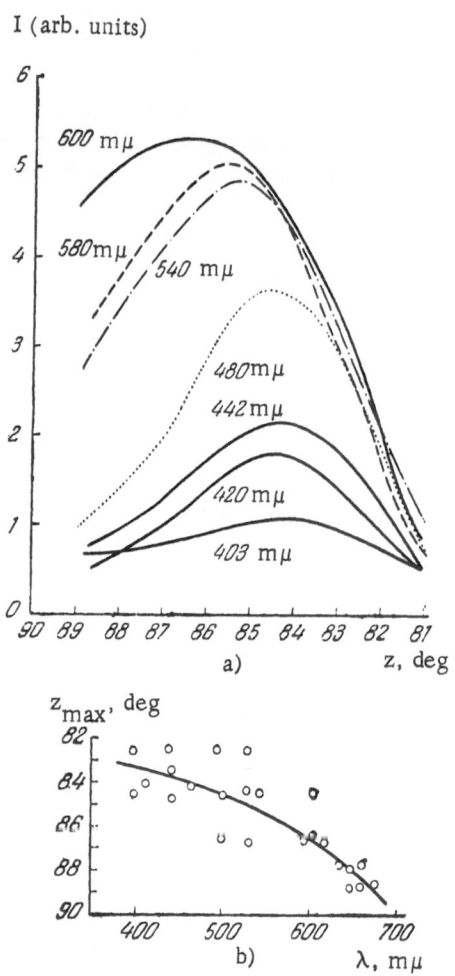

Fig. 22. a) Brightness distribution in the twilight glow along the solar vertical for selected spectral regions; b) zenith distance z_{max} of maximum brightness as a function of wavelength; $\zeta = 90°$.

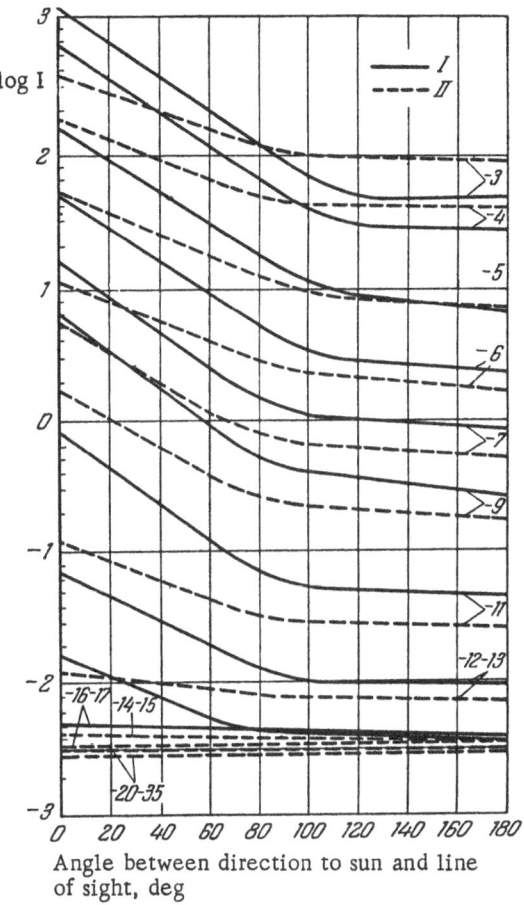

Angle between direction to sun and line
of sight, deg

Fig. 23. Distribution of twilight sky brightness
along the horizon (I) and at z = 30° (II) for various solar altitudes 90° − ζ.

Measurements of the twilight sky brightness along various almucantars
[104, 105, 118] indicate that the brightness varies monotonically in azimuth
and much more slowly than in altitude. This situation is illustrated by Fig. 23,
taken from a paper by Barteneva and Boyarova [118], who employed visual photometry without filters over a relatively wide field of view (10° in diameter).
Divari [105] has investigated the sky-brightness distribution for large ζ (at the
boundary of night) at various azimuths near the horizon.

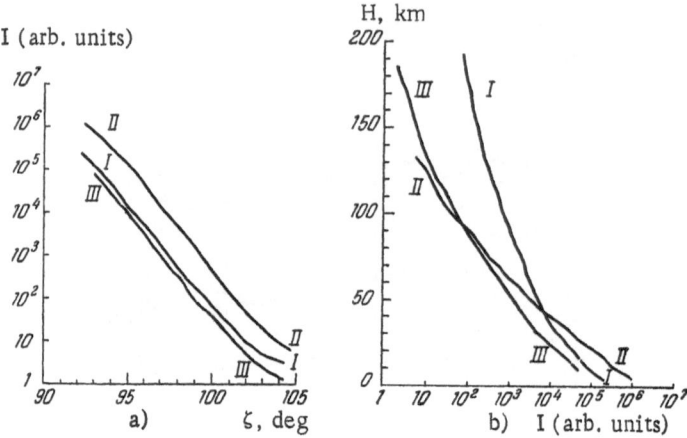

Fig. 24. Sky brightness observed simultaneously in three directions as a function (a) of the sun's zenith distance ζ, and (b) of
the height H of the earth's geometric shadow in the direction of
observation. I) $z = -70°$, away from the sun; II) $z = +70°$, toward
the sun; III) zenith.

One important result that studies of the twilight sky have established is
the finding that, so far as its main features are concerned, the twilight phenomenon advances uniformly in all parts of the sky, providing evidence that
it is formed under identical conditions throughout. Nevertheless, a regular
and very appreciable phase shift is observed: the twilight process experiences
a lag along the solar vertical compared with the antisolar segment. One therefore cannot subclassify the twilight sky as a whole in terms of simultaneous
stages depending only on the changing zenith distance of the sun, as was done
in §2 (Fig. 11). A classification based on the illuminance at the ground would
correspond more or less to the conditions obtaining at the zenith (Fig. 15), but
would not be appropriate for other regions of the sky. As a result, twilights
must be classified individually for each point of the sky. In §§3 and 7 of Chapter IV, we shall develop criteria for subdividing the twilight process by phase.

Again, although the phenomenon proceeds identically in its gross aspects at all points of the sky, much of the fine detail depends on the direction
of the line of sight. This is because the relative contribution of the various
parameters depends on geometric circumstances. And consequently it is just
these differences in detail that take on leading importance when twilight observations are used to study the optical structure of the atmosphere.

The varying behavior of the sky brightness as a function of the sun's
zenith distance for different points along the solar meridian has already been

illustrated in Fig. 19. The effect appears more clearly in Fig. 24, which T. G. Megrelishvili has kindly allowed the author to reproduce here. Figure 24a shows how the sky brightness I varies with the sun's zenith distance for $\lambda_{eff} = 0.52\ \mu$, as observed by Megrelishvili on a single evening from simultaneous measurements at three points of the solar meridian: at the zenith, at $z = +70°$ (in the solar vertical), and at $z = -70°$ (in the antisolar vertical). In Fig. 24b, the same sky-brightness observations are represented as a function of the height H of the earth's geometric shadow in the direction of observation. We notice at once that the slope of the H(I) curves increases as the sun recedes from the point of observation, and also that the sky brightness for a given height of the earth's geometric shadow is much greater in the antisolar vertical than at the zenith or in the solar vertical when the shadow has reached a height of at least 50-60 km above the ground. As first pointed out by V. G. Fesenkov, this behavior evidently arises from the superposition of secondary scattering of sunlight, and so opens the way to an experimental study of the scattering process. Indeed, for a given solar zenith distance the brightness of the rescattered light depends comparatively weakly on the direction of observation. But in the antisolar vertical the height of the earth's shadow is much greater, and hence the brightness of the singly scattered light is much lower, than in the solar vertical.

Figure 24 also shows rather well that the regular trend of the curves is accompanied by appreciable individual variations which, incidentally, undergo strong changes from day to day. These departures from smoothly varying curves are a highly distinctive property of the twilight phenomenon; we shall see later that they reflect the characteristic optical structure of the atmosphere, including even the upper layers, and their variability is clearly due to varying atmospheric conditions as well as horizontal inhomogeneity in the air mass. All authors have, in particular, pointed out that the purple light is subject to strong intensity variations with a definite brightness maximum occurring in the autumn, and with further sharp enhancements after powerful volcanic eruptions.

We turn next to a more detailed review of the color variations in the twilight sky. The first and most distinctive sign of the approach of twilight is the reddening of the sky, an effect which always takes place but whose intensity depends on the position in the celestial sphere and on weather conditions. Gruner [71] and Dorno [7] have carried out photometry of the purple light in different parts of the sky, and they have found that the reddening of the sky changes to a blue coloration when the sun is depressed by about 4-4°.5. This is illustrated by Fig. 25, which is taken from [71]. Link [84] later pointed out that the sky becomes increasingly blue at the zenith also as the sun drops farther below the horizon. Assuming that the blueness was due to an increased contribution of secondary scattering, he attempted by measurements with filters to estimate the relative intensity of the rescattered light. However, this

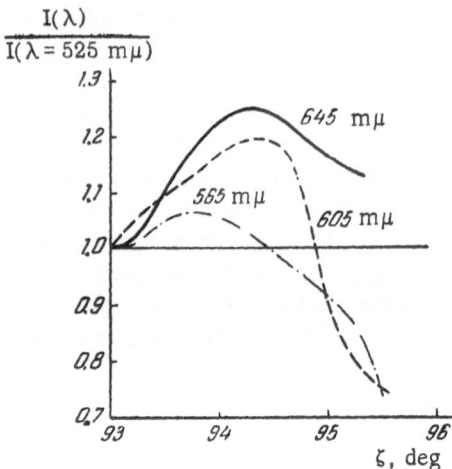

Fig. 25. Color of the sky in the purple-
light region as a function of ζ for small
solar depressions $\xi = \zeta - 90°$.

effort rested on primitive considerations that do not apply to the real atmosphere. That this is the case has, in particular, been shown by Khvostikov,
Magid, and Shubin [93], who confirmed that the twilight sky becomes increasingly blue, but established that the effect does not arise from the appearance
of a λ^{-4} factor as Link had essentially assumed. We shall find in Chapter IV
that the enhanced blueness of the sky has an entirely different explanation.

Megrelishvili [97, 101] began to investigate systematically the color of
the twilight sky in 1942, and she has continued the program until quite recently [121]. These observations have demonstrated that when the sun has
dropped to 9-11° below the horizon, the blue coloration again gives way to a
reddening. Figure 26 shows how the color index at the zenith of the twilight
sky, averaged over each season, depends on the sun's zenith distance, according to extensive measurements in 1945 [101]. The ordinate here is the color
index $CE_{\lambda_2}^{\lambda}$, defined as

$$CE_{\lambda_2}^{\lambda_1} = -2.5\log\frac{I(\lambda_1)}{I(\lambda_2)},$$

(I.3, 1)

where $\lambda_1 = 379$ mμ and $\lambda_2 = 527$ mμ. The measurements were made with filters but the results are expressed in relative units since the spectral sensitivity

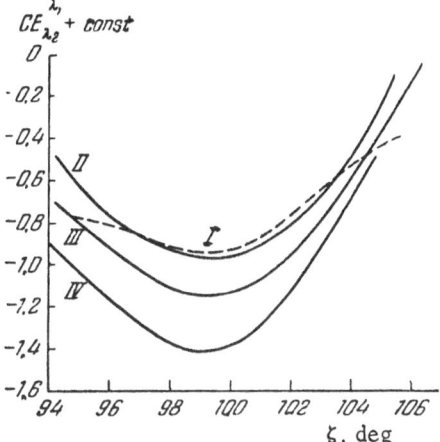

Fig. 26. Color index at the zenith of the twilight sky as a function of ζ for large solar depressions $\xi = \zeta - 90°$. Mean curves for: I) summer; II) autumn; III) winter; IV) spring.

of the instrument as a whole was not determined. As a result, there is a constant term of uncertain value in the color index for all the measurements. Similar results have been obtained more recently by other investigators [107, 113, 116, 130] involving, in particular, a large variety of wavelengths as far as 1 μ.

In Fig. 27 we present the ζ-dependence of color indices defined for several pairs of wavelengths in the direction of observation $z = 70°$, $A = 0°$, according to measurements by Volz and Goody [124] during two twilights.

Figure 26 indicates first of all that the reddening of the sky at the zenith when $\zeta > 99$-$100°$ is a regular occurrence, and in fact by the close of twilight the sky is normally considerably redder even than at the time of its maximum reddening for a solar depression angle of 4-5°. (At $z = 70°$, $A = 0°$, the reddening begins at a later stage — for $\zeta > 102$-$103°$; see Fig. 27.) The reddening is, however, subject to very strong variations, both seasonal (Fig. 26), a direct consequence of meteorological conditions, and from day to day. The color index usually has a more complicated ζ-dependence on an individual day than is shown by the mean curves. Examples of this behavior appear in Fig. 27 and in Fig. 28, which has been prepared [130] on the basis of 1958 observations in the Crimea [122] with filters having $\lambda_{eff} = 448$ and 528 mμ [19].

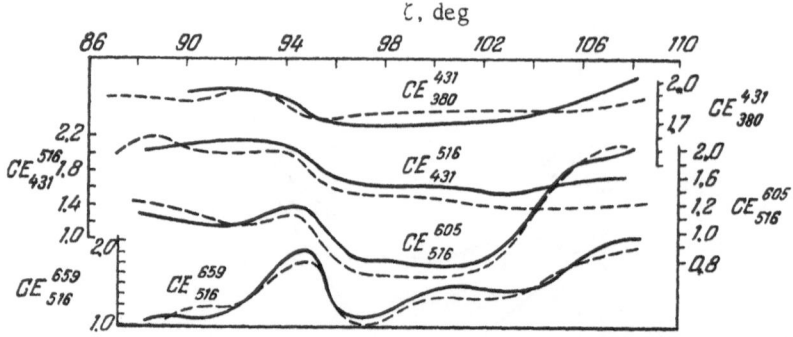

Fig. 27. Color of the twilight sky as a function of ζ for observations in the solar vertical at zenith distance $z = 70°$. Solid and broken curves, measurements on two separate days.

By comparing Fig. 26 with Fig. 16 and Fig. 27 with Fig. 17, we immediately see that the sky is reddened at the zenith for $\zeta \gtrless 100°$ and at the solar horizon for $\zeta \gtrless 103°$. At solar depression angles corresponding to approximately these values (depending on the wavelength), the fall in the sky brightness in the long-wave portion of the spectrum is delayed, while in the short-wave region the brightness continues to drop as swiftly as before. In subsequent chapters, we shall examine in detail the physical processes that might be responsible for this phenomenon. At this point we only wish to remark that the effect also occurs in the spectral region lying beyond the Chappuis absorption bands of ozone, in the interval $0.7–1.0\ \mu$.

That the absorption of light by atmospheric ozone affects the color of the sky has been known for a long time. This was first discovered by Chappuis [131], who even suggested that the presence of ozone is wholly responsible for the azure color of the daytime sky. The role of ozone in connection with twilight was pointed out clearly as early as 1935 by Gauzit [132], who gave a detailed description of the general character of the twilight spectrum. This matter has been considered more recently by Hulburt [133]. Colorimetric investigations by Gadsden [114] qualitatively confirm that absorption of light by atmospheric ozone has an important effect on the color of the twilight sky. Finally, from extensive measurements with filters Divari [116] has quantitatively evaluated the contribution of the Chappuis bands to the light of the twilight sky.

When this book was being prepared for publication, R. M. Goody kindly furnished the author with the manuscript of a forthcoming paper [282] that he

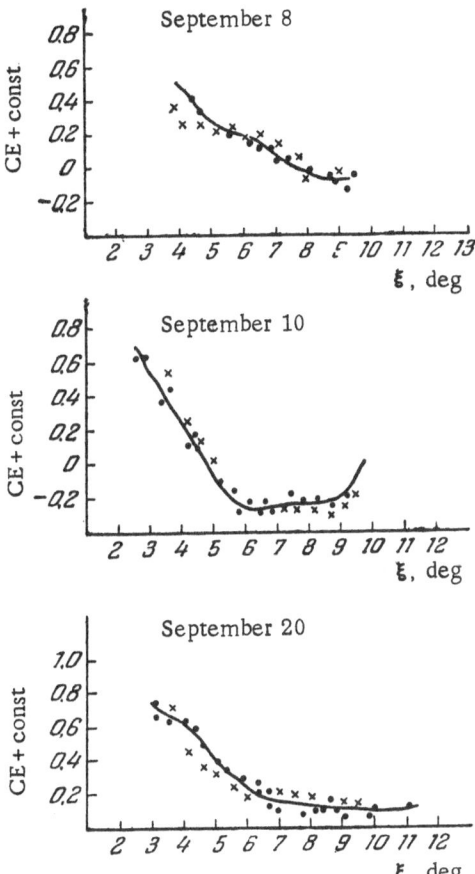

Fig. 28. Color index at the zenith of the twilight sky as a function of ζ for three separate days of observation. Dots, measured points; crosses, computed values (Chapter IV, §5).

and F. E. Volz had completed. These authors have, in particular, computed the spectral variation of the sky brightness at the point z = 70°, A = 0° for ζ = 90° (allowing for single scattering only) in the region of the Chappuis bands for a purely molecular atmosphere R, and for an atmosphere D containing aerosols, under certain assumptions regarding the distribution of the atmosphere with height and the particle distribution with respect to size (Fig. 29).

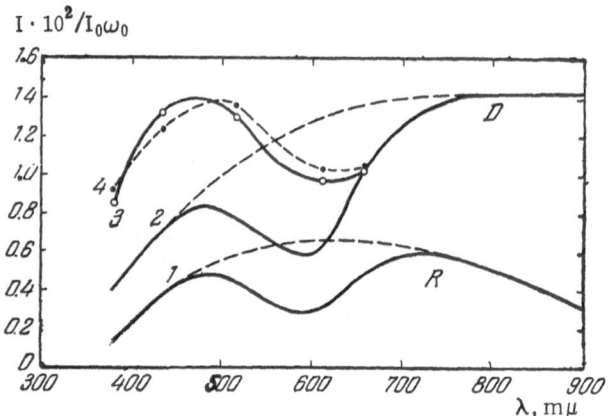

Fig. 29. The influence of the Chappuis bands of ozone on the spectrum of the twilight sky at the point z = 70°, A = 0°, for ζ = 90°. Curves 1 and 2, theoretical (broken curves, no correction for the effect of ozone); curves 3 and 4, observed.

Figure 29 also shows the results of their measurements of the sky brightness with interference filters on two separate days (curves 3 and 4), for ζ = 90°. Similar curves may be found in Divari's paper [116], and may also be derived from the observations of Link [84] and Ljunghall [67]. What Goody and Volz have shown is that by measuring the defect in the twilight-sky brightness in the vicinity of the Chappuis bands, one can estimate quite reliably the total ozone content of the atmosphere, although the procedure they have adopted is very rough, as the authors themselves point out.

So far we have considered only observations with color filters. This by no means reflects a lack of spectra of the twilight sky (see, for example, [86, 134]), but signifies rather that the wealth of such spectra, as obtained by many investigators, have as a rule served only to help explain the intrinsic radiation of the atmosphere (polar aurorae, night-sky emission); the spectra have not been applied to an analysis of the behavior (with respect to ζ and the direction of observation) of the background against which the radiation appears. Moreover, the spectra have been recorded photographically with long exposures, a method which essentially averages them over a long range of values of the sun's zenith distance. As a result, the very abundant spectroscopic material for the twilight sky is practically useless for an understanding of the laws by which the twilight process itself operates. There is one exception, however — an extensive group of studies of the so-called "twilight flash" in the night-sky emission; we discuss this phenomenon briefly below. The only investigations of importance to us here, however, are those specifically devoted to a study of the character and variations of the twilight-sky spectrum at that stage of the process where the intrinsic radiation of the atmosphere still makes no significant contribution.

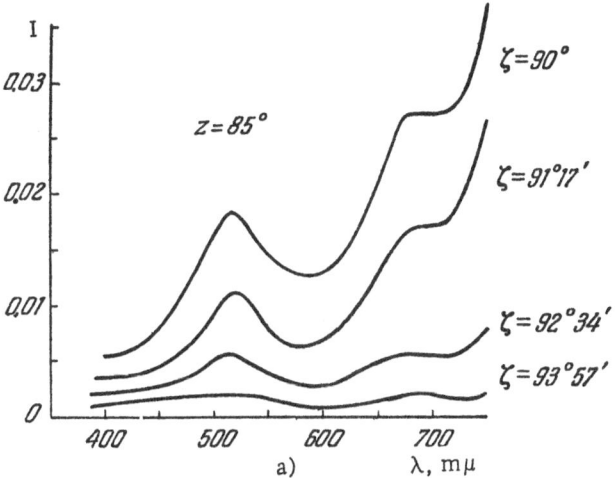

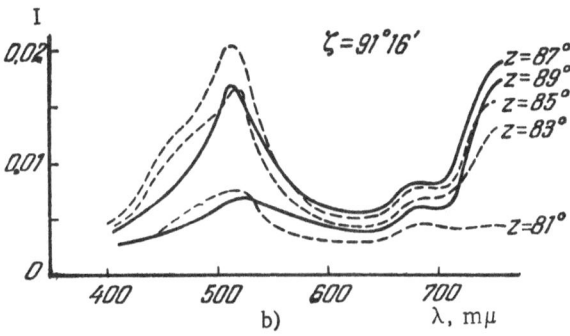

Fig. 30. Examples of the wavelength dependence of the sky brightness (in units of the daytime illuminance) in the twilight glow (a) for selected zenith distances ζ of the sun, and (b) for selected zenith distances z of the line of sight.

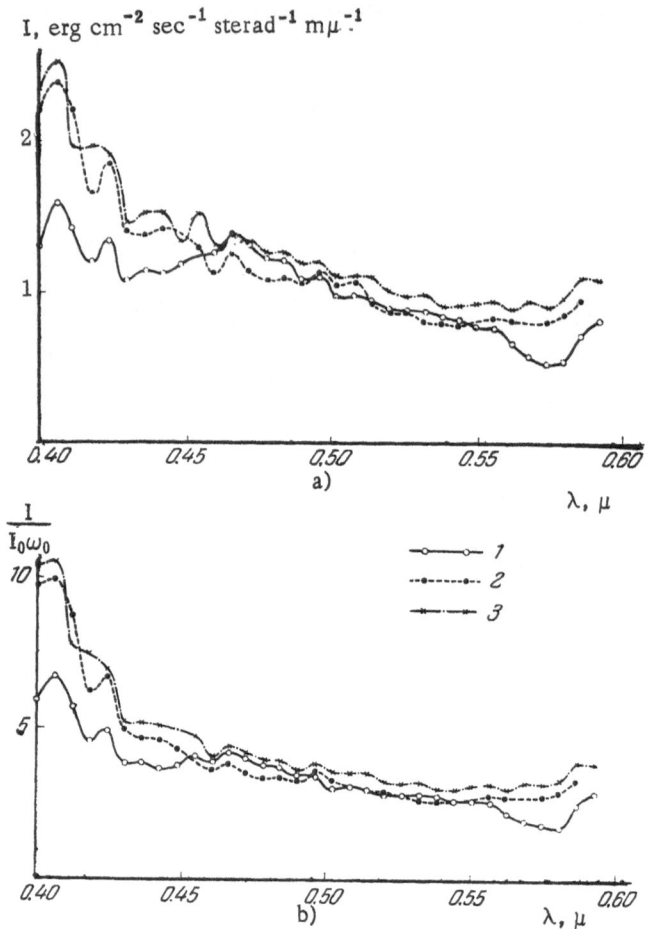

Fig. 31. Spectra of the light at the zenith of the twilight sky for three solar zenith distances ζ (a) in absolute units, and (b) in units of the sun's brightness. 1) $\zeta = 91°35'$; 2) $\zeta = 95°19'$; 3) $\zeta = 97°22'$. The numbers along the axes of ordinates are to be multiplied by:

Figure	Curve		
	1	2	3
31a	10^{-5}	10^{-7}	10^{-8}
31b	10^{-2}	10^{-4}	10^{-5}

Darchiya [120] has recently published a series of spectra of the twilight sky referring to this stage. Her spectra were also obtained photographically, but with short exposures, made possible by restriction to very small values of $\xi = \zeta - 90°$ and to the brightest region of the sky, essentially the twilight-glow segment near the horizon.

The spectra exhibit great variations, as is perfectly natural since they represent observations near the very horizon, where even insignificant fluctuations in the atmospheric extinction will be much exaggerated. We have therefore selected for illustration only two series of relatively well-behaved spectra, and have smoothed them out somewhat so as to remove their less authentic features (Fig. 30). However, to emphasize the variety of the spectra, we have selected cases having considerably different variations of brightness with wavelength in illustrating the dependence of the spectra on the sun's zenith distance ζ and the zenith distance z of the line of sight. All the measurements were recorded in units of the daytime illuminance on a horizontal white screen whose spectrum will, in general, differ appreciably from the solar spectrum because of the presence of scattered radiation from the daytime sky (see, for example, [135]). In particular, this results in a substantial underestimate, progressing rapidly with decreasing wavelength, of the intensity of the short-wave portion of the spectrum. We must regard the sharpness of the brightness maximum near $\lambda = 500$ mμ as due in part to the choice of units, accentuating the decline in intensity toward short wavelengths (compare Fig. 29). The main reason for the decline is, however, the wavelength dependence of the atmospheric transparency, a behavior which reaches a peak at large z. At the same time the spectra clearly reveal the Chappuis absorption bands of ozone between 500 and 700 mμ, with a strong day-to-day variation. The left-hand edge of the bands forms the steep long-wave slope of the green brightness maximum. Notice also the minimum near 700 mμ, apparently due to oxygen and water vapor, and the sharply varying long-wave wing of the spectrum beyond the Chappuis bands. We shall inquire into the reasons for these last fluctuations below.

While this volume was in preparation A. Ya. Driving, O. A. Zigel', I. M. Mikhailin, G. V. Rozenberg, and G. D. Turkin obtained some spectra of the twilight sky at the zenith for solar zenith distances as great as $\zeta = 100°$, using a recording spectrometer equipped with a diffraction grating. Because of the faintness of the radiation from the twilight sky, the measurements were taken with a wide slit, yielding a spectral resolution of 70 A. The scanning time for each spectrum was 1 minute. A Franck-Ritter prism was placed before the spectrometer slit to isolate a component polarized linearly at a 45° angle to the solar vertical. Figure 31 shows some of the spectra that we obtained, expressed in absolute units and in units of the sun's brightness. The

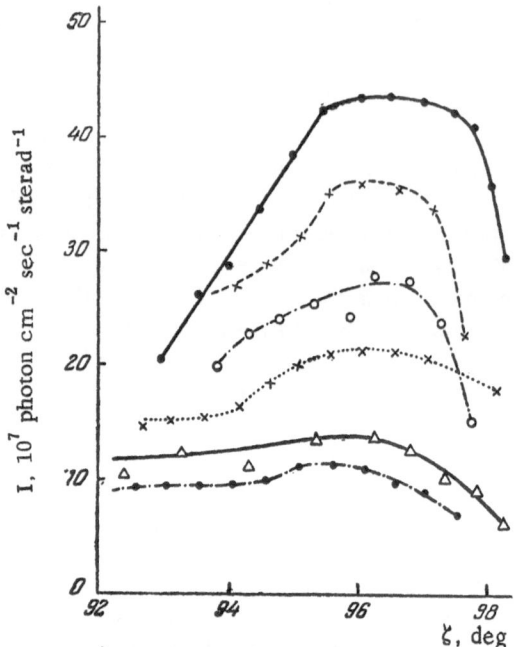

Fig. 32. Intensity of the sodium resonance lines
as a function of the sun's zenith distance on several different days.

instrument was calibrated with a white screen illuminated by the sun; the illumination of the screen by scattered light from the sky was eliminated by
the "sun—shadow" method, and corrections were applied for the spectral transparency of the atmosphere by the so-called long Bouguer method. A comparison of Figs. 29-31 shows that the Chappuis bands appear brightest when
observed close to the horizon.

In 1933 Slipher [136] pointed out for the first time that the negative system of nightglow emission bands of ionized nitrogen, N_2^+, is enhanced during
the period of astronomical twilight. Three years later, Garrigue [137] found
that when the sun is depressed 9° below the horizon, the twilight-sky spectrum
exhibits the red $\lambda 6300$ oxygen line, but that the line weakens as the zenith distance of the sun continues to increase. In the same year, Chernyaev and Vuks
[86] found the sodium doublet in the twilight-sky spectrum; it appears when
the earth's shadow reaches a height of 50-60 km, but dies out as the shadow
rises. Subsequently the twilight-flash and other emission features of the night
sky were discovered: the $\lambda 5200$ line of N I [138], the atmospheric system of

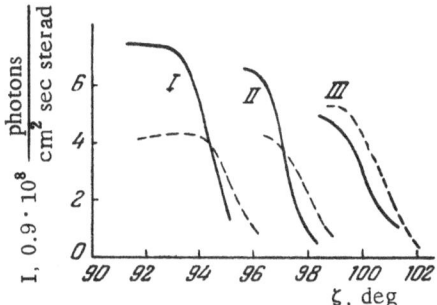

Fig. 33. Intensity of the sodium resonance lines as a function of the sun's zenith distance for three directions of observation, on two different days.

O_2 bands [139, 140], the OH bands [140], the H and K lines of Ca II [141], and the resonance lines of lithium [142] and helium [143]. Two properties are characteristic for all these twilight radiations: 1) they appear against the continuous background spectrum of scattered light from the twilight sky when the sky brightness has become too low to mask the fainter intrinsic emissions of the atmosphere; and 2) their intensity weakens as the sun continues to drop beneath the horizon. We cannot enter here into a discussion of the origin of these radiations (see, for example,[125, 144]), which is inseparable from the general problem of the origin of the night-sky glow (see, for example,[125, 143]), but shall pause only to consider the emission of atmospheric sodium, which has played an important part in the development of the twilight method of atmospheric research.

An extensive literature has been accorded the problem of the twilight flash of the sodium resonance lines [125, 145-162]. No doubt remains today that the effect is due to the resonance fluorescence of sodium vapor formed in a rather well-defined layer high in the atmosphere and irradiated by the light of the setting sun. When the sun is depressed only slightly below the horizon, the fluorescence is lost in the bright scattered light from the sky. As the zenith distance of the sun increases and the twilight brightness falls, the effect becomes observable, only to be suppressed again as soon as the sodium layer falls into the earth's shadow. However, this simple pattern is distorted by the attenuation of the direct sunlight as it passes through the sodium layer, so that its intensity is reduced at small depressions of the sun below the horizon. As a result, the curves for the intensity I of the twilight emission of atmospheric sodium as a function of the sun's zenith distance show a characteristic "plateau" whose form depends on the optical thickness and structure of the sodium layer. Figure 32 illustrates this situation clearly [151]. In Fig. 33, we reproduce examples of the ζ-dependence of I for observations of the twilight sky in three directions, according to measurements by Blamont, Donahue, and Stull [151]. In each case the zenith distance z of the line of sight was 75°, while the azimuths A relative to the sun were (I) 180°, (II) 90°, and (III) 0°. The solid curves refer to the morning of January 3, 1956, the broken curves to the evening of January 5. Note the strong phase displacement as the direction of observation varies, as well as the considerable differences in the shape of the curves and in the absolute values of the brightness on different days and for different azimuths. Moreover, as ζ varies the intensity ratio of the two lines of the doublet is also observed to change.

§4.   Polarimetry of the Twilight Sky

One can gain additional and completely independent information about the twilight process by measuring the polarization of the light from the twilight sky.

The earliest measurements [7, 8, 47, 50, 80, 88, 163-166] refer mainly to the positions of the neutral points (§1), which begin to display a pronounced mobility during the twilight period. Even at the beginning of twilight, at $\zeta = 83°.5$, the Arago and Babinet points behave differently (the Brewster point then falls at the horizon). The distance of the Arago point from the antisolar point decreases smoothly until $\zeta = 91-93°$, whereupon it increases rapidly again, in fact with a discrete jump, according to Dorno.

Wegener [163] has observed that when $\zeta = 98°$, the Arago point is 24-27° from the antisolar point. Photographic measurements regularly yield larger separations than visual observations, no doubt because the neutral points have different positions in different spectral regions. At $\zeta = 100°$, Graff [80] finds a 30° distance for the Arago point.

The Babinet point does not seem to behave in so regular a fashion, for the results of various authors differ considerably with one another. All authors agree that the distance of the Babinet point from the sun rises as ζ increases to 90°. Graff [80] and Roggenkamp [47] claim that this rise continues beyond $\zeta = 90°$. According to the data of Wegener [163], Jensen [165], and Dorno [7, 8], however, a fairly sharp decrease in the separation is observed between $\zeta = 90°$ and 95°, but Wegener and Dorno find that the increase has resumed at $\zeta = 95°$. In particular, Dorno [8] states that at $\zeta = 101°$ both the Arago and the Babinet points have risen high above the horizon (46° and 59°, respectively), while according to Wegener's measurements the distance of the Babinet point from the sun at $\zeta = 100°$ is only 15°.

The differing behavior of the Babinet and Arago points during twilight hours reflects the differing conditions of illumination of the bright and dark horizons, and especially, it would appear, the differing contribution of multiple scattering to the brightness in various regions of the twilight sky (compare [28, 39]). But at the same time the observational data for the neutral points indicate how sensitive the polarization of the light from the sky is to varying atmospheric conditions.

Aside from the positions of the neutral points, measurements have also been made of the degree of polarization in different parts of the sky, but these measurements have been confined to the bright portion of twilight, up to ζ = 95-96° only.

The principal result here has been that the degree of polarization of the light from the twilight sky first increases with ζ, and then decreases. At the zenith, in particular, maximum polarization occurs when ζ = 92-94°, the degree of polarization then having the mean value p = 0.7-0.8.

All the polarization studies of the twilight sky that we have cited so far have been designed primarily as a search for direct connection with the weather in the hope of facilitating weather forecasting. However, for reasons that will become clear in Chapters III and IV, these efforts have not met with success.

I. A. Khvostikov was the first to apply polarization measurements for analyzing the behavior of the twilight phenomenon itself, as well as for studying the structure of the upper atmosphere and the processes occurring there. With these purposes in mind, Khvostikov and his colleagues, in an observing program lasting several years, concentrated on the region of the sky near the zenith, an area where the polarization effects are particularly well defined and where, as was thought at the time, the results could be most simply interpreted. It is noteworthy that subsequent investigators, with few exceptions, have followed in this tradition, and as a result there are hardly any polarization data for other parts of the sky.

The first measurements in the Khvostikov program [167], carried out on Mt. Elbrus in the summer of 1936, showed that beginning at $\zeta \approx 97°$ the degree of polarization p falls off quite rapidly with ζ, but that there is a certain range of ζ between about 98° and 102° where the curves for p as a function of ζ have one or two deep minima. A single curve was also obtained, covering the entire dark portion of twilight to the boundary of night, and showing an extremely deep and broad polarization minimum (p = 0.15) at ζ = 106°.5.

If we convert the solar depression angle into the height of the earth's shadow at the zenith, and note that the effective level at which scattering occurs during twilight is some 20 km above the height of the geometric shadow

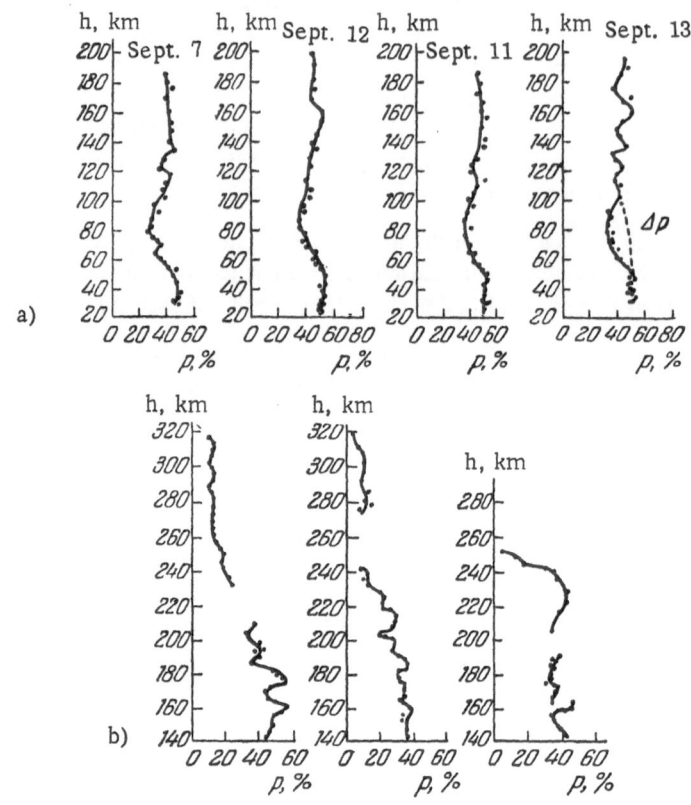

Fig. 34. Examples of the relation between the degree of polarization of the light received from the zenith of the twilight sky and the effective height of the scattering layer. a) Visual measurements; b) photoelectric measurements.

(Chapter III), then we find that the positions of the minima correspond to effective scattering heights of about 100 and 300 km. The presence of more or less well-defined minima of highly variable shape in this height range has been confirmed by later measurements of Khvostikov [168]. The fact that these heights correspond approximately to the levels of the ionospheric E and F layers has led Khvostikov to suggest [167, 168] that the degree of depolarization of the light from the twilight sky is somehow related to the degree of ionization in the upper layers of the atmosphere, and that polarization measurements during twilight hours might provide a direct "cross section" of the ionosphere.

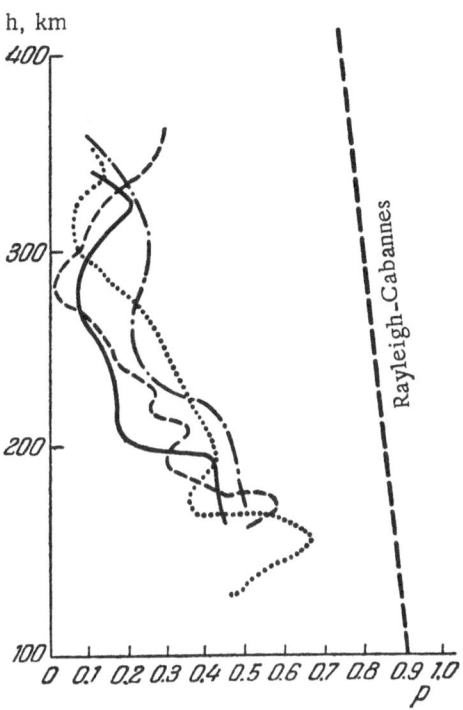

Fig. 35. Examples of the relation between the degree of polarization of the light received from the zenith of the twilight sky and the effective height of the scattering layer (photographic measurements).

At the Atmospheric Optics Laboratory, then under the direction of Professor Khvostikov, the research was pursued in an attempt to test the ionospheric hypothesis. In 1939, a special ionospheric-optics expedition went to the Crimea to conduct polarization measurements by two methods, visually and photoelectrically, with concomitant measurements of the critical frequencies and heights of radio-wave reflection from the ionospheric E and F layers [169]. The optical measurements were made without filters in a broad spectral band limited only by the sensitivity of the receiver. The resulting curves varied greatly in form (Fig. 34), indicating that there are profound differences in the conditions under which the twilight phenomenon develops on different days. Nevertheless, all the curves contain certain features pointed

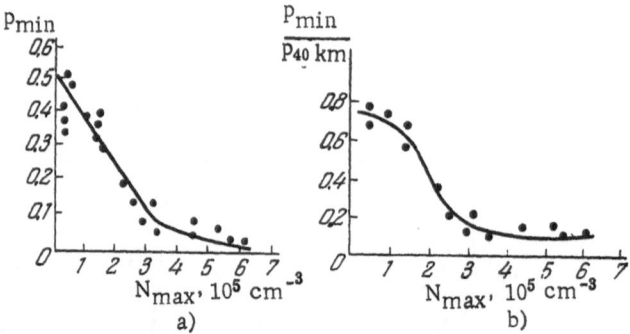

Fig. 36. Depolarization, in regions of minimum polarization, as a function of the maximum electron concentration
in the ionospheric layers.

out in [167, 168]: minima in the polarization occur at effective scattering
heights of about 80 and 250-290 km, and there is a general tendency for the
degree of polarization to fall off sharply as ζ increases.

The same features also stand out clearly in polarization measurements
of the twilight sky performed at Moscow in the summer of 1940 by photographic techniques [95, 96]. In the earlier measurements, the degree of polarization was derived from the intensities of two orthogonally polarized components, with the orientation of the planes of polarization being held invariant
throughout the twilight (one plane coincided with the mean position of the
solar vertical for the twilight). Our measurements, however, employed three
polarizing prisms rotated by 60° angles from one another. This arrangement
made it possible to determine both the direction of polarization and its actual
value. Figure 35 shows some of the resulting curves for p as a function of the
effective height h of the scattering layer; for comparison, we also give the
relation p(h) that would be expected for single molecular scattering.

A comparison of the amount of depolarization, both for an effective
scattering height h ≈ 80 km and for h ≈ 270-300 km, with the critical frequencies of the corresponding ionospheric layers revealed a comparatively
weak and somewhat uncertain correlation: as the critical frequency increased,
the depth of the minimum increased, but the minimum degree of polarization
decreased (Fig. 36) [96, 169-171]. In §§6 and 7 of Chapter IV, we shall discuss some possible explanations for the correlation.

Returning to Figs. 34 and 35, we wish to point out two other circumstances. First, the degree of polarization during twilight is much smaller than

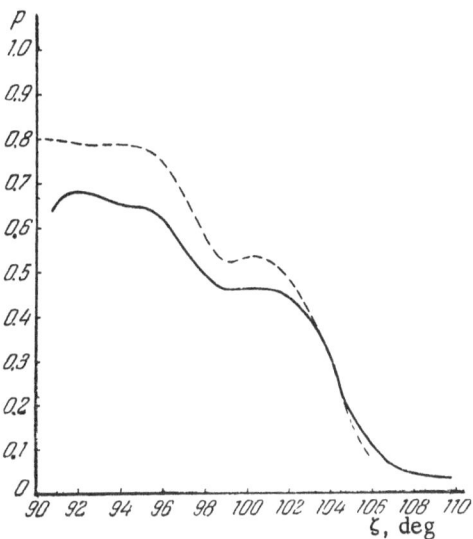

Fig. 37. Examples of the mean degree of polarization of the light received from the zenith of
the twilight sky as a function of the sun's zenith
distance.

would have been expected if only single molecular scattering were operative.
Secondly, random short-period variations are observed in the degree of polarization, covering very small ranges in zenith distance or height of the shadow.
Both of these properties will be treated later, and we shall find that the latter
is purely a time variation. We finally note that the polarization increases
for $\zeta \gtrsim 107°$.

More recently, there have been numerous measurements of the polarization of the light from the zenith region of the twilight sky, and unlike the
earlier observations, these have usually employed color filters [4, 19, 110, 113,
121-123] or spectrographs [172]. In their main aspects the later observations
have confirmed the general behavior established by Khvostikov and his colleagues. Figure 37 shows two samples of the relation between the polarization p at the zenith and ζ. The broken curve was derived visually by Robley
[172], the solid curve photoelectrically (in the spectral range 4000-5700 A)
by Dave and Ramanathan [110], and in both cases a number of days have been
averaged together so that the random (short-period included) variations have
been removed. Averages taken over other groups of measurements or other
spectral intervals yield different curves [110].

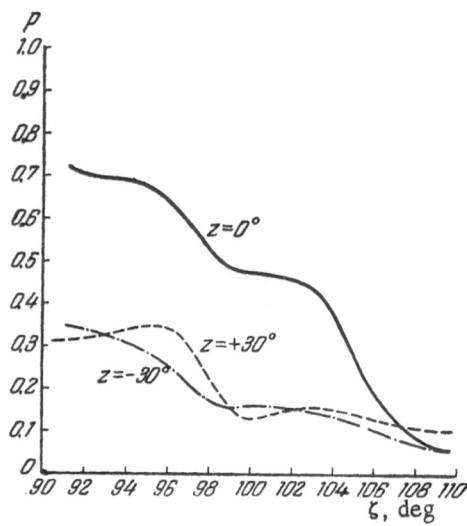

Fig. 38.   Mean ζ-dependence of p for selected
lines of sight.

Measurements of the polarization of the light from the twilight sky at
points other than the zenith have, so far as we know, been carried out only by
Megrelishvili [121] and Dave and Ramanathan [110]. In Fig. 38 we compare
the averages over several days of the ζ-dependence of p for three points on
the solar meridian at zenith distances $z = 0°$, $\pm 30°$ (the minus sign refers to
the antisolar vertical). The measurements were made in the spectral range
4350-4850 A. In practice the intensities of two orthogonally polarized components were measured, one of which ($I_{\parallel}$) corresponded to the vibrations of the
electric vector of the light wave in the plane of the solar meridian. Figure 39
shows the relation between the intensities of the two components (averaged
over several days of observation) and the zenith distance of the sun. We see
that the weaker component $I_{\parallel}$ reaches its limiting (night) value significantly
earlier than the stronger component $I_{\perp}$, for which the electric vector vibrates
normally to the scattering plane. The tail of the $I_{\parallel}(\zeta)$ curve begins at $\zeta \approx$
$103°.5$, while the tail of the $I_{\perp}(\zeta)$ curve does not set in until $\zeta \approx 105°$. This
gives us some insight into the meaning of the drop in polarization between
$\zeta = 103-104°$ and $107-108°$, corresponding to effective scattering levels from
160-200 km to 300-350 km (see Figs. 34, 35, 37). The drop in the function
$p(\zeta)$ will be discussed further in §§6 and 7 of Chapter IV.

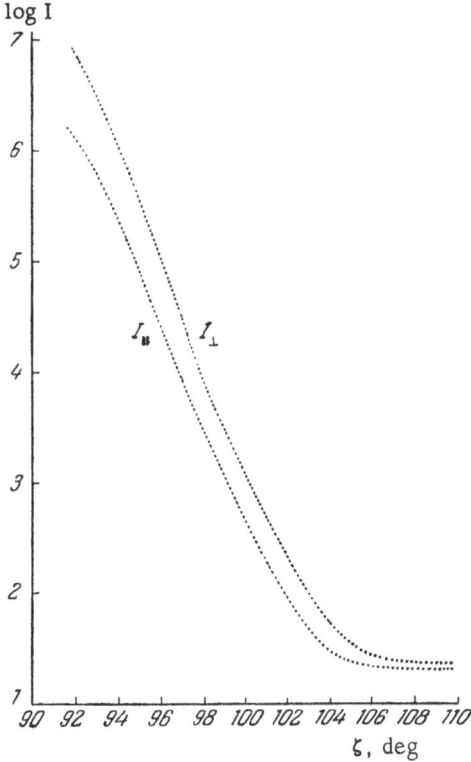

Fig. 39. Mean ζ-dependence of the logarithmic intensities of two orthogonally polarized components of the light received from the zenith of the twilight sky.

We turn now to the dependence of the $p(\zeta)$ relation upon the wavelength of the light. Figure 40 presents some $p(\zeta)$ curves as measured spectrographically on a single day by Robley [172]. For comparison, Fig. 41 shows analogous results obtained by Divari [113] with color filters (the mean for two days).

We must first call attention to the differing behavior of the curves for different wavelengths. The reasons for these variations will again be discussed in §§6 and 7 of Chapter IV. Here we need only say that the effect invalidates any detailed analysis based only on measurements made over wide spectral ranges. It may also be noted that the polarization minimum at $\zeta \approx 97$-$101°$ and its decline for $\zeta > 102$-$103°$ appear to a varying degree at the different wavelengths, and they experience some marked displacements along the axis

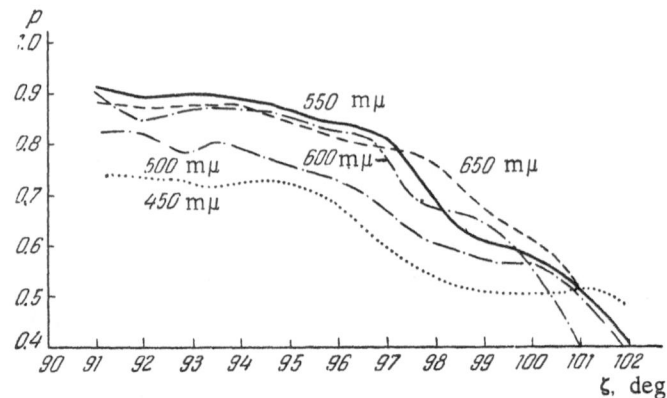

Fig. 40. Polarization curves p(ζ) for selected wavelengths (photographic photometry of twilight-sky spectra).

of solar zenith distances. Moreover, for $ζ \gtrsim 100°$ the light from the sky is polarized considerably more strongly in the short-wave region of the spectrum
than in the yellow—red.

Divari [113] has found a correlation between the value of the atmospheric transparency and the degree of polarization at nearby wavelengths for
$ζ \approx 99$-$108°$, whereas at other ζ the correlation deteriorates. The statistics
covered only 7-8 days, however, and the transparency was determined with
different filters at another time (during the day), rendering the correlation
analysis less convincing.

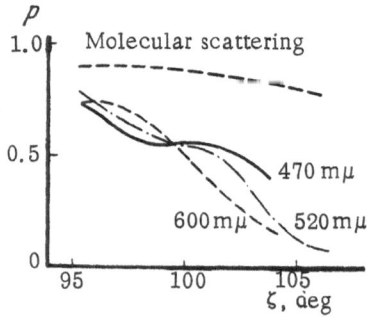

Fig. 41. Polarization curves p(ζ) for selected wavelength bands (photoelectric measurements with filters).

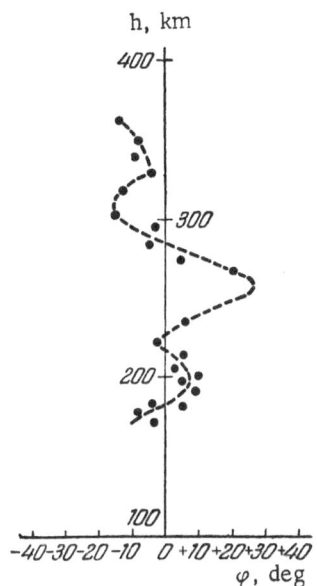

Fig. 42. Example of the relation between the orientation of the plane of polarization of the light from the twilight sky and the effective height of the scattering layer, for large depressions of the sun below the horizon.

Almost all the measurements of the degree of polarization of the light coming from the zenith region of the twilight sky have been performed, as we have seen, by determining the intensity of the light traversing an analyzer (a polaroid or a polarizing prism) oriented in two mutually orthogonal positions. In some cases the instrument has been guided to follow the sun's azimuthal motion, so that one position of the analyzer would coincide with the solar meridian. In other cases the instrument has been held in a constant position, with one orientation of the analyzer coinciding with the solar meridian at some mean epoch during the twilight. In either event it has been assumed a priori that the scattered light from the sky is always polarized strictly in the scattering plane, or strictly perpendicular to it. It has therefore become necessary to make a special check on this assumption, especially since cases are known where it definitely is false. We have in fact seen in § 1 that such cases are observed for the polarized light from the daytime sky. Furthermore, in observations of polarization near the sun during the total solar eclipses of 1932-1934 (the conditions during total eclipses are similar in many respects to twilight conditions), Cohn [173] found a sharp increase in the degree of polarization 8° from the sun during eclipse, corresponding to a deviation of several degrees in the plane of polarization from its orientation outside eclipse. Polarization observations of the solar corona during the eclipse of September 21, 1941, have also revealed sharp deflections, by tens of degrees, of the polarization plane from the radial direction, and in different spectral regions the deflections differed in value and even in sign [174].

The first measurements of the position of the plane of polarization for the light coming from the zenith region of the twilight sky were performed by Rozenberg [95, 96, 171]. The findings were unexpected. The position of the

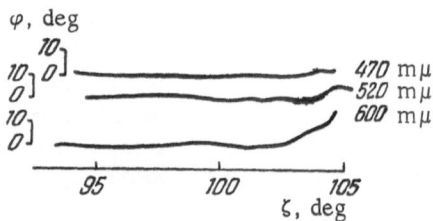

Fig. 43. Position of the plane of polarization of the light from the twilight sky as a
function of ζ at small solar depressions $\xi =$
$\zeta - 90°$, for selected wavelength regions.

plane of polarization turned out to be subject to substantial variations. Beginning with $\zeta \geqslant 104°$, its deviation from the solar meridian can be very considerable, 20-30° or even more. Figure 42 illustrates a sample relation between the deflection angle φ of the plane of polarization from the solar vertical and the effective scattering level, for a single day of observation.

More recently Divari [113] has conducted similar measurements, but
now with filters. His results appear in Fig. 43. We see that if any deviations
of the plane of polarization from the solar vertical exist for $\zeta \leqslant 100°$, they are
small and are random in character. This conclusion is confirmed by measurements by Rozenberg and Turikova [19, 122] with blue (I) and green (II) filters

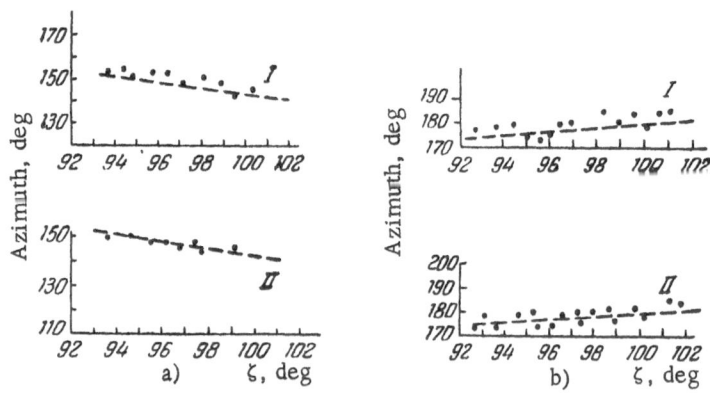

Fig. 44. Azimuth of the plane of polarization as a function of the
sun's zenith distance. Broken lines, ζ-dependence of the sun's azimuth. a) Morning; b) evening.

at relatively small ζ, as shown in Fig. 44, where we present the ζ-dependence of the solar azimuth (broken lines) and the azimuth of the plane of polarization (dots) for the two spectral regions according to observations on two separate days. When $\zeta \gtrsim 101°$, on the other hand, the deviation of the plane of polarization begins to increase, until for solar zenith distances $\zeta = 103\text{-}104°$ it may reach the considerable value of $\approx 10°$ or more, depending on the wavelength (Fig. 43). Unfortunately, Divari's measurements extended only to $\zeta = 105°$, corresponding to an effective scattering height of ≈ 220 km. Thus the strong deflection of the polarization plane from the scattering plane that the author has found at large ζ still awaits verification from further measurements.

One should recognize that measurements of the deviation of the polarization plane are of low accuracy at large ζ because of the weak degree of polarization (see Figs. 35, 37). For this reason one should use with caution the values of φ derived in [95, 96]. The possiblity also remains that the strong deflections of the polarization plane might arise from certain special meteorological conditions (such as the presence of faint noctilucent clouds, or a strong horizontal inhomogeneity in the atmosphere), or from some effect of the night-sky glow. One should recognize further that in the event of horizontal inhomogeneity, secondary-scattering effects might possibly produce a deviation in the plane of polarization, but a much smaller one than is shown in Figs. 42 and 43 [96, 235]. It is also questionable whether any effects of the night-sky glow would explain it satisfactorily [259]. Yet the polarization plane of light scattered by noctilucent clouds might deviate strongly from the scattering plane (see [19]). At any rate, a repetition of the measurements would be desirable here, with filters or spectroscopically if possible. Meanwhile we should view with reserve any quantitative data on the degree of polarization of the light from the twilight sky for $\zeta \gtrsim 104°$, if they have not been corrected for possible rotations of the plane of polarization.

CHAPTER II

STRUCTURE AND OPTICS OF
THE ATMOSPHERE

§ 1. Some Structural Properties of the Atmosphere

Among the latest achievements of science has been an increased understanding of the earth's atmosphere — its structure and the processes operating within it. Still far from complete, these concepts have been worked out just in the past quarter century (see, for example, [175]). Scientists of the elder generation well recall the time when the atmosphere above 10-12 km was considered a dreary wasteland where there was nothing but crystal clear air at a temperature falling off smoothly with height, and a good many observations that did not harmonize with this scheme were thought to be erroneous. But modern research, at first performed indirectly through observation from the ground, and subsequently by direct study from aircraft, stratospheric balloons, rockets, and most recently from artificial earth satellites, has completely overturned the early ideas. Today the pace with which data are accumulating on the physical conditions prevailing at various heights is rapid without precedent, and this applies not only to numerical findings but to many hypotheses of a qualitative character which only recently seemed questionable. Nevertheless, we already have a fairly confident understanding of the atmosphere in basic outline. We must pause to review this topic here, for otherwise no analysis of the twilight phenomenon would be meaningful.

We can provide here only a very brief survey, however, and only of those aspects that are important for the theory of twilight and the twilight method of upper-atmosphere research.

The results of rocket experiments, and observations of polar aurorae, the night-sky glow, meteor trains, and other phenomena have established that the molecular composition of the air remains unchanged to a height of at least 90 km [176]. It is only at this level that we begin to detect faint traces of the separation of gases by diffusion, and if diffusive equilibrium is ever achieved it is not below about 200 km. For this reason the molecular composition of the atmospheric layers near the ground is pertinent (Table 1) [177].

TABLE 1. Molecular Composition of the Air at the Ground

Substance	Density relative to air	Percentage content by volume	Remarks
N_2	0.967	78.09	99.97% by volume combined
O_2	1.105	20.95	
A	1.379	0.93	
H_2O	0.621	10^{-3} to 1	Highly variable
CO_2	1.529	$2.6 \cdot 10^{-2}$	Slightly variable next to soil cover
Ne	0.695	$1.8 \cdot 10^{-3}$	
He	0.138	$5.24 \cdot 10^{-4}$	
CH_4	0.558	$1.7 \cdot 10^{-4}$	
Kr	2.868	$\sim 10^{-4}$	
H_2	0.070	$(5 \cdot 10^{-5})$	Presence doubtful
NO_2	1.529	$3.5 \cdot 10^{-5}$	Variable next to soil cover
CO	0.967	$2 \cdot 10^{-5}$	Variable
Xe	4.526	$8 \cdot 10^{-6}$	
O_3	1.624	10^{-6}	Highly variable
Rn	7.68	$6 \cdot 10^{-18}$	Radioactive, 4-day halflife

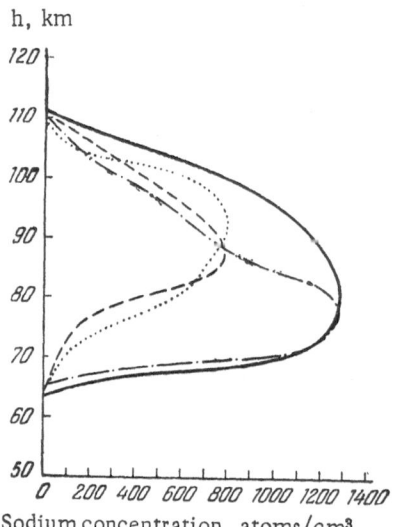

h, km

Sodium concentration, atoms/cm³

Fig. 45. Height distribution of atomic sodium on several representative days.

Beginning at about 60 km one encounters noticeable ionization of the air, increasing steadily with height. Atmospheric compounds and radicals that are unstable near the surface appear simultaneously with the ionization. Somewhere near 90 km a partial dissociation of oxygen sets in, the degree increasing with height. Oxygen appears to be almost entirely in the atomic state above 150 km [178]. Nitrogen, on the other hand, remains essentially undissociated to very great heights, at least up to about 800 km, although traces of dissociation have been detected at heights as low as 150-200 km [178, 179].

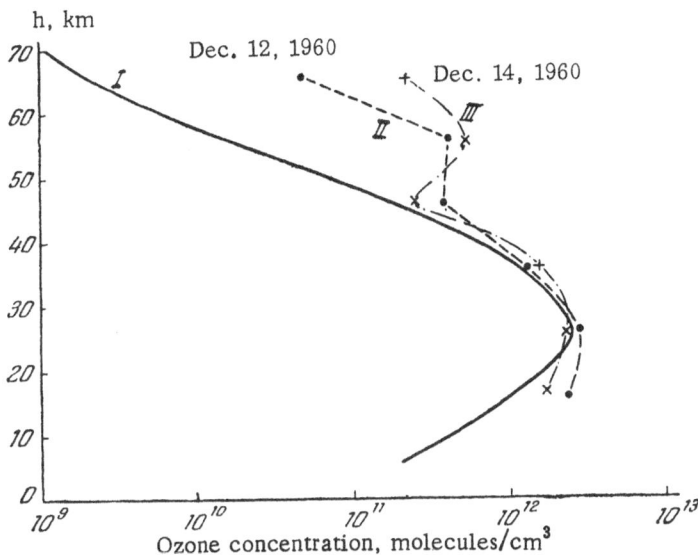

Fig. 46. Rocket determinations of the mean distribution of ozone with height (I), and the distribution obtained on two days by measuring the brightness of an artificial satellite (II, III).

NO ions occur in relatively great abundance in the height range 90-200 km [180], as well as the OH radicals whose emission constitutes one of the chief components of the night-sky glow.

Twilight measurements have revealed that atomic sodium is present at heights of about 50-120 km in a well-defined layer with maximum concentration at about 70-90 km; both the sodium abundance and the height of the layer are highly variable and undergo, in particular, strong seasonal variations [160, 161]. Figure 45 shows sample distributions of sodium with height on different days. The total sodium content in a 1-cm^2 column ranges between $2 \cdot 10^9$ and $2 \cdot 10^{10}$ atoms. There are serious grounds for believing that the sodium layer has a cloudlike structure.

The most important constituent of air from the standpoint of optical effects is ozone. It too occurs as a well-defined but much diffused layer, with maximum concentration at a height of about 25 km. The total ozone content of the earth's atmosphere is approximately 0.3 atmo-cm, * varying between

---

*It is customary to express the abundance of an atmospheric gas as the thickness of an equivalent (or reduced) layer, the layer that the gas would occupy if it were in a pure form and at the standard conditions of 0°C temperature and 760 mm Hg pressure.

about 0.2 and 0.4 atmo-cm with definite latitude and seasonal terms and also a dependence on meteorological conditions [177, 181-183]. The height distribution of ozone has been the subject of long and persistent study, but only with rocket measurements have accurate, reliable results been attained. In Fig. 46 we show the mean height distribution of ozone as established from analysis of rocket data on the absorption of ozone by direct sunlight at various heights. The figure also includes the results of two measurements conducted recently by Venkateswaran et al. [184], who observed the brightness of an artificial satellite of the Echo type during its entry into the earth's shadow. Insufficient study has so far been devoted to the behavior of the very strong fluctuations in the distribution of ozone with height.

Still more variable is the total content and height distribution of water vapor. The processes of condensation, evaporation, and sublimation of water that unceasingly take place in the atmosphere render this constituent of air particularly unstable and are responsible for its strong dependence on temperature fluctuations, and on horizontal and vertical displacements of air masses. Most of the water-vapor mass is concentrated in the bottom layer of the atmosphere, the troposphere. In this layer, whose thickness ranges from 7-9 km in the polar regions to 15-18 km at the equator, the specific humidity of the air on the average hardly varies at all with height, experiencing only the usual ground-layer fluctuations of one order of magnitude because of changes in the weather. But the absolute humidity or water-vapor concentration drops rapidly with height above the earth's surface, following an approximate barometric law.

The situation changes sharply, however, beginning with the tropopause — the upper boundary of the troposphere, separating it from the stratosphere. The drop in absolute humidity with height now considerably outruns the barometric law because of the rapidly falling relative humidity of the air: at a height of 3-4 km above the tropopause it has dropped to 2-5%, corresponding to a specific humidity of $\approx 10^{-3}$ g/kg. Only fragmentary data of relatively low reliability are available on the humidity at higher levels [285]. There is every reason to believe that in the stratosphere and mesosphere the humidity undergoes very substantial variations, greatly exceeding its variability in the troposphere. In particular, above 15-16 km the specific humidity sometimes starts to rise markedly with height, reaching values of $\approx 10^{-1}$ g/kg at 30-35 km. On the other hand, cases are known where the water vapor is mixed quite homogeneously over the entire region above the tropopause with a specific humidity of 0.01-0.05 g/kg. Figure 47, from [285], illustrates examples of the height dependence of the specific humidity q (in g/kg) according to measurements by several different authors. The existence of nacreous and noctilucent clouds (see below) may well indicate an increased (probably sporadically) humidity near the 25- and 80-km levels.

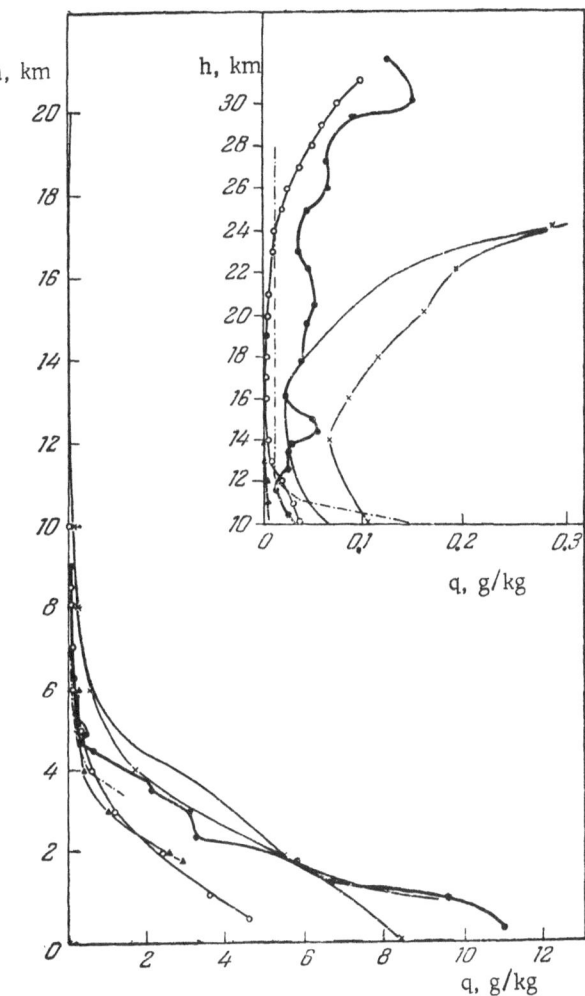

Fig. 47. Height distribution of the specific humidity according to several representative measurements.

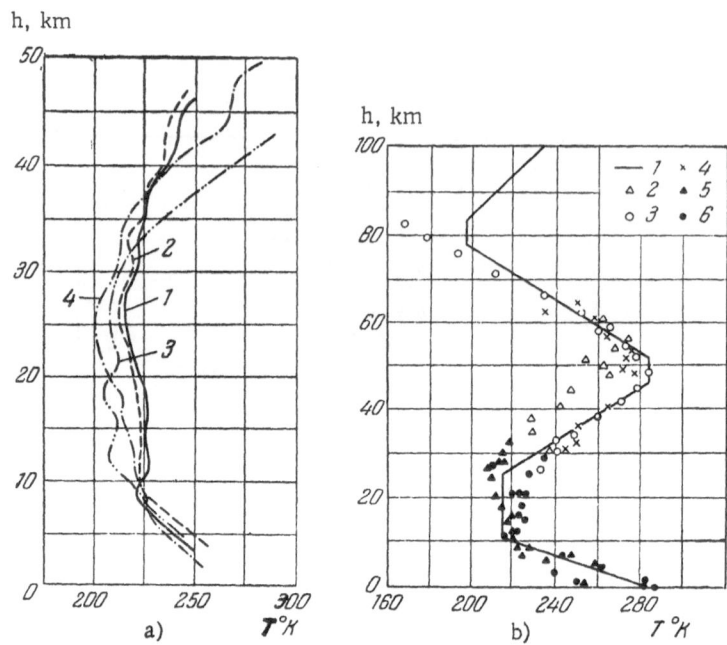

Fig. 48. Sample height distributions of the temperature under various conditions. a) Rocket flights: 1) October, 1957; 2) November, 1957; 3) December, 1957; 4) January, 1958. b) Acoustic survey: 1) NASA standard mean atmosphere; 2) winter; 3, 4) summer; 5, 6) radiosonde observations.

For carbon dioxide, there is every likelihood that the specific concentration remains practically constant up to a height of at least 90 km.

There have been numerous investigations, by a wide variety of techniques, of the distribution of air pressure, density, and temperature with height in the atmosphere (see, for example,[185]). Ground observations have been made of meteors and meteor trains, aurorae, the night-sky glow, twilight and ionospheric phenomena; the structure of the ozone absorption bands has been studied; searchlight beams and acoustic methods have probed the atmosphere. Theoretical considerations of a general character have also been invoked. With the advent of rockets and satellites, new opportunities for experiment have at once been exploited, and we are again finding a very wide range of methods in use for measuring temperatures T, densities ρ, and pressures p (see, for example,[185-187]). One should note, however, that even with the advance from ground observations to instruments carried by rockets and satellites,

T, p, and ρ are still measured by indirect methods, for the motion of a spacecraft relative to the surrounding medium introduces a series of new factors that
radically affect the conduct of the experiment (see, for example,[186]). Thus
we need not be surprised that the accuracy of even the latest data on the temperature, density, and pressure of the upper atmosphere is none too high, and
that the availability of rockets and satellites does not supersede the urgent
need to develop and refine techniques based on observations carried out from
the earth's surface.

A comparison of results obtained by different methods and at different
times and places inescapably demonstrates that the pressure, density, and temperature in the upper levels of the atmosphere undergo substantial fluctuations
of both a local and a global character. It has now become possible to distinguish seasonal, latitude, and even diurnal variations, such as those produced
by atmospheric tides (the analogue of ocean tides). In other words, there is no
longer any question but that a variety of meteorological processes are at work
in the upper atmosphere and certainly in the mesosphere, the height range from
30-35 to 80-90 km: we have the same right to speak of mesospheric weather
as of weather in the troposphere or stratosphere. As an illustration, Fig. 48
presents a series of temperature distributions with height, as measured during
successive months of 1957-1958 on rocket flights above Heiss Island in Franz
Josef Land in the Soviet Arctic [188], together with the results of an acoustic
probe carried out at Churchill, Manitoba, Canada [189].

Despite the large number of measurements by different methods that
have now accumulated, the existence of diverse variations in T, p, and ρ and
the so far unavoidably large uncertainties in their evaluation have made the
measurements inadequate for the reliable formulation of a standard mean atmosphere, or for identifying the scope and character of the latitude, seasonal,
and other fluctuations. As a result we now find standard atmospheres being
quoted in several variants, differing slightly from one another in a number of
details, but all agreeing in rough outline for the basic properties that have become quite well established (see, for example,[185]). We wish to present here
just one of these (Table 2), an atmosphere constructed by Miller [190] from a
thorough analysis of the entire complex of data then available on the upper
layers of the atmosphere. To illustrate the extent of the mutual differences
among models of this type and the real atmosphere, we present in Fig. 49 a
comparison of the height distributions of temperature and air pressure as given
by NASA, by the Rocket Panel, and by Kuiper in the United States, with the
results of rocket flights carried out by the Institute of Applied Geophysics in
the Soviet Union under the direction of Mikhnevich [191]. Similarly, Fig. 50
compares data on the air density in the upper atmosphere as obtained by various methods. Figure 51 shows the height dependence of the concentration of

TABLE 2. Particle Concentration n, Density ρ, Pressure p, Temperature T, and Mean Molecular Weight M as a Function of Height in the Atmosphere (Miller)

Height h, km	Total n, cm^{-3}	Total ρ, g/cm^3	Pressure p, dyn/cm^2	Temperature T, $^\circ$K	Molecular weight M, g/mole
0	$2.55 \cdot 10^{19}$	$1.23 \cdot 10^{-3}$	$1.013 \cdot 10^6$	288.16	
10	$8.60 \cdot 10^{18}$	$4.14 \cdot 10^{-4}$	$2.649 \cdot 10^5$	223.26	
20	$1.85 \cdot 10^{18}$	$8.89 \cdot 10^{-5}$	$5.528 \cdot 10^4$	216.66	
30	$3.71 \cdot 10^{17}$	$1.79 \cdot 10^{-5}$	$1.185 \cdot 10^4$	231.24	28.966,
40	$8.32 \cdot 10^{16}$	$4.00 \cdot 10^{-6}$	$2.997 \cdot 10^3$	260.91	constant
50	$2.25 \cdot 10^{16}$	$1.08 \cdot 10^{-6}$	$8.784 \cdot 10^2$	282.66	to 90-km
60	$7.26 \cdot 10^{15}$	$3.49 \cdot 10^{-7}$	$2.581 \cdot 10^2$	257.55	level
70	$2.09 \cdot 10^{15}$	$1.00 \cdot 10^{-7}$	$6.320 \cdot 10^1$	219.33	
80	$4.50 \cdot 10^{14}$	$2.17 \cdot 10^{-8}$	$1.223 \cdot 10^1$	196.86	
90	$8.31 \cdot 10^{13}$	$4.00 \cdot 10^{-9}$	$2.257 \cdot 10^0$	196.86	28.966
100	$1.65 \cdot 10^{13}$	$7.12 \cdot 10^{-10}$	$4.624 \cdot 10^{-1}$	203.06	25.975
110	$3.83 \cdot 10^{12}$	$1.59 \cdot 10^{-10}$	$1.187 \cdot 10^{-1}$	224.28	24.957
120	$1.05 \cdot 10^{12}$	$4.28 \cdot 10^{-11}$	$3.609 \cdot 10^{-2}$	250.09	24.632
130	$3.20 \cdot 10^{11}$	$1.30 \cdot 10^{-11}$	$1.260 \cdot 10^{-2}$	285.13	24.511
140	$1.06 \cdot 10^{11}$	$4.30 \cdot 10^{-12}$	$5.343 \cdot 10^{-3}$	364.19	24.378
150	$4.42 \cdot 10^{10}$	$1.80 \cdot 10^{-12}$	$2.697 \cdot 10^{-3}$	441.79	24.225
160	$2.14 \cdot 10^{10}$	$8.56 \cdot 10^{-13}$	$1.531 \cdot 10^{-3}$	517.94	24.062
170	$1.15 \cdot 10^{10}$	$4.57 \cdot 10^{-13}$	$9.429 \cdot 10^{-4}$	593.87	23.942
180	$6.69 \cdot 10^9$	$2.65 \cdot 10^{-13}$	$6.178 \cdot 10^{-4}$	669.37	23.842
190	$4.28 \cdot 10^9$	$1.65 \cdot 10^{-13}$	$4.110 \cdot 10^{-4}$	695.89	23.229
200	$2.85 \cdot 10^9$	$1.07 \cdot 10^{-13}$	$2.831 \cdot 10^{-4}$	720.97	22.636
210	$1.95 \cdot 10^9$	$7.16 \cdot 10^{-14}$	$2.007 \cdot 10^{-4}$	744.29	22.072
220	$1.38 \cdot 10^9$	$4.94 \cdot 10^{-14}$	$1.461 \cdot 10^{-4}$	766.76	21.537
230	$1.00 \cdot 10^9$	$3.50 \cdot 10^{-14}$	$1.089 \cdot 10^{-4}$	788.33	21.038
240	$7.41 \cdot 10^8$	$2.53 \cdot 10^{-14}$	$8.281 \cdot 10^{-5}$	809.16	20.574
250	$5.60 \cdot 10^8$	$1.87 \cdot 10^{-14}$	$6.417 \cdot 10^{-5}$	829.75	20.146
260	$4.31 \cdot 10^8$	$1.41 \cdot 10^{-14}$	$5.055 \cdot 10^{-5}$	850.03	19.753
270	$3.37 \cdot 10^8$	$1.08 \cdot 10^{-14}$	$4.061 \cdot 10^{-5}$	874.20	19.395
280	$2.67 \cdot 10^8$	$8.44 \cdot 10^{-15}$	$3.278 \cdot 10^{-5}$	890.87	19.067
290	$2.14 \cdot 10^8$	$6.66 \cdot 10^{-15}$	$2.698 \cdot 10^{-5}$	911.37	18.769
300	$1.73 \cdot 10^8$	$5.33 \cdot 10^{-15}$	$2.230 \cdot 10^{-5}$	931.92	18.498
320	$1.17 \cdot 10^8$	$3.51 \cdot 10^{-15}$	$1.576 \cdot 10^{-5}$	974.1	18.031
340	$8.22 \cdot 10^7$	$2.41 \cdot 10^{-15}$	$1.154 \cdot 10^{-5}$	1017.0	17.644
360	$5.93 \cdot 10^7$	$1.71 \cdot 10^{-15}$	$8.68 \cdot 10^{-6}$	1061	17.324
380	$4.38 \cdot 10^7$	$1.24 \cdot 10^{-15}$	$6.68 \cdot 10^{-6}$	1106	17.059
400	$3.31 \cdot 10^7$	$9.25 \cdot 10^{-16}$	$5.26 \cdot 10^{-6}$	1151	16.838
420	$2.54 \cdot 10^7$	$7.03 \cdot 10^{-16}$	$4.20 \cdot 10^{-6}$	1198	16.653
440	$1.88 \cdot 10^7$	$5.44 \cdot 10^{-16}$	$3.41 \cdot 10^{-6}$	1244	16.496
460	$1.57 \cdot 10^7$	$4.27 \cdot 10^{-16}$	$2.80 \cdot 10^{-6}$	1291	16.364
480	$1.26 \cdot 10^7$	$3.40 \cdot 10^{-16}$	$2.33 \cdot 10^{-6}$	1340	16.265
500	$1.02 \cdot 10^7$	$2.74 \cdot 10^{-16}$	$1.96 \cdot 10^{-6}$	1387	16.155
520	$8.37 \cdot 10^6$	$2.23 \cdot 10^{-16}$	$1.66 \cdot 10^{-6}$	1435	16.070
540	$6.91 \cdot 10^6$	$1.84 \cdot 10^{-16}$	$1.41 \cdot 10^{-6}$	1483	15.988
560	$5.81 \cdot 10^6$	$1.52 \cdot 10^{-16}$	$1.22 \cdot 10^{-6}$	1519	15.809
580	$4.88 \cdot 10^6$	$1.27 \cdot 10^{-16}$	$1.04 \cdot 10^{-6}$	1565	15.737
600	$4.13 \cdot 10^6$	$1.07 \cdot 10^{-16}$	$9.18 \cdot 10^{-7}$	1611	15.670

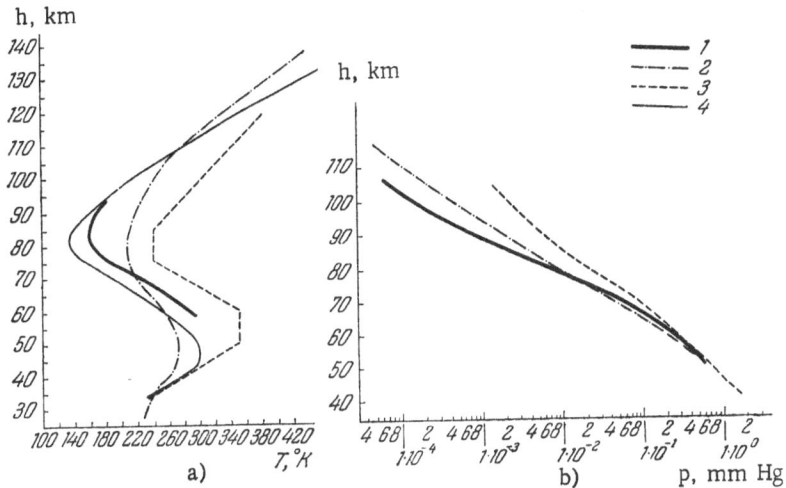

Fig. 49. a) Temperature, and b) pressure as a function of height according to measurements by the Institute of Applied Geophysics and three unified models. 1) Inst. Appl. Geophys.; 2) Rocket Panel; 3) NASA; 4) Kuiper.

neutral molecules according to the results of several experiments. Finally, in Fig. 52 we reproduce the mean temperature, density, and pressure variation at low and intermediate heights according to findings obtained from rocket flights in the temperate zone [186]. Note especially the temperature minima in the stratospheric region (10-35 km) and at about 75-80 km, the temperature maximum near the 50-km level, which coincides approximately with the maximum specific concentration (relative to air) of ozone, and the characteristic bends in the pressure and density curves in the 15-60 km range, a departure from the exponential decline. Incidentally, the behavior of the temperature at a height of about 75-80 km is still not well understood. It seems to be subject to considerable variation: some data indicate that at times it may drop to 150-160°K, or even lower [192].

The data on the variation of the temperature with height in the atmosphere further allow one to compute the height dependence of the pressure E of saturated water vapor. Figure 53 shows the corresponding results obtained by Khvostikov [192]. We shall find it important to note that E > p in the height range from 35 to 65 km, condensation being impossible under these conditions. Above and below this layer condensation is possible if there is a sufficient concentration of water vapor.

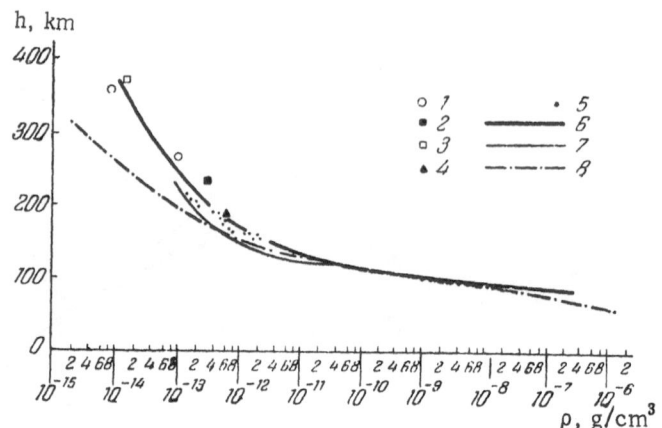

Fig. 50. Density of the upper atmosphere according to rocket and satellite measurements. Satellites: 1) third Soviet (manometers); 2) first, second, and third Soviet (deceleration); 3, 4) Sterne's data for 1958α and 1958γ (deceleration). Rockets: 5) USSR, February, 1958; 6) USSR, containers; 7) US, Viking 7; 8) NASA.

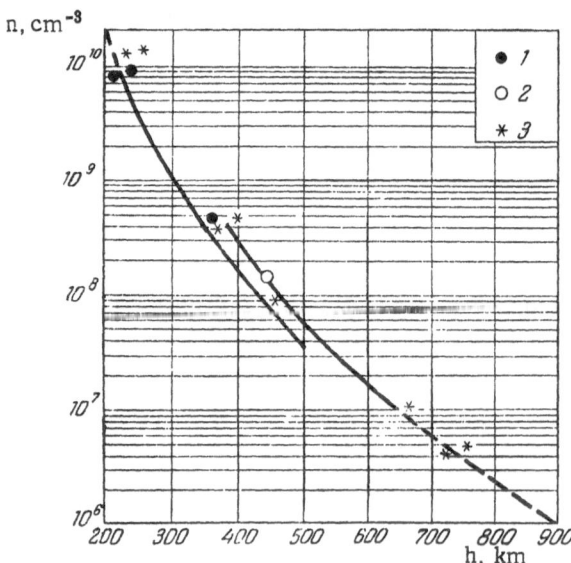

Fig. 51. Neutral-molecule concentration as a function of height according to the results of various methods. 1) Deceleration of Soviet satellites; 2) diffusion of a so-dium-vapor cloud; 3) foreign data on deceleration of Soviet satellites.

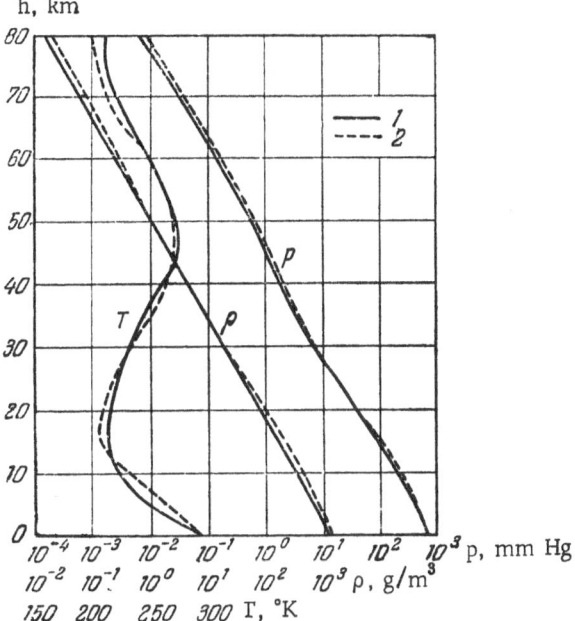

h, km

Fig. 52. Mean distribution of T, p, and ρ with height in the temperate zone. 1) Rocket measurements, USSR; 2) NASA.

So far we have considered exclusively the gaseous component of the atmosphere. But an aerosol component of very distinct origin is also present in the atmosphere at all times and, as we now recognize, at all heights. Some of the aerosol constituents are moisture condensation products, water droplets and ice crystals, occasionally forming such prominent structures as clouds and fog. Other constituents include mineral and volcanic dust, products of human activity such as soot and other combustion products, microorganisms, spores, and pollen from plants, meteoritic fragments decelerated by the atmosphere, and marine salts.

From the optical standpoint, the aerosols form the most active component of the atmosphere. Quite apart from the overwhelming majority of the phenomena of atmospheric optics (rainbows, aureoles, haloes, the glory) for which aerosols are responsible, it is this component which essentially governs both the radiant and the energy conditions of the atmosphere, certainly in the lower layers and perhaps in the mesosphere too. Not only is the concentration of the particles important in this regard, but also their size, shape, and char-

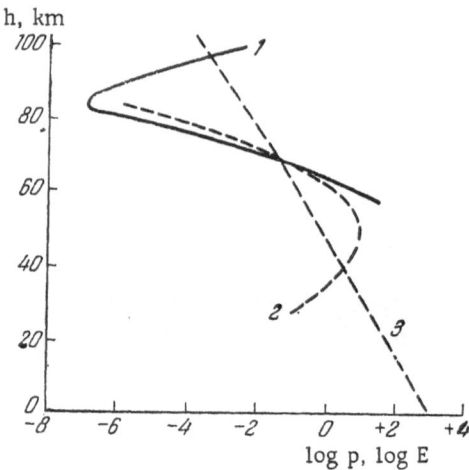

Fig. 53. Air pressure and saturated water-vapor pressure as a function of height. 1) Water-vapor pressure, Mikhnevich's data; 2) water-vapor pressure, rocket measurements above Churchill, Manitoba; 3) air pressure.

acter. The aerosol component is the most variable part of the atmosphere, as can well be imagined by noting, for instance, the great variation that is possible in visibility conditions, or simply the appearance of the daytime sky.

Yet the aerosols still remain the least studied component of the free air. The reasons for this situation lie chiefly in the diverse nature and character of the aerosol particles — particularly the differing ways that they are caught by measuring devices, in the dependence of the parameters of the aerosol component (such as the distribution of particles with respect to size) on meteorological circumstances, and finally in the extremely complicated relations between the microstructure of an aerosol particle and its optical properties. This last factor is especially important, for optical methods form one of the most refined and effective means for studying aerosols; but we are still far from being able to interpret all the information that we can obtain in this way. Since we shall only be interested in aerosols in connection with the twilight-sky problem as one of the factors affecting the progress of twilight, and as a subject of investigation through twilight probing of the atmosphere, we shall limit the present discussion to those properties relevant to the aerosol fractions most efficient in the optical sense, and to the relatively stable, clear weather in which one can conduct twilight measurements.

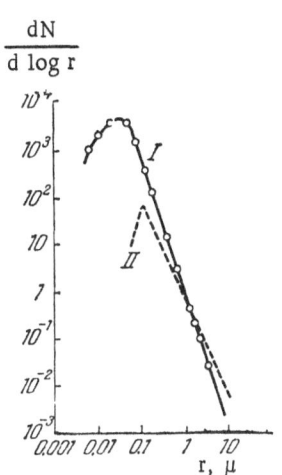

Fig. 54. Numerical distribution of natural aerosols with respect to particle size. I) In ground layer; II) in stratosphere.

In rough outline, aerosol particles may be classified according to their size, taking as a parameter the effective radius r, which is defined as the radius of a spherical particle equivalent in some particular respect — in volume, geometric cross section, scattering coefficient, rate of precipitation, or the like. Junge [193] has suggested dividing the entire spectrum of atmospheric aerosol particles into three groups. The first group consists of particles of size $r < 0.1$ μ (the so-called Aitken condensation nuclei), which are so small that they are still not very different in their optical properties from molecules. They may be studied most effectively by means of expansion chambers. Research in the past few years has shown that these particles hardly ever serve as condensation nuclei in the real atmosphere, since the air is never supersaturated with water vapor to the degree that would be necessary.

The second group consists of "large particles" from 0.1 μ to 1 μ, the most active particles optically. They are notable for a very strong dependence of their optical properties on their size and shape [17, 18].

Finally, the third group, with $r > 1$ μ, comprises the "giant particles" that may be studied effectively with all manner of trapping devices, as well as by certain optical effects (aureoles, rainbows). This group consists mainly of cloud and fog particles and the principal fraction of mineral dust.

Both the large and the giant particles act as true condensation nuclei in meteorological processes. Hygroscopic particles will experience distinct changes in their optical properties even at very low relative humidity, while for nonhygroscopic particles the condensation processes will only become perceptible at a relative humidity in excess of about 70%.

Reliable statistics on the size distribution of the large and giant particles have only been obtained for the lowest layers of the atmosphere, from the ground level up to a height of about 3 km. Junge [193, 194] has found that at the surface the distribution is described quite well by the relation

$$\frac{dN}{d\log r} = \frac{c}{r^3} \, ,$$

(II. 1, 1)

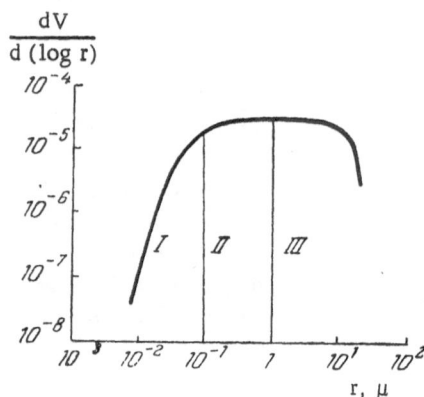

Fig. 55. Volume distribution of natural aerosols in the ground layer with respect to particle size. I) Nuclei; II) large particles; III) giant particles. Unit of relative volume, cm^3 of aerosols per m^3 of air.

where N(r) is the concentration of particles of effective radius greater than r and c is a constant. The solid curve in Fig. 54 represents a sample distribution as determined by Junge from measurements in the Alps. The particle concentration increases rapidly with decreasing particle size, and has a maximum in the Aitken-nucleus region at a point where the optical properties of the particles have become practically indistinguishable from the properties of molecules. An entirely different behavior is found if one considers the distribution of the particles with respect to the volume they occupy (Fig. 55). The total particle concentration by volume is almost invariant over a range in size from 0.1 μ to 10 μ, and falls off rapidly at both ends of this interval. The data of Figs. 54 and 55 and Eq. (II.1, 1) refer to dry aerosols under conditions of low humidity.

Aerosol particles are transported from the ground layer up into the free atmosphere by convective flow, a process accompanied by simultaneous gravitational settling. Hence not only the total concentration but also the spectrum of the particles varies with height. Laktionov [195] has found that at heights below 3 km the concentration of particles larger in size than r is well described by the relation

$$N(r) = \frac{A}{r^\alpha} ,$$

(II.1, 2)

where A and α are functions of height. At the ground α is approximately equal to 3, corresponding to Junge's distribution, but it fluctuates between 2.8 and 4 for the giant particles and between 2 and 3.8 for the large particles. More-over, at a height of 1000 m the value of α increases to 3.1-6.5 for the giant particles and 2.2-4.1 for the large particles. It is noteworthy that Laktionov's data yield different values for α for the large- and giant-particle groups.

Because of the scarcity of observational data it is difficult to tell whether Junge's distribution of particle types is maintained at greater heights. The only results available so far are a few measurements carried out by Junge and his colleagues [286] with impact traps carried by balloons to heights between 12 and 30 km. The data indicate that the particle distribution with size retains a power behavior at these heights, corresponding to $\alpha = 2$ (broken curve in Fig. 54). For particles larger than 1 μ, the value of α seems to exceed 2. There is no clue as to how the particles are distributed with size above 30 km, where aerosol constituents of meteoritic origin should become increasingly im-portant. In all likelihood the distribution of types found by Junge should essen-tially be valid throughout the troposphere in clear, settled, high-pressure weather, but it remains to be seen how far the distribution can be extrapo-lated into the upper regions of the mesosphere, where the convective processes apparently have an entirely different character, and where one has already be-gun to encounter a considerable representation of meteoritic aerosol.

We consider next the variation of the total aerosol content with height. Near the ground the aerosol concentration, at least for the giant particles, varies over a very wide range: from 20,000 particles/cm^3 over large industrial centers to 10-14 particles/cm^3 over forested areas, and only 1-10 particles/liter above the sea. As one rises above the earth's surface the giant-particle con-centration declines rapidly, reaching a value of 10-100 particles/liter at a height of about 1 km above dry land and 1-5 particles/liter above the sea; nevertheless, there is still a strong dependence on the character of the under-lying surface and the convective behavior of the atmosphere [195]. It is note-worthy that measurements of the atmospheric transparency also indicate a ra-pid decline in the aerosol concentration in a layer about 1 km thick at the ground, as illustrated by Fig. 56, which is due to Faraponova [196]. The fig-ure further shows that the height range from 1 to 3-4 km represents a fairly stable layer of enhanced atmospheric turbidity; the presence of this layer and its variability are confirmed by direct measurements of the aerosol concentra-tion [195].

Above 3-4 km the scattering coefficient of the air already approaches the scattering coefficient of pure air, but still exceeds the latter by a factor of 1.5-3 (in the visible region of the spectrum).

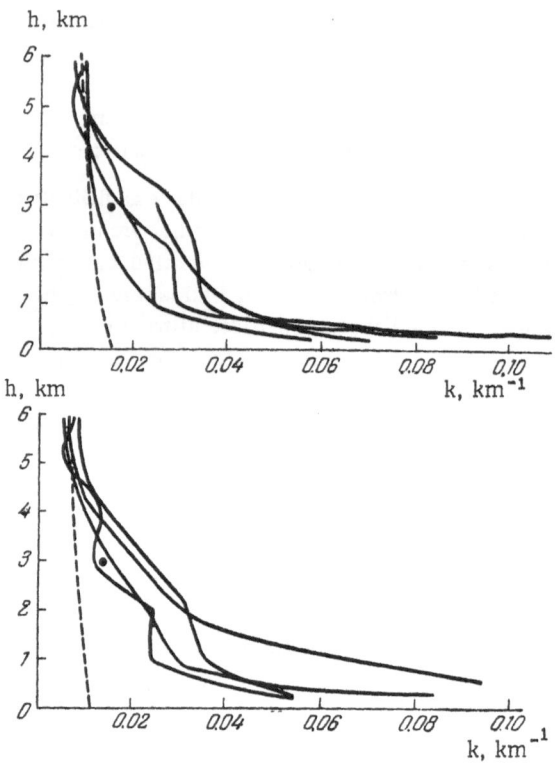

Fig. 56. Sample variations with height of the atmospheric extinction coefficient k, corresponding to the
height distribution of the aerosol content (broken
curves, a molecular atmosphere). Upper panel, $\lambda =$
420 mμ; lower panel, $\lambda = 452$ mμ.

In the stratospheric region, the data of Junge et al. [286] imply that the
aerosol concentration first increases with height, reaches a maximum at about
the 20-km level, and then decreases. The concentration of particles of size
between 0.1 and 1 μ is $\approx 10^{-1}$ cm^{-3} at a height of 20 km, but it fluctuates over
one order of magnitude.

Aside from Junge's measurements, the only data available above 6 km
pertain to the conditions for scattering of light. Among these are measurements of the intensity and polarization of the radiation scattered from a searchlight beam [4, 19], measurements of the daytime sky brightness and the scat-

tering diagram by means of stratospheric balloons [197], and twilight measure-ments. All these findings taken together demonstrate convincingly that aero-sols remain the primary factor governing the scattering conditions throughout the entire range in height, at least up to 80-90 km. In the short-wave portion of the visible spectrum, the scattering power of the aerosols nowhere seems to fall significantly below the scattering power of the gaseous component of the atmosphere, and in fact is considerably greater in the long-wave region. More-over, one finds that the aerosol scattering is highly variable both in time and with height. In particular, well-defined aerosol layers have been discovered, apparently with a cloudlike structure. These layers are most often concen-trated in the tropopause region (at 9-12 km in temperate latitudes), at a level of about 25 km (the nacreous clouds), and at the 80-km level (the noctilucent clouds). Aerosol layers have, however, sometimes been observed at other heights also, especially near the 40-45 km level [19]. It has been established that they often form within temperature inversions [19, 196, 197], a category which, in particular, includes the tropopause and the 80-km level.

In most cases the nature, concentration, and size of the particles in the aerosol layers cannot be determined reliably from the few observations avail-able. We do know that in the tropopause region the layers usually form as condensation products and become ordinary tropospheric clouds. The nacreous or mother-of-pearl clouds that sometimes appear near the 25-km level are understood (from Störmer's observations of halo phenomena and from the po-larization of scattered searchlight beams [19]) to consist of particles of size close to $r = 0.75$ μ with a concentration of several particles/cm^3. Both the op-tical data and an analysis of the synoptic situations at times when nacreous clouds have been observed suggest that these clouds likewise form by conden-sation of water vapor, and consist of supercooled water droplets.

The nature of the noctilucent clouds is still a continuing subject of de-bate. On the one hand, Khvostikov [192] offers some convincing evidence that they consist of ice crystals. But Ludlam [198], for example, suggests that they are agglomerations of dust grains, mainly of meteoritic origin. In any event the size of the particles in the noctilucent clouds is definitely close to the up-per limit for Aitken nuclei (hundredths of a micron), and their concentration is probably of the order of 10^2 particles/liter. (Noctilucent clouds have an op-tical thickness of $\approx 2 \cdot 10^{-6}$.) In this connection we may recall that according to Fig. 53, water vapor cannot condense in the height range from about 40 to 70 km, so that aerosol layers occurring at these heights are definitely com-posed of dust. Moreover, the formation of condensation products at other levels also requires the presence of a sufficient number of solid condensation nuclei — dust particles.

Finally, above about 90 km the air density becomes so low that no particle, whatever its size, will be able to remain in a suspended state. Thus there is no floating aerosol in the ionospheric region. We merely find a stream of meteoritic dust rushing through the ionosphere, although this alone is sufficiently abundant to be almost wholly responsible for scattering light at these heights. We note that some 10^8 tons of meteoritic matter settles to the earth's surface every day. These particles begin to be decelerated efficiently somewhere near the base of the ionospheric layer because of the marked increase in the degree of ionization at that level. The decelerated particles then accumulate in the inversion region near the 80-90 km level. Perhaps it is here that we should seek an explanation for the correlation between the critical frequency for radio-wave reflection and the amount of depolarization of the light from the twilight sky as noted by Khvostikov et al. (Chap. I, §4). Indeed, as we shall see, one consequence of an increasing aerosol concentration should in general be a decrease (or more accurately a change) in the degree of depolarization of the light scattered by the free air.

As early as 1936, an analysis of twilight measurements led Shtaude [85] to suggest the presence of an aerosol layer at heights of 100-150 km. More recently, Link [287] has analyzed observations of the structure of the earth's shadow during lunar eclipses, and he too concludes that such a layer exists. Link finds an optical thickness $\tau \approx 10^{-5}$ for the layer, a value which, as Volz and Goody point out [282], is 10 times that for the noctilucent clouds. From an analysis of their own measurements, Volz and Goody conclude that the layer in fact does not exist, for it would have completely altered the character (and even the scale) of twilight phenomena. Agreeing entirely with this view, we wish to add that the presence of such an aerosol layer would also conflict with current ideas regarding the density of the upper layers of the atmosphere and the processes of sedimentation and deceleration of meteoritic matter taking place there. Furthermore, we should point out that measurements of the twilight-sky brightness seem to leave no place for a correlation between atmospheric transparency and meteor streams, first reported by Kalitin [288] and again observed by Zacharov [289]. The amount of meteoritic dust that would be required in the stratosphere for this purpose would disastrously alter the character of twilight phenomena, as Volz and Goody [282] point out in passing. Nevertheless, the effect of meteor streams on the brightness of the twilight sky remains an open question. While Zacharov [281] finds a distinct long-term increase in the twilight-sky brightness following the Perseid shower, Volz and Goody [282] believe there is no correlation whatever between meteor streams and the twilight sky brightness. The latter authors consider that no such correlation should be expected because the contribution of meteor showers to the total influx of meteoritic matter into the atmosphere is not very large. However, if we recall the effect of meteor showers on the sporadic E layer, we can hardly agree with this objection.

§2. Basic Optical Properties of a Turbid Medium

From the optical standpoint the atmosphere is a turbid medium — a medium in which scattering processes have an important effect on the way light is propagated. Before describing the optical properties of the atmosphere itself, we must therefore pause to summarize the optical properties of scattering media in general.

A narrow pencil of light within an elementary solid angle $d\omega$ may be characterized energetically by its b r i g h t n e s s or i n t e n s i t y I (we shall use both terms interchangeably). The (specific) intensity is defined as the luminous flux per unit solid angle crossing unit area normal to the direction of the pencil. The corresponding i l l u m i n a n c e E on a surface inclined to the pencil by the angle ϑ is defined as (Fig. 57)

$$E = I \cdot \cos \vartheta \, d\omega. \tag{II.2, 1}$$

A medium through which a directed pencil of light is propagated will, in particular, deprive the beam of some of its energy. From this point of view it makes no difference what the subsequent fate of the removed photons is. Whether they are absorbed by the medium itself and transformed into other kinds of energy, or are scattered in other directions, the beam will be attenuated by the same amount. For monochromatic radiation in an isotropic medium, the luminous flux lost by a pencil of light along an element dl of its length will be proportional to the flux in the pencil itself. Since the angular structure of the pencil remains invariant in this process, we have

$$dI = -kI \, dl, \tag{II.2, 2}$$

where k is the a t t e n u a t i o n or e x t i n c t i o n c o e f f i c i e n t of the medium, depending on the properties of the medium, the wavelength λ of the light, and in the case of an inhomogeneous medium on the coordinates x, y, and z also.

The t r a n s p a r e n c y T of a layer of the medium of thickness l is defined by the ratio

$$T = \frac{I_l}{I_0} , \tag{II.2, 3}$$

where I_0, I_l are, respectively, the intensity of the beam upon entry into the layer and departure from it. The reciprocal T^{-1} of the transparency of a layer is

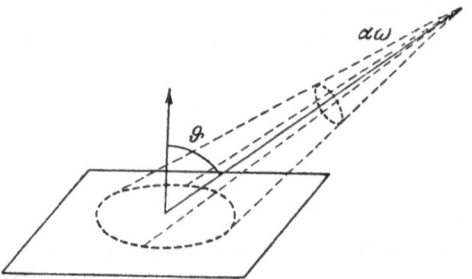

Fig. 57. Illuminance on a surface inclined to
a pencil of light.

called its opacity and the common logarithm of the opacity is the
optical density of the layer:

$$D = -\log T.$$

(II.2, 4)

Upon integrating Eq. (II.2, 2) over the entire thickness of the layer,
we obtain

$$T = \exp\left[- \int_0^l k(\lambda, x, y, z)dl \right],$$

(II.2, 5)

the well-known Bouguer law. The quantity

$$\tau = \int_0^l k(\lambda, x, y, z)dl$$

(II.2, 6)

in the exponent is called the optical thickness of the layer along the
path of the pencil of light. Thus

$$\tau(\lambda) = -\ln T \approx 2.3D.$$

(II.2, 7)

Vavilov [199] has shown that the Bouguer law holds strictly under the
following conditions:

1) intrinsic radiation is absent from the medium throughout the relevant
wavelength range;

2) long-lived excitation states are absent, and thereby also induced emission in the medium (strictly speaking, the Bouguer law would remain valid, but the attenuation coefficient would become negative and would depend on I);

3) the light is strictly monochromatic, or the attenuation coefficient is invariant over the relevant spectral range.

In the case of a scattering medium, we must add the following to the restrictions given by Vavilov:

4) the effects of multiply scattered light are small.

For the earth's atmosphere, the first two conditions certainly hold in the visible spectral region and to a short distance beyond. They fail only in the upper-atmosphere region, above about 90 km, but even here the departures are relatively so slight that they would not have any important effect on the validity of the Bouguer law. However, a special investigation is needed each time to determine whether the other two conditions are satisfied.

If the receiving device has the spectral sensitivity $G(\lambda)$ and if the spectral composition of the pencil of light upon incidence at the layer is $I_0(\lambda)$, then the effective transparency of the layer will evidently be equal to

$$T_{\text{eff}} = \frac{\int G(\lambda) I_0(\lambda) e^{-\tau(\lambda)} d\lambda}{\int G(\lambda) I_0(\lambda) d\lambda} .$$

(II.2, 8)

Thus, for a polychromatic pencil the dependence of the transparency T on the optical thickness τ of the layer will not, in general, be exponential, and the relation (II.2, 7) will no longer hold. However, if the attenuation coefficient k depends only weakly on wavelength in the spectral interval covered by the measurements, then by appealing to the theorem of the mean or by developing $e^{-\tau}$ in series we can remove the exponential term from the integrand in Eq. (II.2, 8). Thus we revert to the Bouguer law, but now with an effective attenuation coefficient k_{eff} depending weakly on the thickness of the layer. In the context of the transparency for the entire thickness of the atmosphere, this departure from the Bouguer law is called the F o r b e s e f f e c t.

However, if a significant amount of selective absorption (or scattering) is present, the situation will change sharply, and will show a relationship to the wavelength-dependent $k(\lambda)$. But in many cases one can nevertheless derive approximate expressions of adequate generality, allowing one to compute the transparency sufficiently accurately. In the long-wave spectral region, for example, where the departures from Bouguer's law mostly arise from the presence of rotational structure in the absorption bands of atmospheric gases

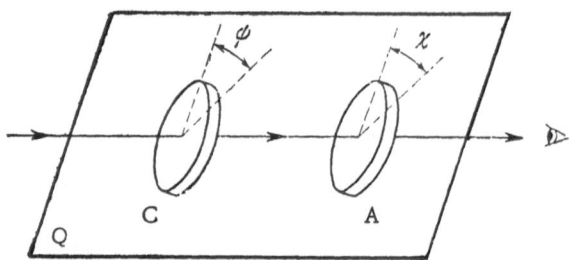

Fig. 58. Definition of the Stokes parameters.

(water vapor, CO_2, ozone), it is useful to introduce schematic models for the structure of the bands [200]. Some of the models yield the following approximate relation over a rather wide range in the thickness of the layer:

$$T \approx 1 - \frac{1}{a} \sqrt{\pi k_0 m} ,$$

(II.2, 9)

where a and k_0 are parameters characterizing the absorption band and m is the mass of the absorbing matter along the trajectory of a ray through the gas layer. In particular, this relation is reasonably accurate under the conditions known to prevail for the water-vapor absorption bands. Since the absorption-line profiles depend, in general, on both pressure and temperature, the transparency shows a corresponding dependence on these quantities. At shorter wavelengths, where the rotational structure of the bands can no longer be resolved, one may use the Bouguer law provided that the working spectral range is sufficiently narrow. The pressure and temperature dependence of the transparency will be preserved, however, and to an extent depending on the particular absorption band.

The fourth condition, that the effects of multiple scattering be small, is reasonably well satisfied for a directed pencil of light if $\tau \lesssim 5$-6, as experiments and theoretical estimates have shown; this boundary is displaced toward greater τ as the absorption increases. We shall return presently to this topic and examine it more fully.

When light is propagated through a scattering medium, not only are the beams attenuated as they travel, but scattered light appears, disseminated in every direction. The fundamental property with which we are concerned here is the great number of mutually noncoherent scattering events, so that a continual mixing of light beams takes place, each beam having its own unique history. Not only does the intensity of the radiation field undergo an angular

redistribution in each scattering event but the polarization also changes. Moreover, the polarization depends on the local properties of the medium, which in
general will not be homogeneous. As a result, the scattered light represents a
noncoherent statistical mixture of beams of widely varying intensity and showing a great variety in type of polarization. It is on the latter effect that the
character and outcome of each scattering event depend most strongly.

In problems concerned with the scattering of radiation and its propagation through a scattering medium, it is therefore not possible to describe the
radiation in terms of the electric- and magnetic-field strengths. We need
other parameters here, additive for noncoherent pencils of light, and providing a description of their state of polarization as well as their energy properties. And we must attempt to describe the medium in terms of optical characteristics that would identify its properties independently of the character of
the radiation field.

We can do so by adopting, as a description of the light beam, the parameters first suggested by Stokes in 1852 — parameters which have since turned out
to be very directly related to the quantum-mechanical matrix for the radiation density [201].

The Stokes parameters can be introduced in their simplest form by ty-
ing them in directly with one of the most convenient procedures for measuring them [96, 201]. Suppose that a compensator C, introducing a quarter-
wave difference, and an analyzer A lie successively along the trajectory of a
beam of light (Fig. 58). Select an arbitrary reference plane Q containing the
direction of the ray. We shall measure the rotation angles ψ of the compensator and χ of the analyzer counterclockwise about the direction of the beam
from the plane Q, looking back toward the ray. We shall then define the
S t o k e s   p a r a m e t e r s   of the beam as the quantities

$$S_1 = I\,(\psi = 0;\ \chi = 0) + I\left(\psi = \frac{\pi}{2}\,;\ \chi = \frac{\pi}{2}\right),$$

$$S_2 = I\,(\psi = 0;\ \chi = 0) - I\left(\psi = \frac{\pi}{2}\,;\ \chi = \frac{\pi}{2}\right),$$ (II.2, 10)

$$S_3 = 2I\left(\psi = \frac{\pi}{4}\,;\ \chi = \frac{\pi}{4}\right) - S_1,$$

$$S_4 = S_1 - 2I\left(\psi = 0;\ \chi = \frac{\pi}{4}\right),$$

where I(ψ, χ) is the intensity of the light passed by both compensator and
analyzer for given values of the angles ψ and χ. One can readily show [201]
that

$$S_1 = I, \quad S_2 = I p \cos 2\psi_0, \quad S_3 = I p \sin 2\psi_0, \quad S_4 = I q, \quad \text{(II.2, 11)}$$

where I is the total intensity of the beam, p is its degree of polarization, q is the "degree of ellipticity" of the polarization of the
beam, and ψ_0 is the angle by which the direction of maximum polarization is rotated with respect to the reference plane Q(Figs. 58 and 59).

In general, an arbitrary partially polarized beam of light of intensity I
may be represented as the sum of two noncoherent beams, a fully (in general,
elliptically) polarized beam of intensity I' = rI and a completely depolarized
beam of intensity I'' = (1 − r)I (Fig. 59). The quantity r = $(p^2 + q^2)^{1/2}$ is called
the amount of polarization or degree of homogeneity of
the beam [96, 201].

The four Stokes parameters S_i (i = 1, 2, 3, 4) may be regarded as the
components of a single Stokes vector-parameter S in four-dimensional func-
tional space, a convention which much simplifies writing the formulas. The
designation of the Stokes parameters by several letters, the usual custom in
the foreign literature, is ill-advised, especially since there is no generally
recognized system of notation.

If we rotate the reference plane counterclockwise (looking back toward
the ray) by the angle ψ', the components of the Stokes vector-parameter S will
change their values to

$$S_i' = \sum_j K_{ij}(\psi') S_j, \quad \text{(II.2, 12)}$$

where the transformation matrix has the form

$$K = \begin{pmatrix} 1 & 0 & 0 & 0 \\ 0 & \cos 2\psi' & \sin 2\psi' & 0 \\ 0 & -\sin 2\psi' & \cos 2\psi' & 0 \\ 0 & 0 & 0 & 1 \end{pmatrix} \quad \text{(II.2, 13)}$$

and the quantities I ≡ S_1, p, q, and r remain invariant relative to the rotation
of the reference plane. Further details of the properties of the Stokes vector-
parameter are described elsewhere [201].

As it traverses a scattering and absorbing medium, a beam of light will
undergo attenuation and, if the medium is anisotropic or if anisotropic particles
are suspended in it, a change in its character of polarization. Either effect may

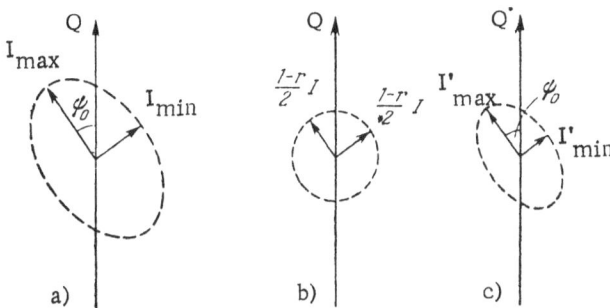

Fig. 59. a) Partially polarized beam of light; $I = I_{max} + I_{min} = I' + I''$, $p = (I_{max} - I_{min})/I = rp'$, $q = rq'$. b) Depolarized component $I'' = (1 - r)I$. c) Fully polarized component $I' = I'_{max} + I'_{min} = rI$, $p = (I'_{max} - I'_{min})/I'$, $q' = \pm 2(I'_{max} - I'_{min})^{1/2}/I'$.

be considered as a change in the Stokes vector-parameter of the beam as it traverses the element dl of length [201]:

$$dS_i = -\sum_j \varkappa_{ij} S_j \, dl, \tag{II.2, 14}$$

where the $\varkappa_{ij}$ (i, j = 1, 2, 3, 4) form the e x t i n c t i o n   m a t r i x. In the case of an isotropic medium the matrix degenerates into the scalar attenuation (extinction) coefficient k which we have discussed above:

$$\varkappa_{ij} = k\delta_{ij}, \tag{II.2, 15}$$

where δ_{ij} is the Kronecker symbol (see [200, 201] for further details).

We turn now to a description of the scattering event itself, in terms of the properties of the scattering medium. Suppose that the volume element dV is irradiated from the direction l^0 by a pencil of light having the Stokes vector-parameter S^0, and consider the pencil of light scattered by the volume element dV in some direction l (Fig. 60). As a reference plane Q for both the incident and the scattered beams we shall take the scattering plane, which contains both the directions l and l^0. Then the linearity of the equations of electrodynamics and the additivity of the Stokes vector-parameter for noncoherent pencils of light imply [201] that the components of the Stokes vector-parameters S of the scattered beam and S^0 of the incident beam stand in the linear relationship

$$dS_i \, d\omega = \frac{1}{r^2} \sum_j D_{ij}(\boldsymbol{l}, \; \boldsymbol{l}^0) \, S_j^0(\boldsymbol{l}^0) \, dV \, d\omega_0,$$

$$(\text{II.2, 16})$$

where r is the distance of the point of observation from the scattering volume element, $d\omega_0$ is an elementary solid angle containing the incident beam, $d\omega = \frac{dV}{dr \cdot r^2}$ is the solid angle subtended by the scattering volume element at the point of observation, and the D_{ij} are the components of a fourth-order matrix characterizing the scattering properties of the medium independently of the state of polarization of the incident beam, and referring to unit volume. We shall call it the s c a t t e r i n g m a t r i x of the medium, and its compo-nent D_{11} the s c a t t e r i n g d i r e c t i v i t y coefficient or the d i f f e r e n t i a l s c a t t e r i n g c r o s s s e c t i o n.

If the scattering medium is isotropic (as with a randomly distributed ori-entation for anisotropic particles), the components of the scattering matrix will not depend on the actual directions $\boldsymbol{l}^0$ of incidence and $\boldsymbol{l}$ of scattering, but only on the scattering angle φ between them (Fig. 60). In this (and only this) case we may introduce [201, 202] the concept of a s c a t t e r i n g c o -e f f i c i e n t σ of the medium (or its total scattering cross section), a quantity depending neither on the state of polarization of the incident beam of radia-tion nor on its direction, and defined (with respect to unit volume) as the frac-tion of the luminous flux striking the volume element dV which is scattered by dV in all directions:

$$\sigma = \oint D_{11} \, d\omega,$$

$$(\text{II.2, 17})$$

with $d\omega$ an elementary solid angle containing the scattered beam. We may now replace the matrix $D_{ij}(\varphi)$ by the n o r m a l i z e d s c a t t e r i n g m a t r i x $f_{ij}(\varphi)$:

$$D_{ij}(\varphi) = \frac{\sigma}{4\pi} f_{ij}(\varphi),$$

$$(\text{II.2, 18})$$

with the component $f_{11}(\varphi)$ satisfying in accordance with Eq. (II.2, 17) the normalization condition

$$\frac{1}{4\pi} \oint f_{11}(\varphi) \, d\omega = 1;$$

$$(\text{II.2, 19})$$

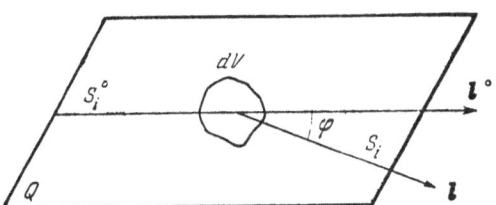

Fig. 60. Specification of the scattering matrix
and scattering function.

this component is usually called the s c a t t e r i n g   f u n c t i o n , and its vector representation, giving the dependence of $f_{11}(\varphi)$ on scattering angle, is
called the s c a t t e r i n g   d i a g r a m   or i n d i c a t r i x .

The scattering diagram therefore shows the angular intensity distribution
of the scattered beams under the condition that the incident beam is depolarized, with $S_2 = S_3 = S_4 = 0$ [Eq. (II.2, 16)]. We shall find below that large and
giant particles yield a highly asymmetric angular distribution, with most of
the light being scattered into the forward hemisphere. Thus it is important to
have a measure of the degree of asymmetry in the scattering. One custom-
arily adopts the d i s s y m m e t r y of the scattering diagram for such a characteristic, as defined by the relation

$$\mathrm{Dis}(\varphi) = \frac{D_{11}(\varphi)}{D_{11}(\pi-\varphi)} = \frac{f_{11}(\varphi)}{f_{11}(\pi-\varphi)} \; ,$$

(II.2, 20)·

where $\varphi \leq 90°$.

The energy that the scattering medium removes from the incident light
beam is not only scattered but also partially absorbed and converted into other
forms of energy. Clearly the fraction of the luminous flux incident upon and
absorbed by the medium, per unit volume of the medium — the a b s o r p t i o n
c o e f f i c i e n t (or cross section) α of the medium — is given by the difference between the coefficients k of .attenuation and σ of scattering:

$$k = \alpha + \sigma,$$

(II.2, 21)

with α in general being a scalar only for an isotropic medium. All the quantities in Eq. (II.2, 21) have the dimensions of cross-sectional area referred to
unit volume, that is, cm^{-1} or, as is more convenient on the atmospheric scale,
km^{-1}, whereas the corresponding quantities referred to a single particle (k_0,
σ_0, α_0) simply have the dimensions of a cross section, or cm^2.

It is very important to note that the quantities α, σ, and k can be expressed in terms of the components D_{ij} of the scattering matrix and $\varkappa_{ij}$ of the extinction matrix, so that the combined spectral and angular behavior of all the components of both matrices exhausts all the information that can be obtained about the properties of the medium from a study of its scattering and absorption of light or other radiations.

Instead of the matrices D_{ij} or f_{ij}, it is sometimes more convenient to use a re d u c e d s c a t t e r i n g m a t r i x, whose components $\widetilde{f}_{ij}$ are given by the relations

$$\widetilde{f}_{ij}(\varphi) = \frac{D_{ij}(\varphi)}{D_{11}(\varphi)} = \frac{f_{ij}(\varphi)}{f_{11}(\varphi)} .$$

(II.2, 22)

The form of the scattering matrix $D_{ij}(\varphi)$ depends essentially on the scattering properties of the medium, in particular its composition, the size, shape, and orientation of the particles suspended in it, and the wavelength of the light. The extent of the anisotropy of the medium or the symmetry of the scattering particles is directly reflected by the number of independent and nonvanishing components in the scattering matrix (see, for example, [18, 201]). However, the exact form of the scattering matrix is known for a very few cases only.

For nonabsorbing gases, with corrections for the anisotropy of the molecules in accordance with the Rayleigh-Cabannes theory of molecular scattering, the normalized scattering matrix has the form [201]

$$f(\varphi) = \frac{3}{4+3d} \begin{pmatrix} 1+\cos^2\varphi+d & -\sin^2\varphi & 0 & 0 \\ -\sin^2\varphi & 1+\cos^2\varphi & 0 & 0 \\ 0 & 0 & 2\cos\varphi & 0 \\ 0 & 0 & 0 & 2\cos\varphi \end{pmatrix},$$

(II.2, 23)

with a scattering coefficient (k = σ because absorption is absent) equal to *

$$\sigma = \frac{8\pi^3(n^2-1)^2}{N\lambda^4} \frac{4+3d}{12-d} ,$$

(II.2, 24)

where N is the number of molecules per unit volume, n is the index of refraction of the medium, $d = 4\Delta/(1-\Delta)$, and Δ is the depolarization of the scattered light for $\varphi = 90°$ and linearly polarized light, with its electric vector per-

---

*A factor of 1/2 was erroneously omitted from Eq. (57) of [201].

pendicular to the scattering plane, being incident on the scattering volume. (Occasionally one uses instead of Δ in the formula the quantity $\rho = 2\Delta/(1+\Delta)$, the depolarization for $\varphi = 90°$ and natural light incident on the scattering volume; sometimes ρ itself is denoted by Δ.) The scattering plane is taken as the plane of reference for both the incident and the scattered beams.

The quantity $n^2 - 1$ in Eq. (II.2, 24) is proportional to N, and it also depends on wavelength because of the dispersion in the index of refraction. Thus the scattering coefficient σ is proportional to the gas density ρ and depends on wavelength like $(n^2 - 1)^2 \lambda^{-4}$. Outside the region of resonance absorption, the λ-dependence of n is weak, and σ is approximately (but not strictly) proportional to λ^{-4}. The depolarization parameter d depends on the structure of the molecules, and for air is about 0.06.

The scattering matrix for gas in a force field has been treated in [203]. In particular, this matrix differs from (II.2, 23) in that the components f_{31} and f_{13} no longer vanish; that is, if the gas is irradiated by natural light we should observe a deviation of the polarization plane from the scattering plane.

In the case of scattering of light by aerosol particles, all 16 components of the scattering matrix can differ, in general. Any symmetry properties of the particles themselves or of their alignment in space will reduce the number of independent components, and some of them will vanish [18]. In particular, for spherical particles (the Mie problem) the scattering matrix will have the form [18, 96, 201]

$$f(\varphi) = \begin{pmatrix} f_1(\varphi) & f_2(\varphi) & 0 & 0 \\ f_2(\varphi) & f_1(\varphi) & 0 & 0 \\ 0 & 0 & f_3(\varphi) & f_4(\varphi) \\ 0 & 0 & -f_4(\varphi) & f_3(\varphi) \end{pmatrix}, \qquad \text{(II.2, 25)}$$

with only four independent components (the scattering plane remains the reference plane for the incident and scattered beams).

The Mie theory enables one in principle to compute the dependence of the scattering and attenuation cross sections σ_0 and k_0, as well as that of the scattering matrix $f_{ik}(\varphi)$, on the wavelength λ and the radius r of the particle. In fact, the radius enters into $f_{ik}(\varphi)$ only through the parameter

$$\rho = \frac{2\pi r}{\lambda}, \qquad \text{(II.2, 26)}$$

and the expressions for k and σ have the form

$$k_0 = K(\rho)\,\pi r^2, \quad \sigma_0 = \Sigma(\rho)\,\pi r^2,$$

(II.2, 27)

where $K(\rho)$ and $\Sigma(\rho)$ are "efficiency factors," the ratios of k_0 and σ_0 to the geometric cross section of the particle. The wavelength enters into $f_{ik}(\varphi)$, $K(\rho)$, and $\Sigma(\rho)$ through the dispersion of the refractive index of the matter of which the particle is composed. However, calculations using the Mie theory are so cumbersome that even though the theory has been available for more than half a century, and computing machines are now at hand, the calculations have so far been carried out for only a small selection of cases. The situation is complicated by the highly capricious behavior of the functions $K(\rho)$, $\Sigma(\rho)$, and $f_{ik}(\varphi)$, which impedes interpolation. Nevertheless, a fairly clear understanding of the basic relationships has been worked out; we cannot discuss the details here, referring the reader instead to the specialized monographs [17, 18]. We only wish to mention some points of a qualitative character that will be of importance for our purposes.

To begin with, the computed data refer exclusively to the quantities σ_0, k_0, $f_1(\varphi)$, and $f_2(\varphi)$; investigators have always ignored the other components of the scattering matrix. One exception is some approximate calculations recently performed at the Atmospheric Optics Laboratory by V. S. Malkova for the ρ- and φ-dependence of the complete scattering matrix, as well as its dependence on the indices of refraction and absorption of the matter comprising very small particles of the Aitken-nucleus type — explicitly, particles with $\rho \le 1$ [204].

For $\rho \ll 1$, the scattering matrix for nonabsorbing particles does not differ from the matrix (II.2, 23) (d = 0 for spherical particles), but the scattering cross section σ_0 is proportional to the sixth power of the radius of the particle (the square of its volume) [17]. As ρ approaches 1, the rise in ρ begins to entail (in addition to an increase in σ_0) a gradual forward elongation of the scattering function in the direction of the incident light beam, with some decrease in the degree of polarization and a weak displacement of the position of maximum polarization. The component $f_4(\varphi)$, which gives the degree of ellipticity of the scattered light, remains very small, reaching a significant value only for $\rho \gtrsim 0.7$-0.8. It is very sensitive, however, to variations in the indices of refraction and absorption of the matter comprising the particle.

In the range $1 \lesssim \rho \lesssim 50$, the sizes which correspond in the visible spectral region to the large particles and a very small fraction of the giant particles, the behavior of σ_0, k_0 and the components $f_1(\varphi)$, $f_2(\varphi)$ of the scattering matrix becomes highly irregular because of interference effects that are very

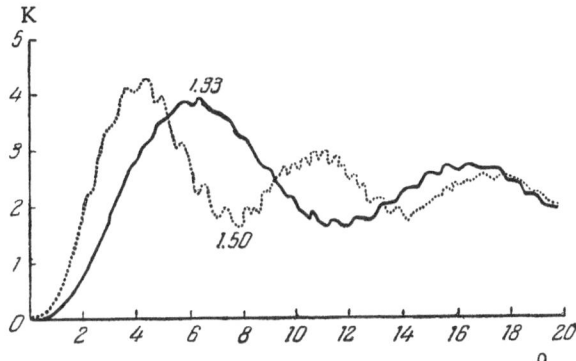

Fig. 61. K(ρ) for spherical particles having refractive
indices of 1.33 (water droplets) and 1.50.

sensitive to variations in the scattering angle and the size of the particle. The
angular functions $f_1(\varphi)$, $f_2(\varphi)$ exhibit numerous and highly mobile extrema
resulting, in particular, in sharp changes in the degree and character of the polarization of the scattered light (the appearance of "negative" polarization,
an excess in the degree of polarization over the values for molecular scattering, and other effects).

We have, moreover, the distinctive diffraction character of the halo and
glory phenomena, whose localization is connected in a regular way with particle size, as well as the progressive forward elongation of the scattering diagram with increasing ρ, and the formation of a diffraction aureole for very
small scattering angles.

When $\rho \gtrsim 50$, f_1 and f_2 exhibit a more gentle behavior; the diagrams
are extended strongly forward and have well-developed,. stable maxima, the
rainbow phenomenon. As ρ increases, the attenuation cross section k_0 approaches the value $k_0 = 2\pi r^2$ [that is, $K(\rho) \to 2$] independently of the nature of
the particle. Figure 61 shows the ρ-dependence of the efficiency factor $K(\rho)$
for spherical particles with refractive indices $n = 1.33$ (water droplets) and
$n = 1.50$ [205].

In addition to scattering by spherical particles, studies have been made
of scattering by particles of ellipsoidal and filamentary shape, but with hardly
any practical computations for these cases [17, 18]. Other particle shapes
have, in general, been inaccessible to theory so far, and information about
them has had to be gleaned entirely from experiment.

To the extent that aerosol particles are suspended in air which itself scatters light, and light is scattered noncoherently by the aerosol particles and the gaseous phase, the components of the scattering matrix $D_{ik}(\varphi)$ are additive for the aerosol and gaseous components of the free air:

$$D_{ik}^{air}(\varphi) = D_{ik}^{gas}(\varphi) + D_{ik}^{aeros}(\varphi).$$

(II.2, 28)

On the other hand, for an aerosol particle itself

$$D_{ik}^{aeros}(\varphi) = \sum_t D_{ik}^{(t)}(\varphi),$$

(II.2, 29)

where t enumerates the particles. Because of the polydispersity and heterogeneous composition of atmospheric aerosols, this last relation may be rewritten as

$$D_{ik}^{aeros}(\varphi) = \sum_s \int_0^\infty n_s(r) D_{ik}^{(s)}(\rho, \varphi) \, dr,$$

(II.2, 30)

where

$$n_s(r) = -\frac{dN_s(r)}{dr}$$

(II.2, 31)

is the density distribution by size of particles of kind s (for particles may differ not only in size but in their species and shape).

If we regard the particles as distributed isotropically in orientation, although as we shall see this does not always obtain, then we may use the relation (II.2, 18) for each kind of particle. Integrating now over all scattering angles and using Eq. (II.2, 19), we have

$$\sigma^{aeros} = \sum_s \int_0^\infty n_s(r)\sigma_0^{(s)}(\rho)dr.$$

(II.2, 32)

If by analogy with Eq. (II.2, 18) we set

$$D_{ik}^{aeros}(\varphi) = \frac{\sigma^{aeros}}{4\pi} f_{ik}^{aeros}(\varphi),$$

(II.2, 33)

then $f_{ik}^{aeros}(\varphi)$ will represent the effective normalized scattering matrix for the aerosol component of the air. An average over the different kinds of particles of different sizes will smooth out the interference properties of the scattering diagrams of the individual particles, and the effective scattering matrices will usually become far milder. We may add that when an experiment is conducted, additional averaging usually takes place over the wavelength bands and scattering angles determined by the design of the instrument, resulting in further smoothing of the function $f_{ik}^{aeros}(\lambda, \varphi)$.

For the Junge distribution (II.1, 2) of a single species of particle with respect to size,

$$n(r) = Cr^{-(\alpha+1)},$$

$$(II.2, 34)$$

where C is a constant, we may use Eqs. (II.2, 26-27) to write the expression (II.2, 32) in the form [206]

$$\sigma^{aeros} = \pi C \int_0^\infty \Sigma(\rho) r^{-(\alpha-1)} dr = \pi C \left(\frac{\lambda}{4\pi}\right)^{-(\alpha-2)} \int_0^\infty \Sigma(\rho) \rho^{-(\alpha-1)} d\rho.$$

$$(II.2, 35)$$

If we neglect the relatively weak λ-dependence of the function $\Sigma(\rho)$ outside the absorption bands due to the dispersion in the index of refraction, then Eq. (II.2, 35) implies that for a Junge-type particle distribution by size,

$$\sigma^{aeros} \sim \lambda^{-(\alpha-2)}$$

$$(II.2, 36)$$

so that for $\alpha = 3$ (corresponding to Junge's distribution for the ground layer of air), $\sigma^{aeros} \sim \lambda^{-1}$, while for $\alpha = 2$ (corresponding to Junge's data for the height range 12-25 km), σ^{aeros} does not depend on λ. However, a distribution of the type (II.2, 34) is valid only over a short range in particle size, and clearly breaks down in the Aitken-nucleus region (see Fig. 54), while the value of α depends on the size, as Laktionov [195] has shown. In the integral of Eq. (II.2, 35) we should therefore replace the limits 0 and ∞ by finite values of r, that is, finite values of ρ depending on λ. This causes the integral to depend on λ, and hence on the form of the function $\Sigma(\rho)$. We have already pointed out that in the real atmosphere observations show that often σ^{aeros} indeed depends approximately exponentially on λ, but the exponent of λ fluctuates over a wide range, and the relation itself is only a rough and idealized one and is usually distorted by a variety of selection effects.

Returning to Eq. (II.2, 28) and again integrating over all directions of scattering, we have

$$\sigma^{air} = \sigma^{gas} + \sigma^{aeros}$$

$$(II.2, 37)$$

and

$$f_{ik}^{air}(\varphi) = \frac{\sigma^{gas} f_{ik}^{gas}(\varphi) + \sigma^{aeros} f_{ik}^{aeros}(\varphi)}{\sigma^{gas} + \sigma^{aeros}}.$$

(II.2, 38)

For the gaseous phase σ^{gas} is proportional to λ^{-4} while for the aerosol phase (large and giant particles) σ^{aeros} depends much more slowly on λ [see Eq. (II.2, 36)]; hence in the short-wave region of the spectrum scattering by the gaseous phase plays the principal role, while at long wavelengths aerosol scattering predominates. Because of the relative stability of the gaseous phase and the strong variability of the aerosol component of the air, light is scattered and attenuated far more stably in the short-wave than in the long-wave part of the spectrum. Moreover, in the long-wave region the aerosol scattering matrix attains great weight, leading to an increase in the dissymmetry of the scattering diagram with increasing wavelength [207]. Finally, because of the sharp forward elongation of the aerosol scattering diagram, the role of aerosol scattering differs for different scattering angles, and is much greater for $\varphi < 90°$ than for $\varphi > 90°$.

We pass now to the effects of multiple scattering, which are important in the atmosphere because of its relatively great optical thickness and comparatively weak absorption of light. The phenomenon involved here consists in light being scattered in some volume element, illuminating the surrounding air, and there experiencing secondary scattering. Higher-order scattering usually does not play a significant part in the atmosphere, becoming important only if there is a marked increase in the optical thickness, such as in the presence of strong turbidity or in the ultraviolet spectral region, as well as under twilight conditions, especially for large depressions of the sun below the horizon.

The propagation of light through a turbid medium with allowance for the effects of multiple scattering is described by the e q u a t i o n o f t r a n s - f e r of the radiation, which takes the following form, including polariza-tion effects:

$$(\mathbf{r}(\theta, \psi) \; \text{grad} \, S_i (\theta, \psi)) = \sum_{j} \left[-\varkappa_{ij} S_j (\theta, \psi) + \right.$$

$$\left. + \frac{\sigma}{4\pi} \oint f_{ij}(\theta, \; \psi, \theta', \psi') \; S_j(\theta', \; \psi') \sin \theta' \; d\theta' \; d\psi' \right].$$

(II.2, 39)

where θ, ψ, and θ', ψ' are the polar angles specifying the directions of the pencils of light, and $\mathbf{r}(\theta, \psi)$ is the unit vector in the direction (θ, ψ). For an isotropic medium the equation simplifies, assuming the form

$$\frac{dS_i(\theta, \psi)}{d\tau} = \sum \left[-S_j(\theta,\psi) + \frac{1}{4\pi(1+\beta)} \oint f_{ij}(\varphi) S_j(\theta', \psi') \times \right.$$

$$\left. \times \sin\theta' \, d\theta' \, d\psi' \right], \qquad \text{(II.2, 40)}$$

where φ is the scattering angle, $d\tau = kdl$ is an element of optical thickness of the medium along the direction (θ, ψ), and

$$\beta = \frac{\alpha}{\sigma} \equiv \frac{k}{\sigma} - 1$$

$$\text{(II.2, 41)}$$

is the specific absorption coefficient of the medium [taken arbitrarily along the direction $\mathbf{r}(\theta, \psi)$]. This equation has a very simple meaning: the Stokes vector-parameter of the beam changes along the path length dl first because of extinction (the first term in the right-hand member of the equation), and secondly because of the scattering by the volume element dV in the direction (θ, ψ) of beams reaching the volume from all initial directions (θ', ψ') (the second term in the right-hand member). Note especially that the contribution of the second term, which incorporates the effects of self-irradiation of the medium, depends in an essential way on the value of the specific absorption β, and decreases as β increases.

Even for a plane-parallel atmosphere (the one-dimensional problem), the solution of the equation of transfer with allowance for the elongation of the scattering diagram and for polarization effects poses an exceptionally complicated task, one which has been surmounted only in the simplest cases [10-16]. Under twilight conditions, where one must take into account the sphericity of the earth's atmosphere (a three-dimensional problem), general methods for solving the equation have not yet been developed. Further difficulties arise from the need to allow for atmospheric refraction and the associated divergence of the solar rays (the corresponding generalization of the equation of transfer has recently been given by Minin [208]). Up to the present, then, all attempts to consider secondary scattering for twilight conditions have involved direct tracing of the successive fortunes of individual beams of light in their propagation through the atmosphere: from the sun into the atmosphere until the first scattering event occurs, thence through the atmosphere to the second scattering event, and once again through the atmosphere to the observer (see Chap. IV, §6) — one of the most laborious ways of solving the equation of transfer, but one of the simplest conceptually.

We must return now to the question of the transparency of a layer within the scattering medium, discussing more fully the fourth criterion for the applicability of the Bouguer law (II.2, 5). This law rests only on the consideration that when a light beam penetrates into the medium it is attenuated by absorption and scattering, corresponding only to the first terms in the right-hand members of the transfer equations (II.2, 39) or (II.2, 40). We have seen, however, that when light is scattered in a medium the attenuation of the beam because of extinction is accompanied by an intensification due to multiple scattering, corresponding to the second term in the right-hand member of the equation of transfer. Moreover, scattered light beams will emerge from the layer in every possible direction, and all will to some extent be received by the measuring device. Thus the relation between the luminous flux collected by the measuring device upon emergence from the layer and the intensity of the beam incident upon the layer will be a fairly complicated one, and strictly speaking can be established theoretically only by solving the equation of transfer with allowance for the individual properties of the medium (such as the scattering matrix) as well as the measuring device (such as the aperture of the receiver). Great caution must be used in applying the Bouguer law in the case of turbid media, and also a certain precision in terminology.

We must first of all discriminate clearly between the concepts of idealized transparency T, total transmissivity Θ, diffuse transmissivity Θ_d, instrumental transparency t, and instrumental transmissivity t_d.

Let us suppose that a layer of a turbid medium (such as a layer of the atmosphere or an extended cloud) is uniformly illuminated by light from a remote source subtending the small solid angle ω_0. Further let us suppose that we are able completely to filter out all the light scattered in the layer, and so distinguish only the light that has not experienced a scattering event. If I_1 denotes the intensity of the light received by a measuring device equipped with an appropriate filter and placed at the point of emergence from the layer, then by analogy with Eq. (II.2, 3) we shall define the idealized transparency of the layer as the ratio

$$T = \frac{I_1}{I_0} ,$$

(II.2, 42)

where I_0 is the intensity of the light beam incident on the layer. In the absence of scattering effects this quantity coincides with the ordinary definition of transparency. But even if multiple scattering is present we could apply the Bouguer law to T provided, of course, that the other three criteria of applicability are satisfied. However, T could then only be evaluated theoretically, for it is not possible in practice to design a filter such as we would require.

We shall here define the total transmissivity Θ of the layer, unlike the definition of Gurevich and Chokhrov [209], as the ratio of the luminous-flux density Φ emergent from the layer to the luminous-flux density Φ_0 incident on the layer:

$$\Theta = \frac{\Phi}{\Phi_0} \, ,$$

(II.2, 43)

where

$$\Phi_0 = \int_0^1 \int_0^{2\pi} I_0(\vartheta, \; \psi) \cos \vartheta \, d(\cos \vartheta) \, d\psi$$

(II.2, 44)

with ϑ the angle between the direction of the incident beam and the normal to the surface of the layer, and ψ the azimuth of the beam. Thus Θ is equivalent to the ratios of the illuminances on surfaces parallel to the boundaries of the layer at the points of emergence and incidence of the luminous flux. Unlike T, this quantity is amenable both to theoretical calculation and to direct experimental measurement, as is readily seen.

Next we define the diffuse transmissivity of the layer as the quantity

$$\Theta_d = \Theta - \frac{\Phi_1}{\Phi_0} \, ,$$

(II.2, 45)

where

$$\Phi_1 = \int_0^1 \int_0^{2\pi} I_1(\vartheta, \; \psi) \cos \vartheta \, d(\cos \vartheta) \, d\psi =$$

$$= \int_0^1 \int_0^{2\pi} T(\vartheta, \; \psi) I_0(\vartheta, \; \psi) \cos \vartheta \, d(\cos \vartheta) \, d\psi$$

(II.2, 46)

is the density upon emergence from the layer of the luminous flux that has not experienced a scattering event. In other words, Θ_d is equal to the ratio of the luminous-flux density emergent from the layer and produced by the scattering of light beams, to the luminous-flux density incident on the layer. Since an experimental separation of direct and scattered light is not possible, this quantity, like the idealized transparency T, can only be determined theoretically.

Now let us assume that behind the layer we have an optical system that can form an image of the source illuminating the layer, and in its focal plane, practically coinciding with the plane of the image (the source is assumed remote), let us place a diaphragm subtending the small solid angle ω.

Evidently the luminous flux transmitted by the diaphragm is given by

$$\Phi_\omega = SI(\vartheta,\ \psi)\,\omega, \qquad\qquad (II.2,\ 47)$$

where S is the area of the entrance pupil of the receiving system and $I(\vartheta,\ \psi)$ is the intensity of the light emerging from the layer in the direction $(\vartheta,\ \psi)$ admitted by the diaphragm.

We define the **instrumental transmissivity** t_d of the layer as the ratio

$$t_d = \frac{1}{S}\,\frac{\Phi_{\omega d}}{\Phi_0}\,, \qquad\qquad (II.2,\ 48)$$

in which $\Phi_{\omega d}$ represents the luminous flux scattered by the layer and incident on the receiver. Clearly

$$I_d(\vartheta,\ \psi) = \frac{1}{\pi}\,g(\vartheta,\ \psi)\,\Theta_d \Phi_0, \qquad\qquad (II.2,\ 49)$$

where $g(\vartheta,\ \psi)$ characterizes the angular distribution of the scattered light upon emergence from the layer, normalized as

$$\frac{1}{\pi}\int\limits_0^1\int\limits_0^{2\pi} g(\vartheta,\ \psi)\cos\vartheta\,d(\cos\vartheta)\,d\psi = 1,$$

so that

$$t_d(\vartheta,\ \psi) = \frac{\omega}{\pi}\,g(\vartheta,\ \psi)\,\Theta_d. \qquad\qquad (II.2,\ 50)$$

If the remote source of light has the small angular size ω_0, and if its image is completely occulted by the diaphragm, then $\Phi_\omega = \Phi_{\omega d}$, since only the light scattered by the layer is incident on the receiver. In this event measurements will evidently give us just the value of the instrumental transmissivity of the layer. However, if the image of the source falls partially or wholly within the aperture of the diaphragm, the situation will change radically, for in addition to the scattered light the receiver will now accept direct unscattered light from the source. Assuming that the center of the diaphragm aperture coincides with the center of the image of the source, we shall define the instrumental transparency t of the layer as the ratio

$$t = \frac{\Phi_\omega}{S I_0 \omega_0}.$$

(II.2, 51)

One can readily see that in this case

$$\Phi_\omega = \Phi_\omega d + \begin{cases} S T I_0 \omega_0, & \text{for} \quad \omega \geqslant \omega_0, \\ S T I_0 \omega, & \text{for} \quad \omega \leqslant \omega_0, \end{cases}$$

(II.2, 52)

and by Eq. (II.2, 44),

$$\Phi_0 = I_0 \omega_0 \cos \vartheta,$$

so that the instrumental transparency of the layer is given by

$$t = t_d \cos \vartheta + \begin{cases} T, & \text{for} \quad \omega \geqslant \omega_0, \\ T \dfrac{\omega}{\omega_0}, & \text{for} \quad \omega \leqslant \omega_0, \end{cases}$$

(II.2, 53)

where t_d is defined by the expression (II.2, 50). It is this quantity that we measure if the instrument is pointed, for example, at the sun.

The first term in the expression (II.2, 53) for the instrumental transparency takes into account all the scattering effects in the layer (apart from extinction). Only in the event that they are small may we unconditionally use the Bouguer law. Consequently, a criterion for the applicability of the Bouguer law would be a small value for the ratio

$$\frac{g(\vartheta, \psi) \Theta_d \cos \vartheta}{T} \omega \quad \text{for} \quad \omega \geqslant \omega_0$$

or

$$\frac{g(\vartheta, \psi) \Theta d \cos \vartheta}{T} \omega_0 \quad \text{for} \quad \omega \leqslant \omega_0.$$

We note first of all that as long as $\omega > \omega_0$, the scattering effects will be small when ω is small. But there would be no point in using apertures ω smaller than ω_0, since no gain would result in freedom from scattering effects, and one would only weaken the total luminous flux being measured. We also remark that the presence of an aureole due to the scattering of light by particles with

$\rho \gg 1$ will sharply increase $g(\vartheta, \psi)$ in the vicinity of the direction toward the source of light, particularly when the optical thickness of the layer and hence the effects of multiple scattering are small.

Experiments by Gurevich and Chokhrov [209] as well as theoretical estimates have shown that the effect of multiple scattering remains negligible for small ω if the optical thickness τ of the layer, measured along the path of the incident beam, is less than about 5-6. If τ is greater than this value, the admixture of multiply scattered light will become appreciable, and its contribution will increase rapidly with τ. For $\tau \approx 7$-9, both terms in Eq. (II.2, 53) will have become comparable in size, while for $\tau \approx 10$-12 the direct beam will be entirely lost in the background fog due to multiple scattering [210, 211].

These quotations refer to a light beam penetrating a semi-infinite medium, and the values will increase somewhat for light traveling through an unbounded layer of finite thickness [211]. One should also observe that multiple scattering is a spatial effect. Hence a reduction in the cross section of the volume penetrated by the beam (such as by confining the beam to a long, narrow channel with absorbing walls) will tend to weaken the effects of multiple scattering, and to displace the optical thicknesses mentioned above toward larger values of τ.

In the general case the calculation of Θ_d meets with considerable difficulties and can be performed only with the aid of high-speed computing machines [13-16, 211]. However, for layers of large optical thickness (measured along the trajectory of the ray) and for $\beta \ll 1$, the following simple approximation may be used [212]:

$$\Theta_d = \frac{g(\vartheta)\,\text{sh}\,y}{\text{sh}\,(x+y) - A\,\text{sh}\,x}, \tag{II.2, 54}$$

where

$$y \approx \eta\sqrt{\beta}, \quad x \approx \frac{\eta}{l}\sqrt{\beta}\,\tau^*, \tag{II.2, 55}$$

τ^* is the optical thickness of the layer along the normal to its surface, η and $l = \eta^2/4$ are constants depending on the form of the scattering matrix, A is the albedo of the underlying surface, and $g(\vartheta)$ is a function depending relatively weakly on φ and not differing very much from unity (for $\tau^* \sec\vartheta \gtrsim 5$, the azimuthal dependence of g vanishes). If $y \ll 1$ and $x \ll 1$, that is, if the specific absorption β of the medium is very small (a reasonably adequate assumption, for example, in the short-wave spectral region in the atmosphere, away from the ozone absorption bands), the expression (II.2, 54) simplifies still further to

$$\Theta_d \approx g(\vartheta) \frac{l}{(1-A)\tau^*+l} ,$$

(II.2, 56)

where for the atmosphere the constant l is of the order of several units($l = 4/3$ for Rayleigh scattering). Moreover, the Bouguer law remains valid for T:

$$T = e^{-\tau^* \sec \vartheta}.$$

(II.2, 57)

Thus, in this case the expression for the instrumental transmissivity of the layer will take the form

$$t_d \approx \frac{\omega}{\pi} g(\vartheta) g(\vartheta') \frac{l}{(1-A)\tau^*+l} ,$$

(II.2, 58)

where ϑ is the angle of incidence of the incoming ray, and ϑ' is the angle of view. The corresponding instrumental transparency for $\omega \geq \omega_0$ and $\vartheta' = \vartheta$ will be

$$t = \frac{\omega}{\pi} g^2(\vartheta) \cos \vartheta \frac{l}{(1-A)\tau^*+l} + e^{-\tau^* \sec \vartheta}.$$

(II.2, 59)

One should, however, remember that this expression is admissible only in the event that $\tau^* \sec \vartheta \gtrsim 5$. Nevertheless, when the layer is illuminated by a diffuse luminous flux rather than a direct beam, the expression (II.2, 54) still may be used, but with g(ϑ) replaced by 1 [212].

§3. Atmospheric Optics

In the first approximation, we may regard the optical properties of the earth's atmosphere in clear settled weather as depending on only one coordinate, elevation above sea level. But although this assumption is certainly justified for the nitrogen—oxygen component of the gaseous phase of the free air, it is true only on the average in the real atmosphere. At any given moment of time one finds a more or less well-defined horizontal inhomogeneity in the atmosphere, because of the nonuniform distribution over its extent of both gaseous impurities (water vapor, ozone, sodium) and particularly aerosols [19]. The latter are usually injected into the atmosphere in the form of cloudlike formations carried by the wind, with a scale depending essentially on the prior history of the accompanying air masses and the nature of their motion. The boundaries of air masses of different origin therefore usually act also as boundaries of regions having different optical properties. In other respects the horizontal inhomogeneities in the optical properties of the atmosphere normally exhibit a highly variable and irregular behavior, appearing superposed on a general uniform background as statistical noise (excluding, of course, large-scale variations, in latitude for example, which are imperceptible to measurement and are eminently local in character compared with the scale of the entire earth). If the turbidity of the atmosphere is low the level of this noise will not usually be very high. Nevertheless, it will not be so low that we can neglect it altogether, especially since like all noise it possesses a very wide frequency spectrum. Measurements have shown [19] that the variations in this term of the atmosphere's vertical optical density, the term associated with the presence of a substantial aerosol component, undergo fluctuations of several times because of variations in the aerosol content, and the spectrum of these fluctuations includes periods of anywhere from a fraction of a second to several days. We point out further that these fluctuations occur not only in the value of the atmospheric transparency, but also in its spectral dependence, and here too their amplitude reaches the same order of magnitude.

A still greater scale is typical for the fluctuations in the local optical properties of the atmosphere, such as the scattering coefficient in a particular restricted volume. There amplitude reaches an order of magnitude in clear settled weather [19]. Whenever measurements are taken (or are used for studying the atmosphere itself), one must therefore consider not only the average values of any of the atmosphere's optical parameters, but also their instantaneous values, which can only be determined through auxiliary measuring programs. Nevertheless, since we shall be discussing the statistical noise we must first specify carefully the mean optical characteristics of the atmosphere.

We shall, then, regard the optical characteristics of the atmosphere, in particular the attenuation coefficient k, as functions of two variables only — the height h and the wavelength λ. We consider first the behavior of the optical thickness τ of the atmosphere as these parameters and the direction l of propagation of the light ray vary.

For a ray traveling at the angle z to the zenith direction, the elementary optical thickness of air that it traverses in the height interval dh is

$$d\tau = k(h, \lambda)\sec z\, dh. \tag{II.3, 1}$$

If we neglect the curvature of the earth's surface and refraction, then the optical thickness traversed by the ray along its trajectory from the level h_1 to the level h_2 will be

$$\tau(h_1,\ h_2) = \sec z \int_{h_1}^{h_2} k(h, \lambda)\, d\lambda = \sec z \cdot \tau_V(h_1,\ h_2), \tag{II.3, 2}$$

where $\tau_V(h_1, h_2)$ is the optical thickness traversed by a ray passing from level h_1 to level h_2 in the vertical direction.

However, the angle z will in fact vary with height, both because of re-fraction and because of the curvature of the earth's surface, corresponding to a varying zenith direction along the trajectory of the ray. Hence the sec z in Eq. (II.3, 2) must be replaced by its integrated mean value

$$m = \overline{\sec z}, \tag{II.3, 3}$$

which depends materially on the vertical structure of the atmosphere. This mean sec z along the trajectory of the ray is called the a i r m a s s m(z). One will note that for the real atmosphere, the relative scale compared to the size of the globe ensures that the air mass will differ significantly from sec z only if the observing site has a large enough value of z, essentially z > 70-80°. With this correction, the relation (II.3, 2) may be written as

$$\tau(h_1, h_2) = m\tau_V(h_1, h_2). \tag{II.3, 4}$$

We therefore find, using Eq. (II.2, 7), that outside regions of selective absorp-tion, and neglecting the Forbes effect, the idealized transparency of the at-mosphere along the direction of the ray is

$$T = P^m, \tag{II.3, 5}$$

where

$$P = e^{-\tau_V} \tag{II.3, 6}$$

is the transparency of the atmosphere in the vertical direction. (In the earth's atmosphere, if $\omega \ll 1$ the instrumental transparency in the vertical direction will differ from the idealized transparency essentially because of compara-tively weak aureole effects.) We shall henceforth affix asterisks to all quan-tities referring to the entire thickness of the atmosphere from the earth's sur-face to infinity (τ^*, P^*, etc.), primes to quantities referring to the entire thickness of the atmosphere above some level h [$\tau'(h)$, $P'(h)$, etc.], and no symbol to quantities referring to the atmospheric layer between the surface and the level h [$\tau(h)$, $P(h)$, etc.]. We therefore have, for arbitrary h,

$$\tau^* = \tau(h) + \tau'(h); \quad P^* = P(h)P'(h). \tag{II.3, 7}$$

Moreover, by Eqs. (II.2,21) and II.2, 37)

$$k^{air} = \sigma^{mol} + \alpha^{sel} + k^{aeros}. \tag{II.3, 8}$$

TABLE 3. Scattering Coefficient σ^{mol} for Pure Air at 0°C, 760 mm Hg [213]

λ, μ	$\sigma, 10^{-3} \text{ km}^{-1}$	λ, μ	$\sigma, 10^{-3} \text{ km}^{-1}$	λ, μ	$\sigma, 10^{-3} \text{ km}^{-1}$
0.30	152.5	0.49	19.66	0.67	5.498
0.31	132.6	0.50	18.10	0.68	5.178
0.32	115.8	0.51	16.69	0.69	4.881
0.33	101.6	0.52	15.42	0.70	4.605
0.34	89.59	0.53	14.26	0.71	4.348
0.35	79.29	0.54	13.21	0.72	4.109
0.36	70.45	0.55	12.26	0.73	3.886
0.37	62.82	0.56	11.39	0.74	3.678
0.38	56.20	0.57	10.60	0.75	3.484
0.39	50.43	0.58	9.876	0.76	3.302
0.40	45.40	0.59	9.212	0.77	3.132
0.41	40.98	0.60	8.604	0.78	2.973
0.42	37.08	0.61	8.045	0.79	2.824
0.43	33.65	0.62	7.531	0.80	2.684
0.44	30.60	0.63	7.057	0.90	1.670
0.45	27.89	0.64	6.620	1.00	1.092
0.46	25.48	0.65	6.217	1.10	0.7447
0.47	23.33	0.66	5.844	1.20	0.5250
0.48	21.40				

The first term refers to the gaseous phase of air without allowance for selec-tive absorption, that is, only to light attenuated by molecular scattering; the second term represents selective absorption by impurities in the gaseous phase, for which $k^{sel} = \alpha^{sel}$; and the third, attenuation of light by atmospheric aerosols.

Correspondingly, we have

$$\tau^{air} = \tau^{mol} + \tau^{gas,sel} + \tau^{aeros}.$$

(II.3, 9)

We consider successively the behavior of all three terms in the attenua-tion of light by the atmosphere.

Penndorf [213] has made comprehensive calculations of the scattering coefficient for the gaseous phase of the free air, on the basis of Eq. (II.2, 24). Table 3 presents his values of σ^{mol} for various wavelengths, referred to 0°C temperature and 760 mm Hg pressure. It is not difficult to extract also the height dependence of σ^{mol} from Eq. (II.2, 24). In fact, if we note that

$$N = N_0 \frac{\rho}{\rho_0} \quad \text{and} \quad (n^2 - 1) = (n_0^2 - 1) \frac{\rho}{\rho_0},$$

(II.3, 10)

TABLE 4. Optical Thickness $\tau *^{mol}$ of an Isothermal Atmosphere
at Sea Level [213]

λ, μ	$\tau *$ mol	λ, μ	$\tau *$ mol	λ, μ	$\tau *$ mol
0.40	0.3630	0.51	0.1335	0.61	0.06433
0.41	0.3277	0.52	0.1233	0.62	0.06022
0.42	0.2966	0.53	0.1140	0.63	0.05643
0.43	0.2691	0.54	0.1057	0.64	0.05294
0.44	0.2447	0.55	0.09805	0.65	0.04971
0.45	0.2231	0.56	0.09110	0.66	0.04673
0.46	0.2038	0.57	0.08477	0.67	0.04397
0.47	0.1865	0.58	0.07897	0.68	0.04141
0.48	0.1711	0.59	0.07367	0.69	0.03903
0.49	0.1572	0.60	0.06880	0.70	0.03682
0.50	0.1447				

where N_0, ρ_0, and n_0 are the values of the molecular concentration, the air density, and the index of refraction under the standard conditions of Table 3, and N, ρ, and n are the values at some height h, then we have

$$\sigma^{mol}(h) = \sigma^{mol}(0) \frac{\rho}{\rho_0} .$$

(II.3, 11)

We can also readily obtain the height dependence of the molecular optical thickness above the level h, that is, the function $\tau'^{mol}(h)$. If we regard the pressure at height h as equal to $p(h) = g \int_{h}^{\infty} \rho \, dh,$ where g is the gravitational acceleration, and use the equation of state for an ideal gas, $\rho = pM/RT$, where M is the molecular weight of air, R the universal gas constant, and T the absolute temperature, then we find from Eq. (II.3, 11) that

$$\tau'^{mol}_v(h) = \sigma^{mol}(0) \frac{p}{p_0} H_0,$$

(II.3, 12)

where $p_0 = 760$ mm Hg and the scale height

$$H_0 = \frac{RT_0}{gM}$$

(II.3, 13)

is the height of a homogeneous atmosphere with $T_0 = 273°K$.

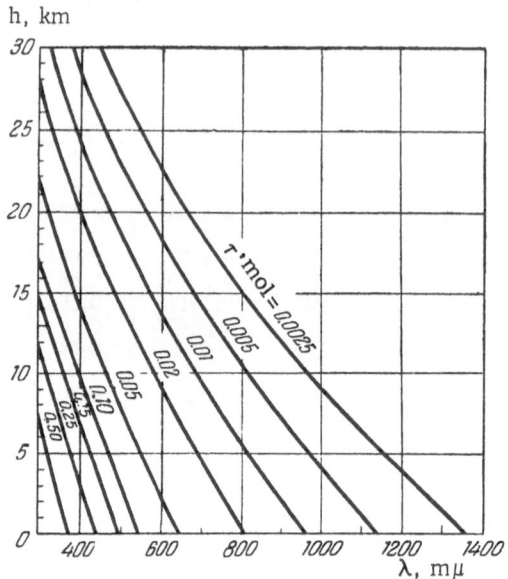

Fig. 62. The vertical optical thickness of pure
air as a function of height and wavelength.

Table 4 presents the wavelength dependence of the vertical optical
thickness at sea level of an isothermal gaseous atmosphere containing no selectively absorbing impurities [213]. The corresponding graph of curves of equal
τ'^{mol} as a function of height and wavelength [28, 290] appears in Fig. 62.
These values refer to mean conditions. Variations in the height distribution
of the temperature, and hence of the pressure, will displace the isolines, and
the displacement can in principle be used to derive the instantaneous behavior
with height of the air pressure and temperature. In particular, this procedure
has been the basis of most attempts to investigate the upper layers of the atmosphere by the twilight method.

We come now to the second term in Eq. (II.3, 8). In the visible region
of the spectrum, selective absorption of light by the gaseous phase is determined mainly by the presence of ozone, water vapor, and oxygen in the
atmosphere.

Oxygen exhibits two relatively strong bands in the visible region, with
a well-resolved structure (Fig. 63). These bands are centered at $\lambda \approx 760$ mμ
(more accurately, 7593-7718 A; band A) and at $\lambda \approx 690$ mμ (more accurately,
6867-7164 A; band B); both bands are comparatively narrow and are responsible

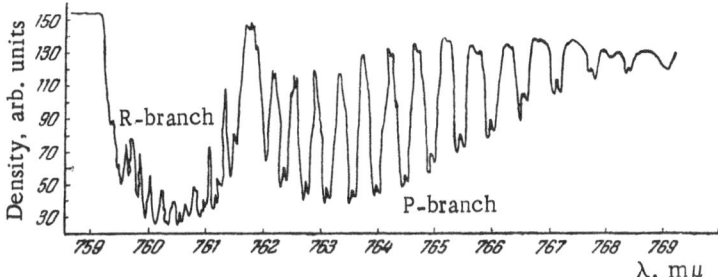

Fig. 63. A tracing of the telluric A-band of oxygen in absorption.

for little integrated absorption [214]. Moreover, there are the very faint bands a (6267-6350 A), a' (5790-5830 A), and a'' (5377-5396 A). At low resolution the A- and B-bands of oxygen stand out distinctly against the background spectrum of the sun or scattered light from the sky (see Fig. 67) only when the path of the light beam through the atmosphere has become sufficiently long; these conditions occur during twilight, especially at large zenith distances of the line of sight. This circumstance may, in particular, be used to determine the optical thickness of the oxygen (and hence of the air also) traversed by a ray of light under twilight conditions.

Atmospheric ozone, apart from its strong absorption bands in the ultraviolet and infrared spectral regions, has in the visible region a faint but widespreading band system, discovered by Chappuis and bearing his name. Figure 64 shows a profile of this system. In Fig. 65 we present an example of the spectral dependence of the vertical transparency of the atmosphere on a particularly clear day, according to measurements by Toropova [215]; it shows a definite dip in the 600-mµ region due to absorption by atmospheric ozone. The Chappuis-band absorption stands out especially well if we represent the vertical optical density $D^* = -\log P^*$ of the atmosphere as a function of λ^{-4}, as is done in Fig. 66 [215]. The component of D^* due to molecular scattering is here shown as the straight line (II.2, 24). The Chappuis bands appear most strikingly in spectra of the twilight sky, where the path of the rays through the atmosphere is considerably longer (see Chap. I, §3, Figs. 29-31).

The absorption of light by water vapor is concentrated mainly in the infrared region of the spectrum. In the spectral region accessible to measurement with available photomultipliers, only very faint high overtones occur. Table 5 gives the position of the centers of these absorption bands, together with their conventional designations.

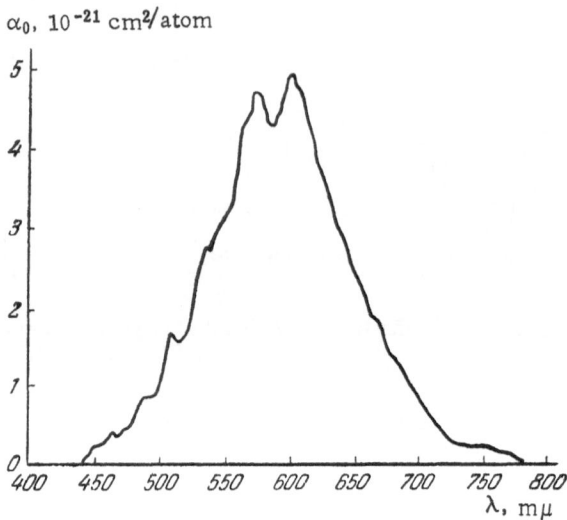

α_0, 10^{-21} cm²/atom

Fig. 64. The Chappuis absorption bands of ozone in the visible spectral region; α_0 is the absorption cross section.

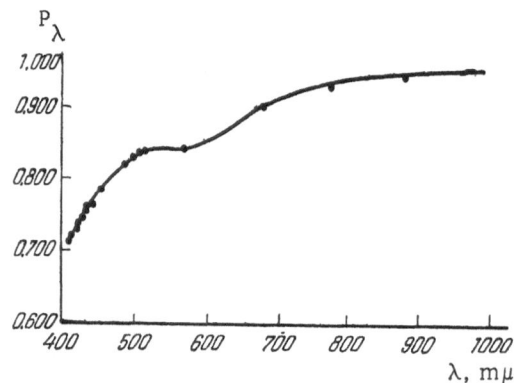

Fig. 65. Sample spectral dependence of the vertical transparency of the atmosphere on a particularly clear day.

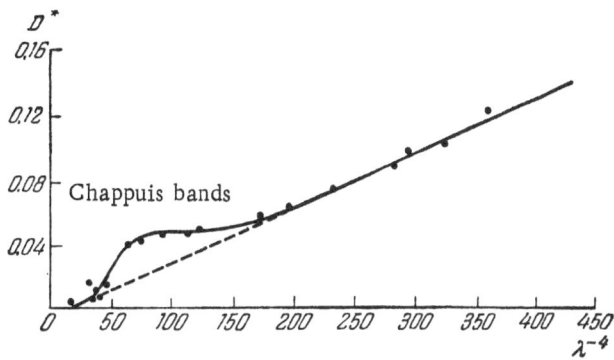

Fig. 66. D^* as a function of λ^{-4} for a day of high transparency.

TABLE 5. Absorption Bands of Water Vapor

Center of absorption band, μ	0.72	0.82	0.93	1.13
Name of band	α	β	$\rho\sigma\tau$	Φ

TABLE 6. Percentage Absorption of Sunlight by Water Vapor

Band	Spectral region, μ	Water-vapor content, g/cm²					
		0.5	1	2	4	6	8
α	0.70-0.74	1.4	3.1	5.2	9.2	12.6	17.4
β	0.79-0.84	2.2	3.9	6.3	10.1	13.9	18.7
$\rho\sigma\tau$	0.86-0.99	8.8	13.2	19.0	27.8	33.0	37.0
Φ	1.03-1.23	9.1	14.0	19.8	27.0	31.2	34.6

I (arb. units)

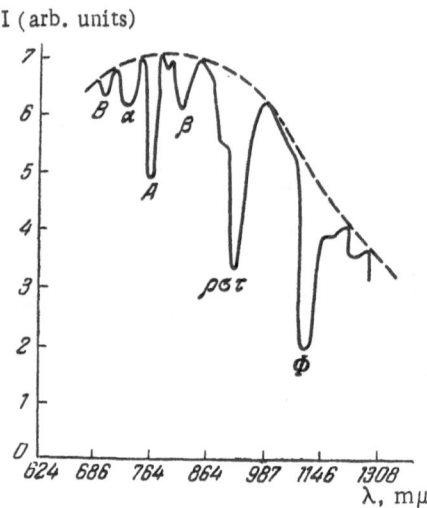

Fig. 67. Long-wave portion of the spectrum of sunlight at the earth's surface. The
absorption bands of water vapor and oxygen
are shown.

Since the Bouguer law definitely does not hold for water vapor, it is not
meaningful to speak of an absorption coefficient for these bands. Such measurements as those of Toropova [215] and the laboratory investigation carried
out at the Atmospheric Optics Laboratory by Elagina [291] have shown that
a square-root law is considerably closer to reality. As a rough guide, Table 6
presents values of the absorption of light over the entire thickness of the atmosphere, as a function of the total water-vapor content along the path of the
ray (in grams per 1-cm^2 column); the data are those of Fowle [216].

The situation is illustrated qualitatively by Fig. 67, which shows the long
wave section of the spectrum of direct sunlight at the earth's surface. The
telluric absorption bands of oxygen (A- and B-bands) and water vapor stand out
distinctly in Fowle's measurements [217]. The fine structure of the solar spectrum is lost in Fig. 67 because of the low resolution. Figure 68, from Rainer's
data [216], provides a somewhat more detailed picture of the energy distribution in the visible region of the solar spectrum (the minima in the energy are
numbered serially). The same figure shows how the spectrum of the sun is de-
formed when its rays are attenuated as they pass through the atmosphere for different values of the air mass m. In Table 7 we list standard data on the relative energy distribution in the ultraviolet and visible regions of the solar spec-

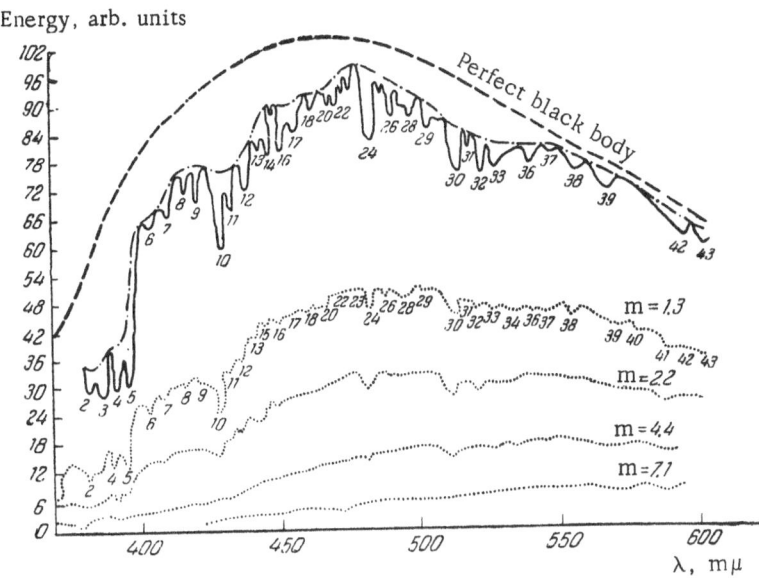

Fig. 68. Energy distribution in the spectrum of direct sunlight beyond
the atmosphere, and at the earth's surface.

trum outside the atmosphere [216], and in Table 8, the corresponding absolute values of the energy (in W/m²) for spectral intervals 10 mµ wide (see [135] for further
details). Figure 69 illustrates the energy spectrum of sunlight in absolute units.

Returning now to the last term in the relation (II.3, 8), the attenuation
of light by the aerosol component, we remark first of all that the value of the
true absorption of light by aerosol particles remains an open question. The
relative absorption of light by clouds and fog is known; in addition to absorption by their water-vapor component, they are also responsible for considerable absorption because of their water droplets or ice crystals. As a result of
diffraction effects, however, this absorption differs materially from absorption
by an equivalent layer of continuous water or ice, both in magnitude and in
its spectral behavior. For spherical water drops the value of the absorption
can be computed from the Mie theory, but so far only a very small range of
such calculations have been performed, and they cannot serve as a basis for
generalization [17, 18]. As for the dry fraction of atmospheric aerosol, here
the situation is still quite unclear. On the one hand, careful measurements of
the attenuation and scattering coefficients for the ground layer of the atmosphere, obtained on a single day by Pritchard and Elliott [218], have shown that
the true absorption of light by aerosol particles is at least comparable in mag-

TABLE 7. Energy Distribution in the Solar Spectrum Beyond
the Atmosphere (relative units)

λ, mμ	Abbot	Pettit	Stand- ard	λ, mμ	Abbot	Pettit	Stand- ard	λ, mμ	Abbot	Stand- ard
292		0.096	0.096	510	1.085		1.085	770	0.642	0.642
295	—	0.17	0.17	520	1.059	—	1.059	780	0.614	0.614
300	—	0.27	0.27	530	1.034	—	1.034	790	0.603	0.603
305	—	0.26	0.26	540	1.017	—	1.017	800	0.594	0.594
310	—	0.322	0.322	550	1.009	—	1.009	810	0.578	0.578
315	—	0.366	0.366	560	1.000	—	1.000	820	0.566	0.566
320	—	0.422	0.422	570	0.991	—	0.991	830	0.556	0.556
325	—	0.463	0.463	580	0.986	—	0.986	840	0.548	0.548
330	—	0.465	0.465	590	0.983	—	0.983	850	0.540	0.540
340	0.530	0.476	0.476	600	0.944	0.95	0.944	860	0.532	0.532
350	0.573	0.466	0.520	610	0.952	—	0.952	870	0.522	0.522
360	0.600	0.530	0.600	620	0.931	—	0.931	880	0.512	0.512
370	0.638	0.540	0.638	630	0.910	—	0.910	890	0.498	0.498
380	0.620	0.497	0.620	640	0.893	—	0.893	900	0.487	0.487
390	0.639	0.644	0.639	650	0.875	—	0.875	910	0.475	0.475
400	0.734	0.852	0.734	660	0.860	—	0.860	920	0.466	0.466
410	0.915	—	0.915	670	0.846	—	0.846	930	0.456	0.456
420	0.970	0.857	0.970	680	0.833	—	0.833	940	0.445	0.445
430	0.969	—	0.969	690	0.814	—	0.814	950	0.430	0.430
440	1.029	—	1.029	700	0.791	0.655	0.791	960	0.418	0.418
450	1.096	1.075	1.096	710	0.768	—	0.768	970	0.406	0.406
460	1.120	—	1.120	720	0.744	—	0.744	980	0.392	0.392
470	1.135	—	1.135	730	0.722	—	0.722	990	0.378	0.378
480	1.136	—	1.136	740	0.702	—	0.702	1000	0.362	0.362
490	1.121	—	1.121	750	0.682	—	0.682			
500	1.107	1.07	1.107	760	0.665	—	0.665			

nitude with the scattering they produce. Yet a detailed study of the radiation balance at various heights, as carried out by Kastrov [219], indicates that although true absorption of light by aerosols does occur in the free atmosphere, it is a minor effect.

Because of the highly variable concentration and nature of the aerosols with respect to time and height, the aerosol component of the atmospheric attenuation of light behaves very erratically. As we have already seen, in the bottom layer near the ground the air is particularly turbid, and even in clear dry weather k^{aeros} will often reach values of several tenths of an inverse kilometer there (see, for example, [19]). The quantity k^{aeros} can vary over a very wide range, from practically 0 up to 1 km^{-1} in dry weather, and up to a few hundred inverse kilometers in thick fog. There is a rapid decline in k^{aeros} with height (see Fig. 56), and somewhere near the 3-5 km level it becomes close to σ^{mol} for typical regions of the visible spectrum. This means, in

TABLE 8. Energy Distribution in the Solar Spectrum Beyond the Atmosphere
(The spectral flux $S_0 \cdot \Delta\lambda$ of solar radiation is expressed in W/m^2 for $\Delta\lambda = 0.01$ μ on the American pyrheliometric scale of 1948.)

λ, μ	$S_0 \cdot \Delta\lambda$	λ, μ	$S_0 \cdot \Delta\lambda$	λ, μ	$S_0 \cdot \Delta\lambda$	λ, μ	$S_0 \cdot \Delta\lambda$	λ, μ	$S_0 \cdot \Delta\lambda$	λ, μ	$S_0 \cdot \Delta\lambda$
0.22	0.13	0,41	17.50	0.60	18.98	0.79	11.80	0.98	7.99	1.85	1.60
0.23	0.36	0.42	17.60	0.61	18.40	0.80	11.54	0.99	7.83	1.90	1.42
0.24	0.42	0.43	16.43	0.62	17.97	0.81	11.36	1.00	7.68	1.95	1.30
0.25	0.56	0.44	19.11	0.63	17.55	0.82	11.13	1.05	7.00	2.00	1.19
0.26	1.33	0.45	20.95	0.64	17.12	0.83	10.91	1.10	6.09	2.05	1.09
0.27	1.71	0.46	21.61	0.65	16.83	0.84	10.68	1.15	5.47	2.10	0.98
0.28	1.52	0.47	21.50	0.66	16.46	0.85	10.46	1.20	5.05	2.15	0.90
0.29	3.58	0.48	21.80	0.67	16.08	0.86	10.24	1.25	4.62	2.20	0.83
0.30	4.14	0.49	20.23	0.68	15.70	0.87	10.04	1.30	4.25	2.25	0.75
0.31	6.14	0.50	20.77	0.69	15.33	0.88	9.83	1.35	3.90	2.30	0.67
0.32	7.19	0.51	20.74	0.70	14.83	0.89	9.63	1.40	3.57	2.35	0.63
0.33	8.77	0.52	19.34	0.71	14.46	0.90	9.43	1.45	3.29	2.40	0.58
0.34	8.64	0.53	20.16	0.72	14.13	0.91	9.24	1.50	2.98	2.45	0.52
0.35	9.14	0.54	20.19	0.73	13.81	0.92	9.05	1.55	2.71	2.50	0.47
0.36	9.42	0.55	19.30	0.74	13.24	0.93	8.87	1.60	2.48	3.00	0.26
0.37	9.41	0.56	20.08	0.75	13.05	0.94	8.68	1.65	2.26	4.00	0.10
0.38	9.29	0.57	19.80	0.76	12.82	0.95	8.50	1.70	2.06	5.00	0.04
0.39	9.88	0.58	19.84	0.77	12.43	0.96	8.33	1.75	1.87	6.00	0.02
0.40	15.46	0.59	19.22	0.78	12.13	0.97	8.16	1.80	1.71	7.00	0.01

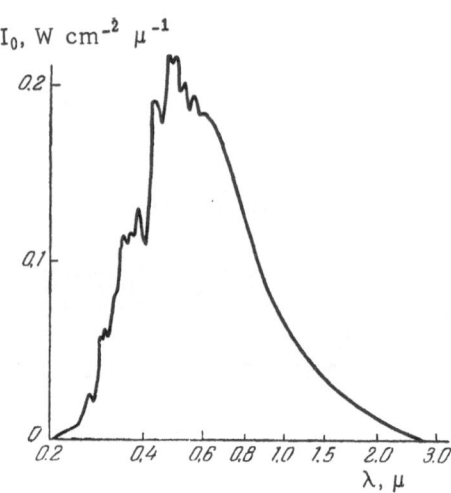

Fig. 69. The energy spectrum of sunlight.

particular, that in the long-wave region aerosol scattering predominates at all heights, while at short wavelengths, molecular scattering is most important beginning at about 5 km. Nevertheless, at greater heights one sometimes encounters aerosol layers with a sharply enhanced scattering power. Thus, in a searchlight probe of the atmosphere Driving [19] observed aerosol layers at a height of 22-25 km with $\sigma^{aeros} \approx 5 \cdot 10^{-2}$ km^{-1}, which is at least an order of magnitude greater than the molecular scattering at this level. Moreover, noctilucent clouds are occasionally observed at heights of about 80 km; estimates of their scattering coefficient lead to computed values on the order of millionths of an inverse kilometer, which once again is several orders of magnitude in excess of the value for molecular scattering. Finally, searchlight data [19] distinctly indicate the presence at still other heights, especially the 40-km level, of aerosol layers whose scattering power exceeds that of pure air by several times.

The spectral behavior of k^{aeros} can vary greatly [19]. Cases have been encountered where k^{aeros} either does not depend on wavelength at all, or increases or decreases smoothly with λ. Sometimes one even finds cases where the function $k^{aeros}(\lambda)$ has a well-defined maximum at some value of λ, particularly in the short-wave portion of the visible spectrum. At times irregular behavior of $k^{aeros}(\lambda)$ is also observed. And the form of the function $k^{aeros}(\lambda)$ frequently is subject to random short-period variations [19].

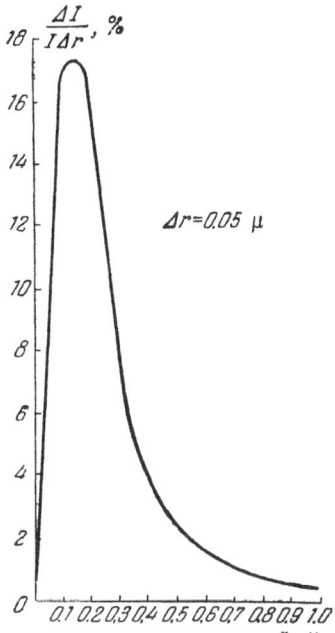

Fig. 70. Scattering efficiency for various fractions of atmospheric aerosol on the Junge distribution

The nature of the scattering and the λ-dependence of σ^{aeros} depend essentially on the nature and size of the scattering particles. From this standpoint it is important to inquire as to what fraction of atmospheric aerosol is the most efficient component in the optical sense. Laktionov [195] has made such a calculation for scattering at a 45° angle, starting with the Junge distribution and an idealized $\sigma_0(\rho)$-dependence. The latter was adopted from the experiments of Ya. I. Kogan and Z. A. Burnasheva [220] and represented visual measurements in white light, so that the function $\sigma_0(\rho)$ was essentially replaced by the relation $\overline{\sigma_0(r)} = \int v(\lambda)\sigma_0(\rho)\,d\lambda$, where $v(\lambda)$ is the visibility function. According to these data, $\overline{\sigma_0(r)} \sim r^6$ for $r < 0.08$ μ, $\overline{\sigma_0(r)} \sim r^3$ for 0.25 $\mu < r < 0.5$ μ, and $\overline{\sigma_0(r)} \sim r^2$ for $r > 0.5$ μ. Laktionov thereby obtained the curve shown in Fig. 70, where the ordinate represents the fraction of the total scattered intensity I received for an interval $\Delta r = 0.05$ μ in particle size. A different dependence will of course result for another particle distribution with size or for scattering at other angles. In particular, for scattering at small angles the contribution of the larger particles will increase sharply because of the high directivity of their scattering diagrams. Nevertheless, Fig. 70 does show that in the optical sense a relatively narrow size range of large particles is the most active in the visible region of the spectrum, and this is just the range in particle size for which interference effects appear especially strongly in the scattering phenomena.

Another important property of the light scattered by atmospheric aerosols is its effective scattering diagram (II.2, 32), corresponding to the effective scattering diagram (II.2, 38) for the real air. These and other diagrams are characterized by a strong forward elongation. Figure 71 shows a typical atmospheric scattering diagram, from Pyaskovskaya-Fesenkova's monograph [9]. Barteneva's extensive study [20] of the form of the scattering diagram for the free air in the ground layer, under different turbidity conditions, has shown,

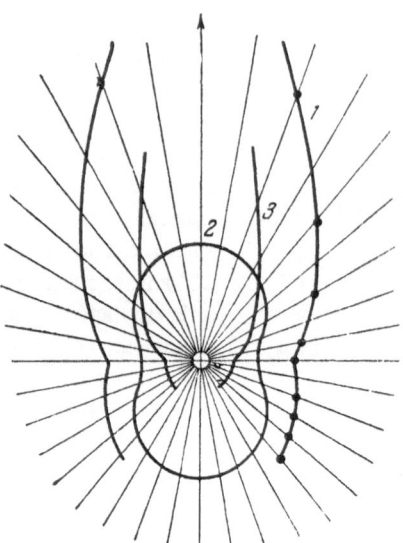

Fig. 71. A typical diagram for the atmospheric coefficient $D_{11}(\varphi)$ of directed
light scattering (relative units). 1) Observed diagram; 2) pure air; 3) aerosol
component.

in accord with the results of other investigators (see, for example, [19]), that
all the various scattering diagrams divide into two distinct classes. One class
is characterized by a smooth, gentle dependence of the scattering function f_{11}
on the scattering angle φ. In the other class, a well-defined minimum is present near $\varphi \approx 100°$ and a sharp maximum in the region $\varphi \approx 140°$ (the first rainbow), with the maximum appearing only in fog (Fig. 72). The absence from
Fig. 72 of any interference effects arises not only from the polydispersity of
the aerosols, but also from the fact that the measurements were taken visually in white light. At the Atmosphere Optics Laboratory, A. Ya. Driving
I. M. Mikhailin, G. D. Turkin, and G. V. Rozenberg have recently conducted
observations with good spectral resolution ($\Delta\lambda = 6$ A), and some well-defined
extrema have been found in the $f_{11}(\varphi)$ curve. Figure 73 shows some examples
of this behavior.

In Fig. 72 we would hardly be justified in extrapolating to very small
scattering angles. Direct measurements [218] show that in the region of very
small angles — the aureole region — the atmospheric scattering diagram is
usually much more strongly elongated. Table 9 illustrates this property [218].

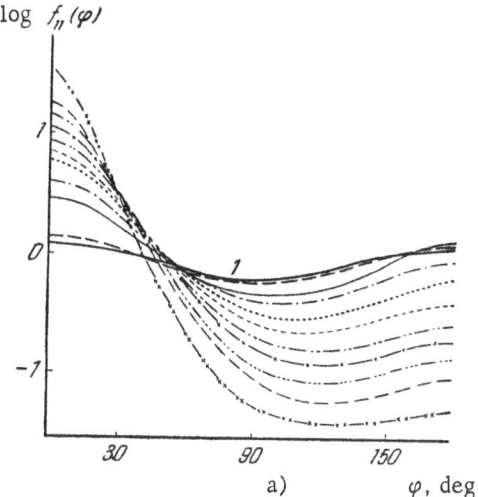

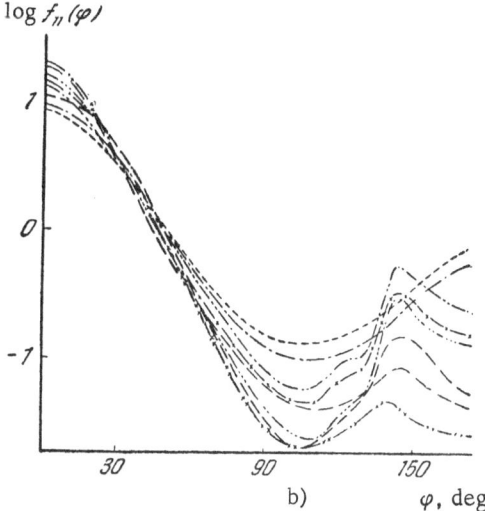

Fig. 72. The two classes of scattering diagram for
the free air (visual measurements in white light). a)
"Gentle" type (1, Rayleigh scattering); b) "peaked"
type.

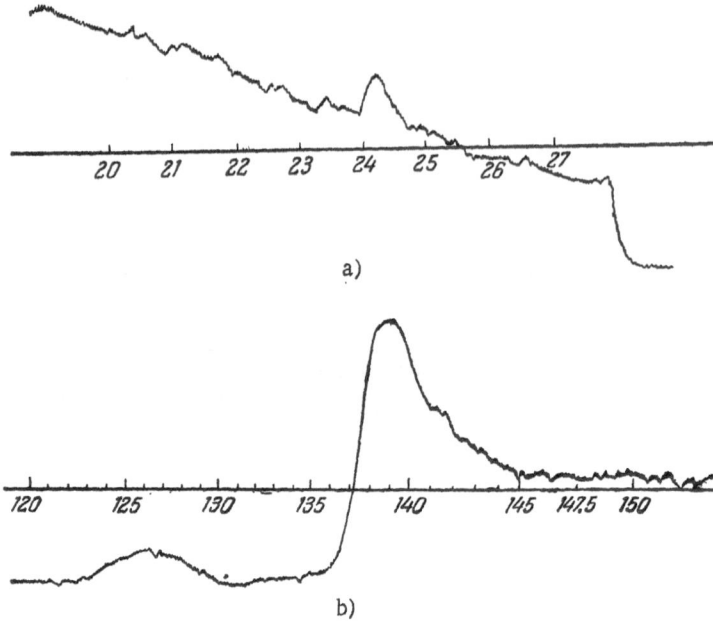

a)

b)

Fig. 73. Sample scattering diagrams measured in polarized light under good monochromatic conditions. a) Halo region of φ; b) rainbow region. The abscissa is the scattering angle.

TABLE 9. Coefficients $D_{11}(\varphi)$ of Directed Light Scattering in the Ground Layer of the Atmosphere on a Clear Night (Unit, 10^{-3} km^{-1})

φ, deg	Spectral region			φ, deg	Spectral region		
	440 mμ	530 mμ	620 mμ		440 mμ	530 mμ	620 mμ
2	650	590	520	90	5.0	3.0	2.4
4	190	170	160	100	4.4	2.7	2.2
6	100	88	82	110	4.2	2.5	2.0
8	76	59	56	120	4.2	2.5	2.0
10	59	46	39	130	4.3	2.6	2.0
15	43	33	27	140	4.5	2.8	2.1
20	33	23	20	150	4.7	2.9	2.1
30	23	16	14	160	4.8	3.0	2.2
40	17	11	9.5				
50	12	8.2	6.6	k	200	160	157 In
60	8.9	6.0	4.9	σ	122	88	69 10^{-3}
70	6.9	4.5	3.7	α	79	76	89 km^{-1}
80	5.7	3.6	2.9				

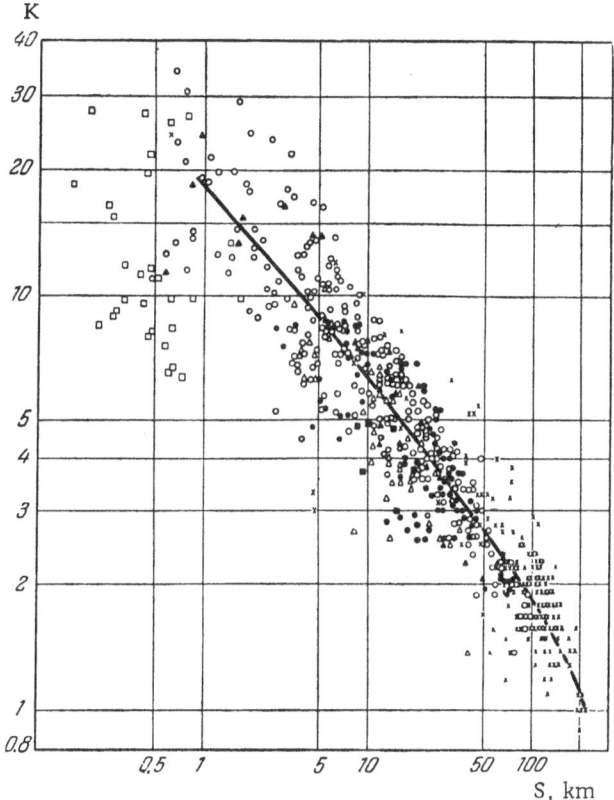

Fig. 74. Correlation between the elongation of the scattering
diagram and the meteorological visibility S.

The values of k, σ, and α for the conditions of measurement are given at the
end of the table. We readily see that on what the authors considered a clear
night, the atmosphere was nevertheless quite turbid (see [19]), and true absorption of light by aerosol particles comprised a considerable part of their
attenuation effect. In the presence of light haze ($\sigma = 0.70$ km^{-1}), 18% of the
entire energy of scattered light is concentrated into the region of angles less
than 2°, and 56% into the region $\varphi < 20°$ [218].

All investigators have pointed out the correlation between the directivity of the scattering diagram and the value of the scattering coefficient (or
reciprocally, the distance of meteorological visibility). This is illustrated in
Fig. 74, which shows the correlation between the ratio K of the light fluxes

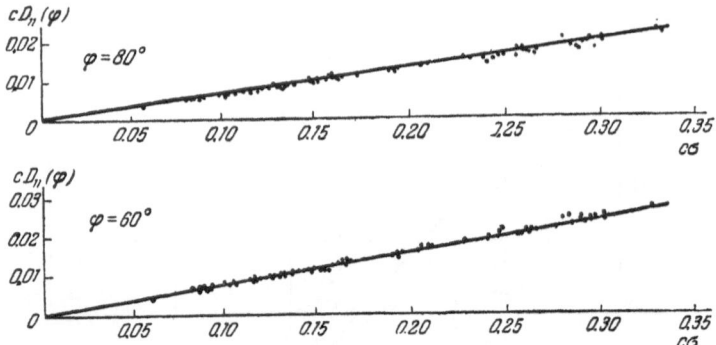

Fig. 75. $D_{11}(\varphi)$ as a function of σ for moderate scattering angles.

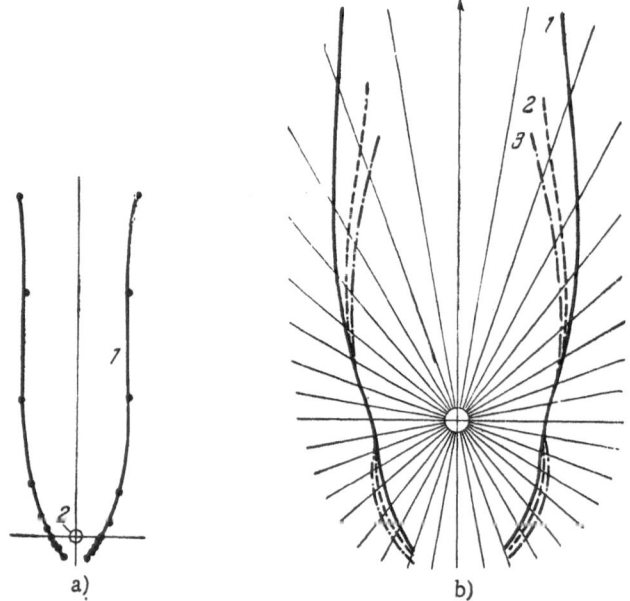

Fig. 76. Wavelength dependence of the scattering diagram.
a) Diagrams of (1) the aerosol and (2) the gaseous component of the coefficient $D_{11}(\varphi)$ of directed scattering in the
near infrared spectral region. b) Mean diagrams of $D_{11}(\varphi)$
for air in different spectral regions: 1) $\lambda = 625$ mμ; 2)
$\lambda = 546$ mμ; 3) $\lambda = 476$ mμ.

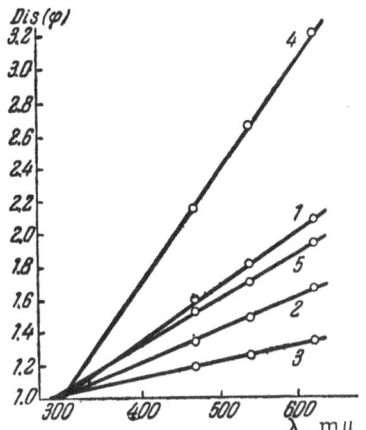

Fig. 77. Dissymmetry Dis(φ) of the scattering function as a function of λ at selected φ for the free air (1-3) and its aerosol component (4, 5). 1) $\varphi = 20°$; 2) $\varphi = 40°$; 3) $\varphi = 60°$; 4) $\varphi = 40°$; 5) $\varphi = 60°$.

scattered in the forward and back-ward directions, and the meteoro-logical visibility S [20]. As would be expected, an increase in the tur-bidity of the air is regularly accom-panied by a forward elongation of the scattering diagram. But in the interval of scattering angles from about 30° to 80°, the form of the scattering diagram depends little on atmospheric conditions [9, 20], so that an extremely strong correlation is set up between $D_{11}(\varphi)$ and σ, as shown for example by Fig. 75 [9]. At small and large φ, however, the form of the scattering diagram is very sensitive to variations in σ.

As one moves into the long-wave portion of the spectrum, the contribution of scattering by the gas-eous phase decreases, as we have seen, and aerosol scattering plays an in-creasingly important role, resulting in further elongation of the scattering dia-gram. This and other effects are illustrated in Fig. 76, adapted from [221] and [9]. Figure 77 shows the λ-dependence of the dissymmetry of the scattering function $f_{11}(\varphi)$ of the free air and its aerosol component [9].

Measurements of the scattering diagram at various heights [197] have established, first, that at all heights, at least up to 19 km, the scattering is es-sentially of the aerosol type, but secondly, that its form changes irregularly with height (Fig. 78). We clearly find here layers with an enhanced elonga-tion in their scattering diagrams; they tend to occur near temperature in-versions [197].

So far we have spoken exclusively of one of the components of the scat-tering matrix for the atmosphere, the scattering function $f_{11}(\varphi)$. The reason is the great scarcity of material on the behavior of the other components. To some extent Fig. 79 may help to provide a graphic idea of the character of the polarization effects when light is scattered by the free air. The figure con-tains a mosaic of photographs of a horizontal searchlight beam as obtained si-multaneously by a series of cameras arranged in a semicircle, for various ori-entations P of the polarizer on the searchlight and A of the analyzer on the

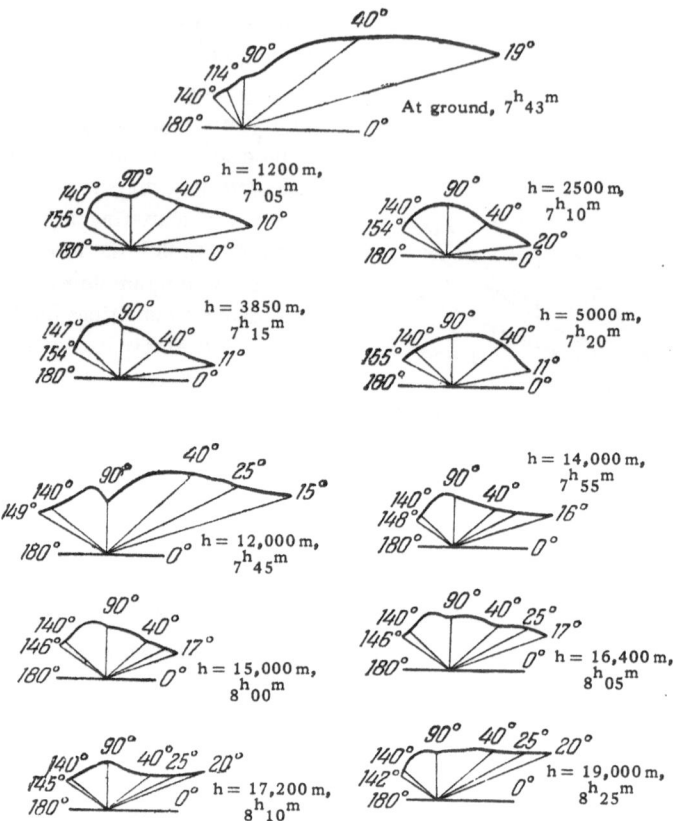

Fig. 78. Scattering diagrams $D_{11}(\varphi)$ at selected heights as measured on a single flight.

cameras. The dependence of the components of the scattering matrix on the
scattering angle as derived from these measurements has been described elsewhere [19]. But because of major uncertainties the measurements should
merely be regarded as provisional. The degree of polarization of light scattered at various angles when the atmosphere is illuminated by an unpolarized
light beam has been measured by Bullrich et al. [222], Toropova [222],
and others.

For several cases, measurements are reported in [218] of the angular
dependence of the scattered-light intensity for various positions of polarizer
and analyzer, as part of a larger program. In one case the complete scattering matrix was computed. The authors here proceeded essentially on the

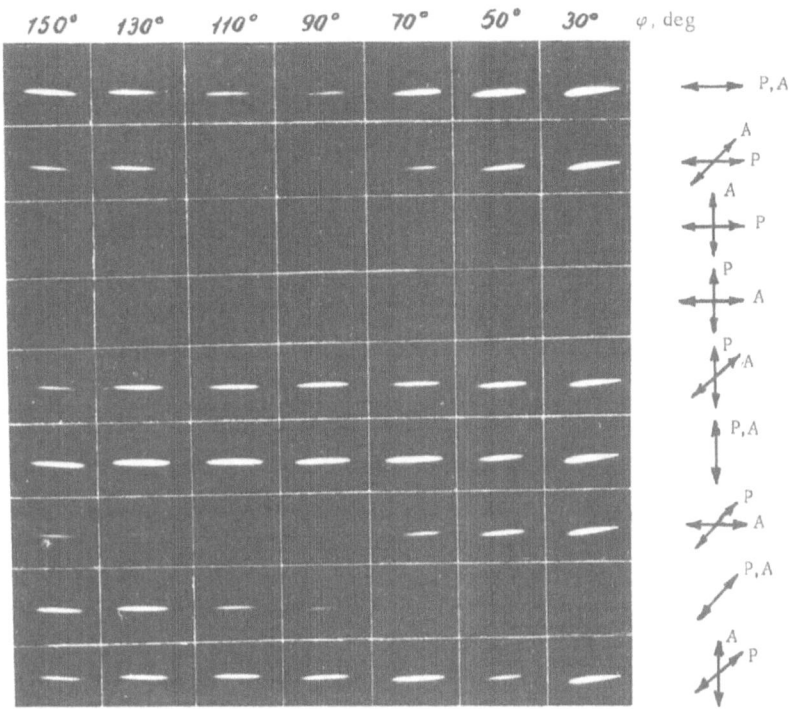

Fig. 79. Sample photographs of a searchlight beam for selected orientations P of the polarizer and A of the analyzer, and selected scattering angles φ.

assumption that the scattering particles were spherical in shape. However, we have shown [19] that the form of the actual scattering matrix conflicts with this assumption.

At the Atmospheric Optics Laboratory, A. Ya. Driving, G. I. Gorchakov, I. M. Mikhailin, G. D. Turkin, G. V. Rozenberg, N. D. Rudometkina, and Yu. Samsonov, as well as G. I. Gorchakov, have recently obtained careful measurements of all 16 components of the scattering matrix of the free air on a series of nights. We have used both a photoelectric photometer with filters [19] and equipped with a polarizing adapter, and a recording spectrophotometer equipped with a polarizing adapter. Figure 80 presents some of the results of this investigation, as secured by Gorchakov with various filters on different nights (see [19] and Eq. (II.2, 25)). These findings demonstrate conclusively that the aerosol particles in the atmosphere are often anisotropic,

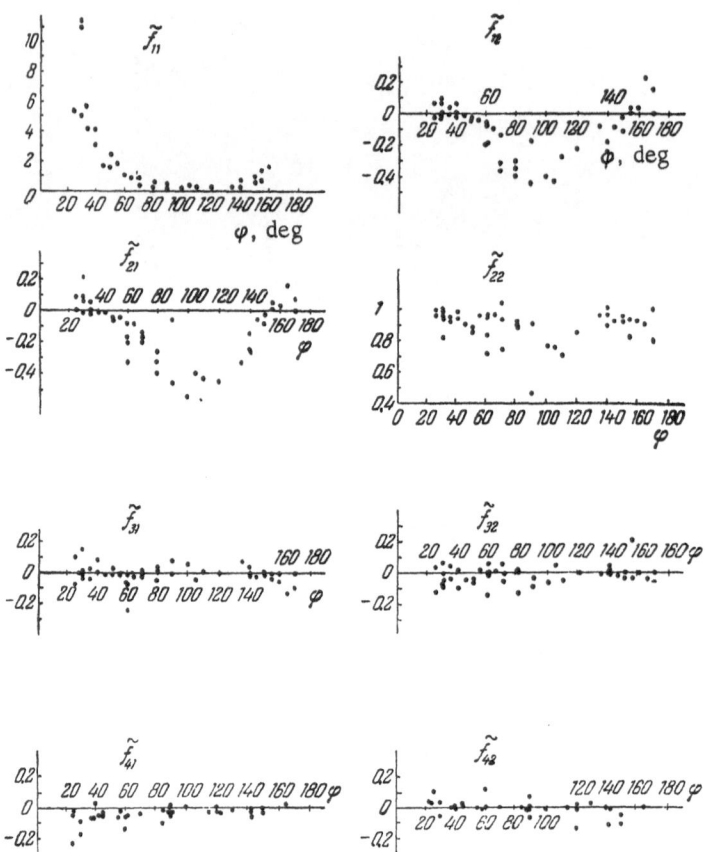

Fig. 80. Examples of the angular dependence of the components $\tilde{f}_{ij} = f_{ij}(\varphi)/f_{11}(\varphi)$ of the reduced scattering matrix [the relation $\tilde{f}_{11} \equiv 1$ is replaced by the function $f_{11}(\varphi)$]. The abscissa is the scattering angle φ in degrees.

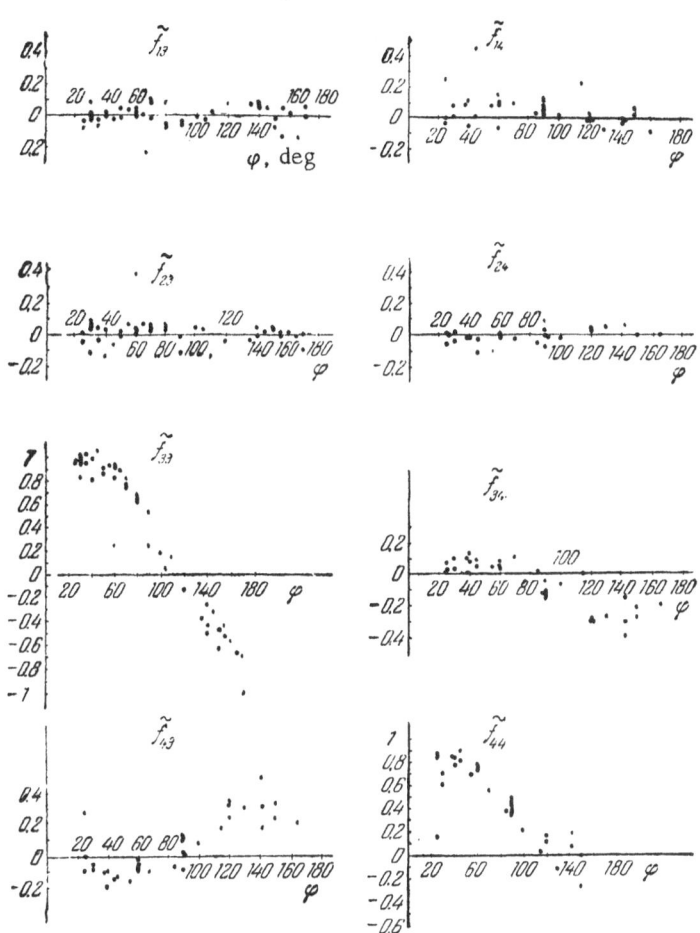

Fig. 80.(Continued)

in particular the particles comprising haze, and perhaps that their alignment distribution is also anisotropic. In this regard, we note that if anisotropic aerosol particles are themselves anisotropically aligned, the effect would be associated with all kinds of halo phenomena and the fall of large raindrops. At any rate, one should regard the development of such an anisotropy as a possibility, and thereby also deviations in the plane of polarization for scattering not only at the earth's surface, but at any height. We may add that measurements of the scattering matrices have shown that not only can scattering entail deviations in the polarization plane, but even elliptical polarization. However, under the conditions prevailing when the atmosphere is illuminated by natural sunlight, elliptical polarization would only be expected if the aerosol component of the air causes secondary scattering, and if the illumination conditions are sufficiently asymmetric. This may, in particular (see [223] and Fig. 80), happen during twilight.

CHAPTER III

PRINCIPLES OF THE
TWILIGHT THEORY

§ 1. The Geometry of Sunlight

Sunlight is both attenuated and scattered as it makes its way through the earth's atmosphere. In the final analysis, it is the relation between these two acts that governs the whole complex of optical phenomena observed whenever the sky is bright, including the twilight period. We could, in principle, derive the full sweep of twilight phenomena just by solving the equation of radiative transfer, Eq. (II.2, 40), under the appropriate boundary conditions. But a general method of solution has yet to be developed for three-dimensional problems of the type applicable to twilight conditions. Even then we could only work out the behavior of the twilight sky if we knew the optical structure of the atmosphere in advance. But it is precisely the converse problem which is of greatest interest: to derive the structure of the atmosphere from the behavior of the twilight sky. Indeed a prime aim of twilight theory, in its present state at any rate, is to discover the basic laws of the phenomenon, and to distinguish the most important factors governing its progress. Clearly we can succeed only by adopting a simplified model for the phenomenon, later refining it by correcting for various side effects. It is not enough for our model to be a good representation of reality in a general way. Our analysis must in addition remain within the scope of the equation of transfer, and eventually lead to its solution. We have already pointed out that one way to proceed is by following the successive fortunes of each light beam through all the vicissitudes it encounters as it travels through the atmosphere. Because of the atmosphere's vertical inhomogeneity and spherical shape, geometric factors become decisive here; a further complication is that the observer essentially has to express them in his own coordinate system. A clear understanding of the geometry of solar rays will therefore be a prerequisite for all the ensuing discussion.

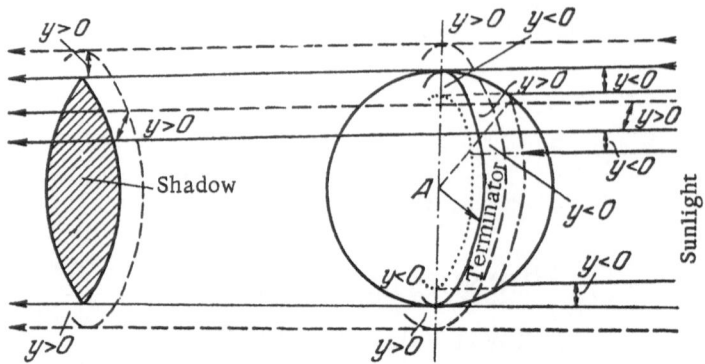

Fig. 81. The earth's geometric shadow.

It is essential that we know just where a given line of sight intersects a solar ray having a specified height at perigee (in particular, the boundary of the earth's shadow), and how the intersection moves as the sun's zenith distance or the line of sight varies. The problem has been solved many times, both by neglecting refraction [85, 96, 224] and by including it under various assumptions as to the structure of the atmosphere [5, 225-233, 292, 293], and even by allowing for the shape of the geoid [234]. However, some of these results have been presented in tabular form, often with superfluous detail, and intended only for solving particular problems, not for qualitative analysis of how phenomena develop; others are given as cumbersome formulas, practically impervious to analysis; and the remainder are in the form of approximate expressions, too restricted in application, and with a rather ill-defined range of validity. We are faced, then, with the need to discuss the matter anew, in the approximate formulation most suited for treating the twilight phenomenon, but with the understanding that many refinements are to be found in the papers cited above.

We shall at first neglect the sun's angular size, thus regarding the solar rays illuminating the earth as strictly parallel. Again, we shall disregard refraction at present, especially since it turns out to have no relevance whatever for several most important problems, as we shall see later. The earth's geometric shadow may then be represented as a cylinder of revolution (Fig. 81) of radius R, the radius of the earth. The generating lines will be the solar rays tangent to the earth's surface (we tentatively neglect the irregularities in the surface, as well as the departures of the earth's figure from a sphere).

Since the atmosphere is spherically symmetric, every ray that passes the earth's surface at the same height above the terminator, with the same height y at perigee, does so under identical conditions. Thus all the solar rays comprise a one-parameter family, with y as a suitable parameter for classification.

If instead of passing by the earth a ray strikes the surface, it will fall within the extension of the earth's cylindrical shadow. The ray may therefore be assigned a negative "perigee height" y, the maximum depression below the terminator of the continuation of the ray inside the globe (Fig. 81).

Suppose now that the observer is stationed at the point N, at height L above the earth's surface, and that the zenith distance of the sun is ζ at the point of observation. Figure 82a represents the case where the observer is located in the night hemisphere, and Fig. 82b the case of the day side. During twilight the earth's rotation induces a gradual change in ζ from about 80° to 110°, and a corresponding displacement of the observer across the terminator from one hemisphere to the other.

We define the direction of the line of sight in terms of its zenith distance z and its azimuth A relative to the solar vertical. The position of an arbitrary point P along the line of sight can be assigned by giving its height h above the surface, as well as z and A. We shall require the height y at perigee of the ray illuminating the point $P(z, A, h)$, the zenith distance $\zeta'(P)$ of the sun at P, and the zenith distance $z'(P)$ of the line of sight at P.

To begin with, we note that the angle φ by which sunlight is scattered from the point $P(z, A, h)$ to the observer is evidently given by the relation

$$\cos \varphi = \cos z \cos \zeta + \sin z \sin \zeta \cos A. \qquad (\text{III.1, 1})$$

Now let ϑ denote the angle between the directions from the center of the earth to the observer N and the observed point P. We then find that

$$\cos \vartheta = \frac{R + L + l \cos z}{R + h}, \quad \sin \vartheta = \frac{l \sin z}{R + h}, \qquad (\text{III.1, 2})$$

where l is the distance of the observed point from the observer, so that we may immediately write

$$(R + L + l \cos z)^2 = (R + h)^2 - l^2 \sin^2 z. \qquad (\text{III.1, 3})$$

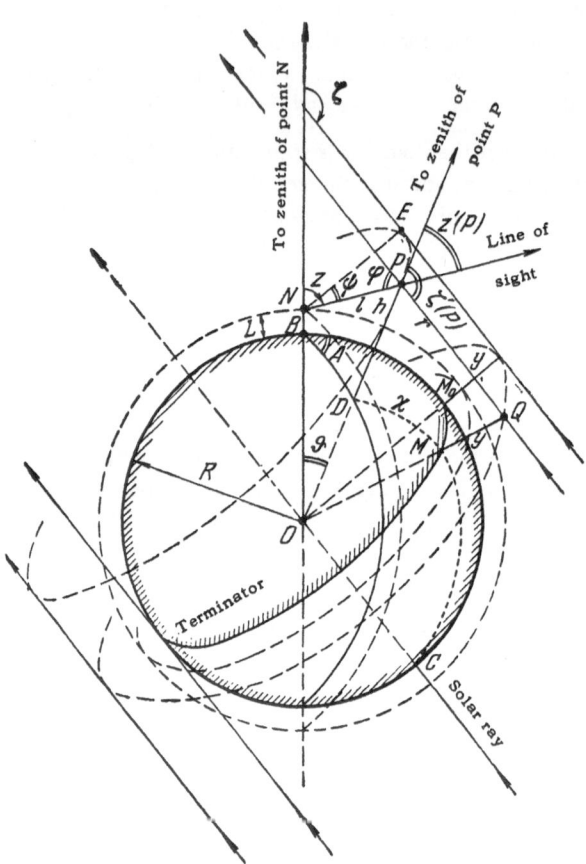

Fig. 82a. The geometry of sunlight. Observer in the
night hemisphere.

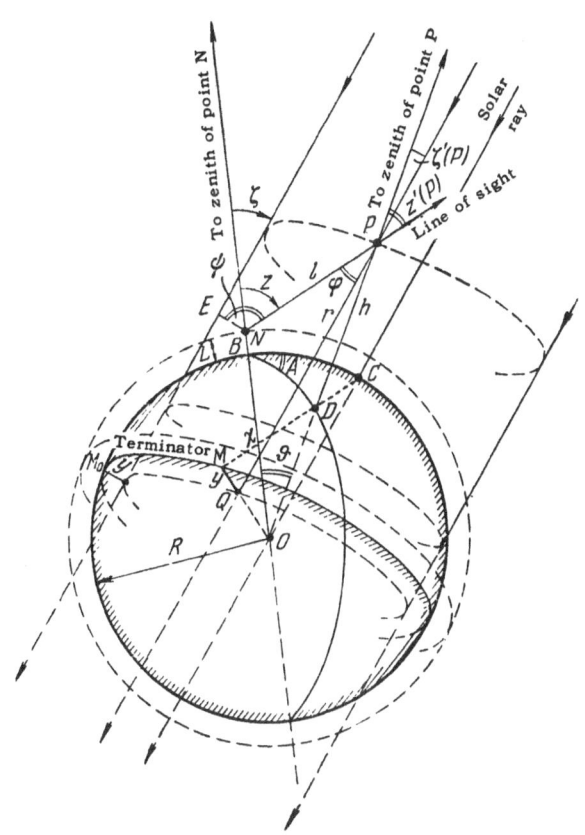

Fig. 82b. The geometry of sunlight. Observer in the
daytime hemisphere.

On the other hand, since the direction OQ from the center of the earth to the perigee Q of the ray illuminating the point P is perpendicular to the ray, the triangle OPQ gives

$$\frac{R+y}{R+h} = \cos \chi, \qquad (III.1, 4)$$

where χ is the angle between the directions OP and OQ.

Next, let C denote the point where the solar ray whose extension passes through the center of the earth intersects the earth's surface, D the point where the line OP intersects the surface, and M the point where the line OQ intersects the surface. The points C, D, and M clearly lie along a great circle, and the angular separation of C and M is $\pi/2$; thus cos (CD) = ± sin χ, where the plus sign refers to the daytime hemisphere (Fig. 82b) and the minus sign to the night hemisphere (Fig. 82a). The spherical triangle BDC now yields

$$\pm \sin \chi = \cos \vartheta \cos \zeta + \sin \vartheta \sin \zeta \cos A, \qquad (III.1, 5)$$

where the choice of sign remains invariant over each hemisphere. If we square the relation (III.1, 4), introduce it into the right-hand member of Eq. (III.1, 5), replace the cos ϑ and sin ϑ there by the expressions (III.1, 2), and make some simple reductions by means of (III.1, 3), we obtain

$$[(R+L) \sin \zeta + l \cos \psi]^2 = (R+y)^2 - l^2 \sin^2 z \sin^2 A, \qquad (III.1, 6)$$

where

$$\cos \psi = \cos z \; \sin \zeta - \sin z \cos \zeta \cos A, \qquad (III.1, 7)$$

so that ψ is the angle between the line of sight and the perpendicular dropped from the observer N to the solar ray intersecting the observer's zenith direction and having the same perigee height y as the ray illuminating the point P (Figs. 82a and b).

Transforming to dimensionless variables, set

$$\left. \begin{array}{l} x = \dfrac{y-L}{R+L}, \quad u = \dfrac{l}{R+L}, \quad w = \dfrac{h-L}{R+L}, \\[2mm] \varepsilon = \sin^2 z \sin^2 A, \quad v = 1 - \sin \zeta + x. \end{array} \right\} \qquad (III.1, 8)$$

Then Eqs. (III.1, 3) and (III.1, 6) assume the form

$$(1 + u \cos z)^2 = (1 + w)^2 - u^2 \sin^2 z, \left. \right\}$$
$$(\sin \zeta + u \cos \psi)^2 = (1 + x)^2 - u^2 \varepsilon. \quad \text{(III.1, 9)}$$

The simultaneous solution of these equations will establish the required relation between h and y, as a function of ζ, z, and A. However, the rigorous solution of the system (III.1, 9) is rather complicated, and is inconvenient for qualitative analysis. We shall find later that only under twilight conditions is it important to know the perigee height of the ray illuminating the observed point. An approximate solution of the system (III.1, 9) is therefore possible; to derive it, we shall estimate the values of the quantities appearing in the equations.

The maximum height h at which scattered light is still perceptible under twilight conditions changes strongly as ζ increases. When $\zeta < 90°$ it does not exceed L + 30 km, but by the time $\zeta \approx 110°$ it may reach 400 km. In any event, heights $h \gtrsim 400$ km need not be considered because of the extreme tenuity of the air. We shall find it necessary to know the perigee heights y of the rays only when they lie between about -100 and $+100$ km. We shall further limit the observer to heights below 400 km. The distance l to the observed point may considerably exceed this value, reaching 2000-3000 km when the line of sight is close to the horizon. Finally, during twilight $\sin \zeta$ will differ little from unity; in any event $1 - \sin \zeta \lesssim 0.06$.

We obtain the following limits from Eqs. (III.1, 8):

$$\left. \begin{array}{c} -0.07 < x < 0.02, \\ -0.06 < w < 0.06, \\ -0.06 < v < 0.12, \\ u \lesssim 0.5. \end{array} \right\} \quad \text{(III.1, 10)}$$

Nothing is lost, then, if in the system (III.1, 9) we neglect the squares of x, v, and w and the fourth powers of u. We may therefore resort to an approximate extraction of the square roots of the right-hand members of Eqs. (III.1, 9), obtaining the expressions

$$\left. \begin{array}{c} u \cos z = w - \dfrac{1}{2} u^2 \sin^2 z, \\ u \cos \psi = v - \dfrac{1}{2} u^2 \varepsilon. \end{array} \right\} \quad \text{(III.1, 11)}$$

From the second equation,

$$u = \frac{\cos \psi}{\varepsilon} \left(\sqrt{1 + \frac{2v\varepsilon}{\cos^2 \psi}} - 1 \right),$$ (III.1, 12)

and substitution of this value of u into the first equation yields

$$w = \frac{\cos \psi}{\varepsilon} \left[\frac{v \sin^2 z}{\cos \psi} + \left(\cos z - \sin^2 z \frac{\cos \psi}{\varepsilon} \right) \left(\sqrt{1 + \frac{2v\varepsilon}{\cos^2 \psi}} - 1 \right) \right].$$ (III.1, 13)

A further simplification is possible if $2v\varepsilon/\cos^2 \psi$ is small compared to unity. In this event, by developing the expression under the square root in series and retaining only the first three terms of the expansion, we find

$$w \approx \frac{v \cos z}{\cos \psi} \left(1 + \frac{v \sin^2 z}{2 \cos z \cos \psi} - \frac{v \sin^2 z \sin^2 A}{2 \cos^2 \psi} \right).$$ (III.1, 14)

The second term in parentheses represents a correction that is significant when the line of sight is close to the horizon, and the last term is a correction important for azimuths away from the solar vertical. It is not hard to show that the approximation (III.1, 14) corresponds to replacing the spherical earth by a cylindrical body with axis normal to the solar meridian.

The approximation (III.1, 14) is satisfactory provided that

$$\frac{2v\varepsilon}{\cos^2 \psi} \lessgtr \frac{1}{2}.$$ (III.1, 15)

For y < L we have x < 0, and if sin ζ > 1 − |x| the condition (III.1, 15) will be satisfied for all lines of sight without exception. If v = 0, that is, if x ≤ 0 and sin ζ = 1 − |x|, the condition will be violated only for lines of sight falling within the plane cos ψ = 0 (see Figs. 82). The same relation (III.1, 15) will hold for arbitrary values of v if ε = 0, that is, for lines of sight in the solar meridian; the directions toward the sun and the antisolar point, for which cos ψ = 0, again are exceptions.

In the remaining cases, there is a region on the sphere wherein the relation (III.1, 15) fails and the approximation (III.1, 14) no longer is realistic. This region extends along the horizon and is inclined to it by an angle equal to the sun's altitude above the horizon. It contracts to a point toward the sun and the antisolar point, while at azimuths A ≈ 90° and 270° a narrow band extends symmetrically on both sides of the horizon.

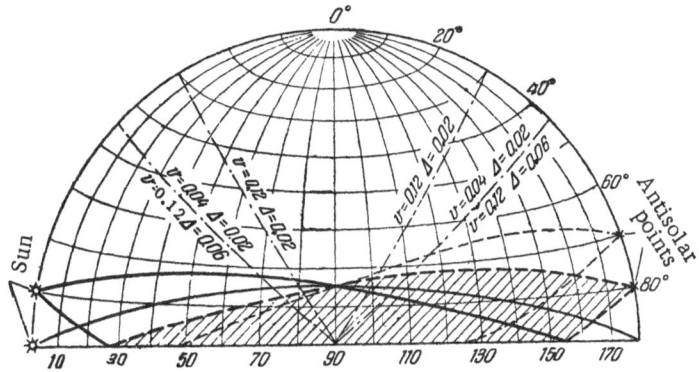

Fig. 83. Boundaries of the region in the upper hemisphere where
the approximation (III.1, 14) fails, and of the region where the
correction terms in this formula may be neglected.

Figure 83 shows the boundaries in the upper hemisphere where the approximation (III.1, 14) is inapplicable, for $v = 0.12$ and the solar zenith distances $\zeta = 80, 90, 100$, and $110°$ (the region for $\zeta = 100°$ is shaded). In the
lower hemisphere, the boundaries are symmetric with respect to replacement
of the solar vertical by the antisolar vertical. We have already mentioned
that under twilight conditions v will not exceed the value 0.12; in fact it is
usually much smaller than 0.12. In this event the region of inapplicability
of our approximation will contract, for according to (III.1, 15) its boundaries
at the widest place, $A = 90°$, are given by the relation $tg\, z = \sin \zeta / \sqrt{v}$. Here
the formula (III.1, 14) may be used throughout the twilight period for lines of
sight covering the overwhelming majority and the most interesting parts of
both the upper and the lower hemisphere.

Returning now to Eq. (III.1, 14), we note that the correction terms in
parentheses are proportional to $\sin^2 z$, and therefore tend to vanish in the zenith
region of the sky. Moreover, at $A = 90°$ or $270°$ they differ only by the small
factor $(1 - \sin \zeta)$, and so almost entirely compensate each other. Consequently
there is a band of the sky at high altitudes, extending across the solar meridian,
within which we may neglect the correction terms. Figure 83 shows the boundaries of this region for selected values of the sum of the two correction terms,
$\Delta = 0.02$ and 0.06, and for the values $v = 0.12$ and 0.04, with $\zeta = 90°$. At other
ζ the boundaries differ very little during the twilight period. The behavior is
entirely analogous for the lower hemisphere.

We see from Fig. 83 that whenever $z \lessgtr 40°$ (or $z \gtrless 140°$), we may neglect the correction terms and set

$$w = \frac{v \cos z}{\cos \psi} .$$

$$(III.1, 16)$$

Moreover, if the observed point lies on the solar meridian, then by Eq. (III.1, 7),

$$\cos \psi = \sin (\zeta - z), \qquad (III.1, 17)$$

where z is considered positive on the solar vertical and negative on the anti-solar vertical. Formula (III.1, 16) then takes the particularly simple form

$$w = \frac{v \cos z}{\sin (\zeta - z)} ,$$

$$(III.1, 18)$$

while the approximation (III.1, 14), which as we have seen is valid over the entire solar meridian, becomes

$$w = \frac{v \cos z}{\sin (\zeta - z)} + \frac{v^2 \sin^2 z}{2 \sin^2 (\zeta - z)} .$$

$$(III.1, 19)$$

In the opposite configuration, where the azimuth of the line of sight is $A = 90°$ or $270°$, Eq. (III.1, 7) yields

$$\cos \psi = \sin \zeta \cos z$$

$$(III.1, 20)$$

whence

$$w = \frac{v}{\sin \zeta} ,$$

$$(III.1, 21)$$

so that w is practically independent of z provided, of course, that the condition (III.1, 15) is satisfied, that is, for $tg\ z \lessgtr (\sin \zeta) / 2 \sqrt{v}$.

Turning now to the true height h of the observed point and the perigee height y of the ray illuminating it, we have, by (III.1, 8),

$$h = w (R + L) + L,$$

$$(III.1, 22)$$

whence for the region of the sky near the zenith,

$$h = \frac{[R (1 - \sin \zeta) + y - L\ tg\ z \cos \zeta\ \cos A] \cos z}{\cos \psi} ;$$

$$(III.1, 23)$$

along the solar meridian,

$$h = \frac{[R\,(1-\sin\zeta)+y-L\,\mathrm{tg}\,z\cos\zeta]\cos z}{\sin(\zeta-z)};$$

(III.1, 24)

and along the A = 90, 270° meridian at right angles to it and at the zenith and nadir,

$$h = \frac{R\,(1-\sin\zeta)+y}{\sin\zeta}.$$

(III.1, 25)

Defining

$$\gamma = \frac{\cos z}{\cos\psi},$$

(III.1, 26)

we obtain for the height of the earth's geometric shadow (the ray with perigee height y = 0) in the direction (z, A):

$$H = [R\,(1-\sin\zeta)-L\,\mathrm{tg}\,z\cos\zeta\cos A]\,\gamma$$

(III.1, 27)

and also

$$h = H + \eta,$$

(III.1, 28)

where

$$\eta = y\gamma$$

(III.1, 29)

is the excess of the height at the point where the line of sight intersects the ray with perigee height y, over the height of the earth's shadow on the same line of sight. In particular, at the observer's zenith and nadir,

$$H_0 = \frac{R\,(1-\sin\zeta)}{\sin\zeta},$$

(III.1, 30)

$$\eta_0 = \frac{y}{\sin\zeta}.$$

Figures 84a and b present families of curves for the height H of the earth's geometric shadow as a function of ζ and z, as computed for the solar meridian from Eqs. (III.1, 22) and (III.1, 19) for an observer located on the surface (L = 0). Figures 85a and b illustrate families of curves for the factor γ as a function of ζ and z for the portion of the solar meridian near the zenith. Finally, Fig. 86 shows the behavior of H(ζ) at selected z and H(z) at selected ζ for an observer at a height of 400 km.

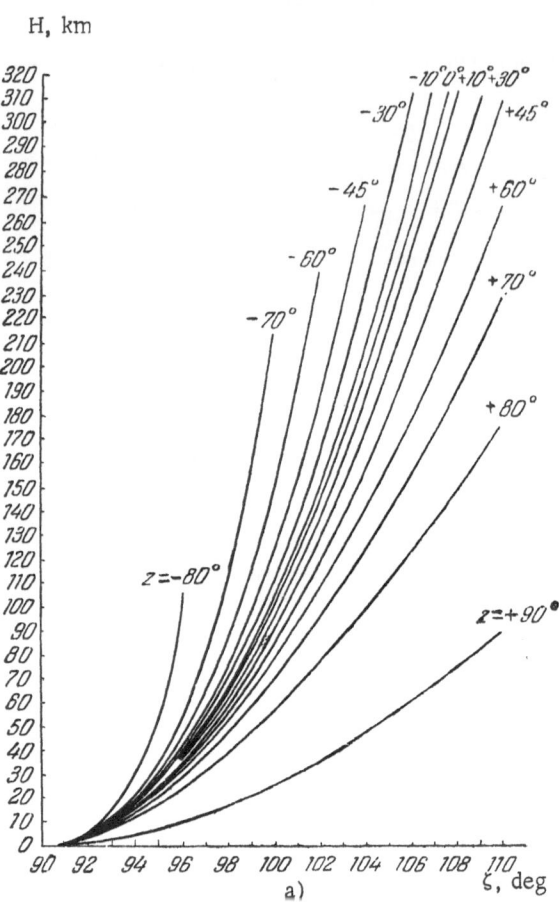

Fig. 84. Height H of the earth's geometric shadow as a function of ζ and z in the solar meridian for an observer on the earth's surface.

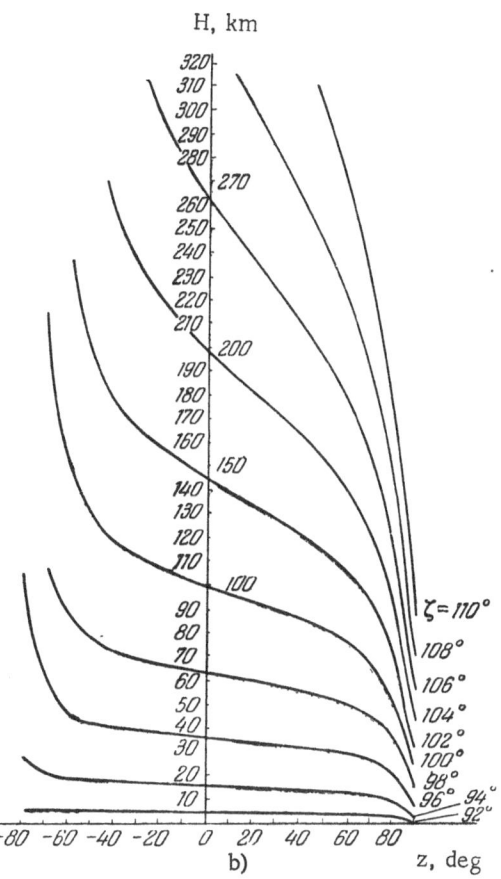

Fig. 84. (Continued)

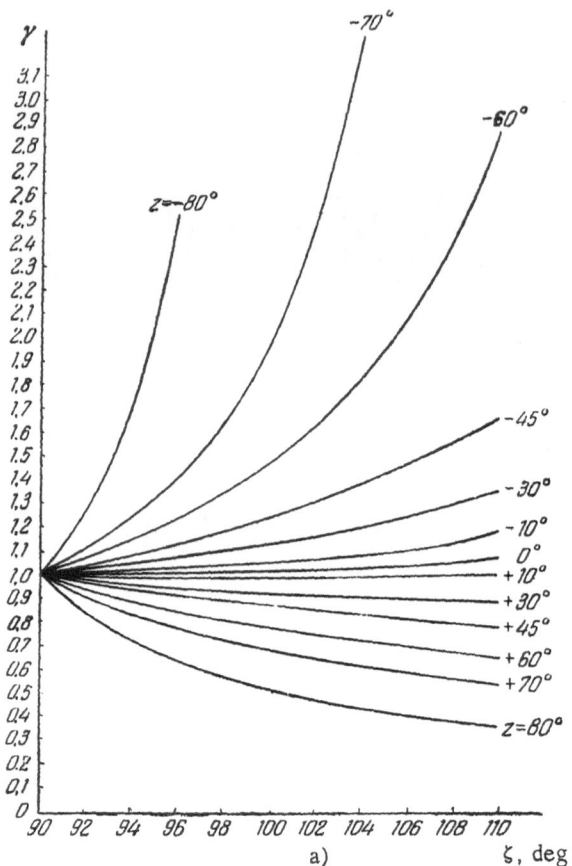

Fig. 85. The factor γ as a function of ζ and z for the
portion of the solar meridian near the zenith.

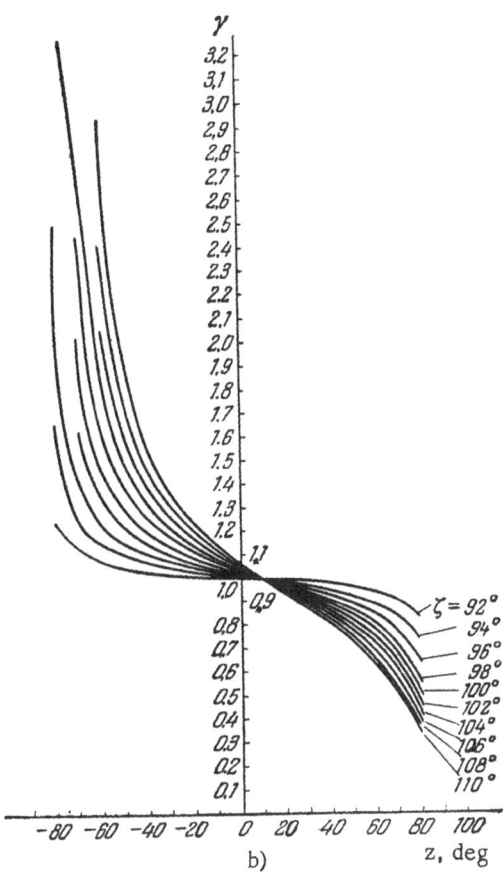

Fig. 85.(Continued)

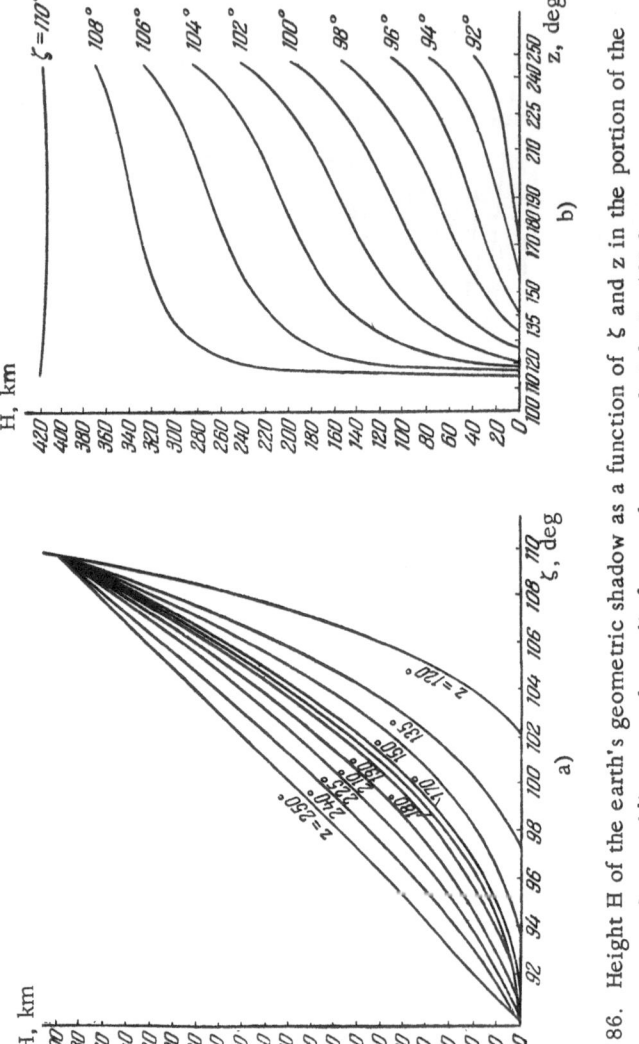

Fig. 86.   Height H of the earth's geometric shadow as a function of ζ and z in the portion of the solar meridian near the nadir for an observer at height L = 400 km.

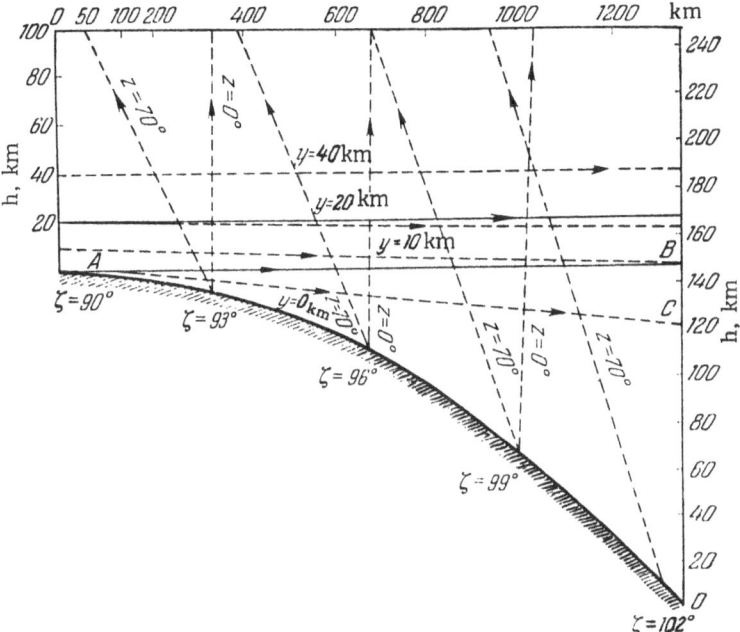

Fig. 87. The trajectory of solar rays of various perigee heights y
with (broken) and without (solid lines) allowance for refraction.
AB) Geometric shadow of the earth; AC) shadow corrected for
refraction. (See p. 144.)

We now proceed to evaluate the zenith distances $\zeta'(P)$ of the sun and
$z'(P)$ of the line of sight at the observed point P. Figures 82 show that

$$z'(P) = z - \vartheta,$$

$$\zeta'(P) = LDC = \begin{cases} \dfrac{\pi}{2} + \chi \text{ in the night hemisphere,} \\ \dfrac{\pi}{2} - \chi \text{ in the daytime hemisphere} \end{cases} \qquad \text{(III.1, 31)}$$

or, with Eq. (III.1, 5),

$$\cos z' = \cos z \cos \vartheta + \sin z \sin \vartheta$$
$$\cos \zeta' = \cos \zeta \cos \vartheta + \sin \zeta \sin \vartheta \cos A. \qquad \text{(III.1, 32)}$$

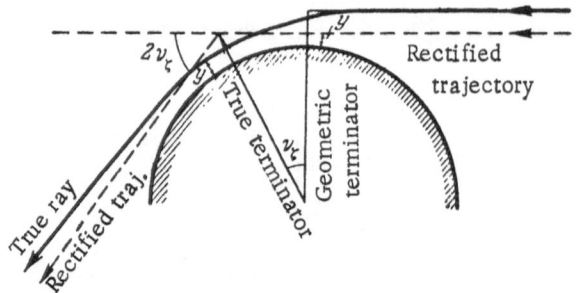

Fig. 88. A schematic trajectory.

Inserting the values (III.1, 2) for cos ϑ and sin ϑ, transforming to the dimensionless variables (III.1, 8), and making use of Eq. (III.1, 1), we find

$$\cos z' = \frac{\cos z + u}{1 + w}, \quad \cos \zeta' = \frac{\cos \zeta + u \cos \varphi}{1 + w},$$

$$\text{(III.1, 33)}$$

where φ is the scattering angle. In view of the condition $w \ll 1$, we may write

$$\cos z' \approx \cos z + u, \quad \cos \zeta' \approx \cos \zeta + u \cos \varphi,$$

$$\text{(III.1, 34)}$$

where u is given by the relation (III.1, 12). The difference between z and z',
as well as between ζ and ζ', becomes noticeable only for large values of u,
when the observed point lies near the horizon. Moreover, for the sun's zenith
distance the correction is important only when cos φ is large, with the line of
sight relatively close to the direction of the sunlight. Since the approximation we have adopted to determine the relation between h and y becomes inapplicable near the horizon at azimuths much different from the azimuth of
the solar meridian, we may limit consideration to the vicinity of the meridian.
Then Eq. (III.1, 12) simplifies to

$$u \approx \frac{v}{\cos \psi},$$

$$\text{(III.1, 35)}$$

so that

$$\cos z' \approx \cos z + \frac{v}{\cos \psi}, \quad \cos \zeta' \approx \cos \zeta + \frac{v \cos \varphi}{\cos \psi}. \quad \text{(III.1, 36)}$$

It remains to consider the e f f e c t of r e f r a c t i o n. The refraction is

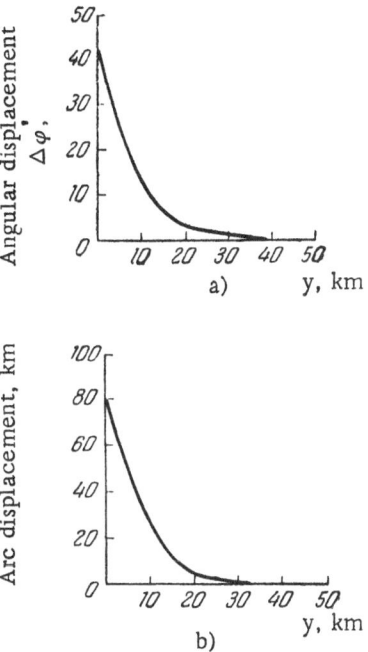

Fig. 89. Displacement of the terminator by refraction as a function of perigee
height. a) In angular measure; b) in
linear measure.

governed by the behavior of the refractive index of air with height, and hence
by the structure of the gaseous component of the atmosphere. The presence
of aerosols does have some influence on the refractive index n of the air, but
the effect is weak enough to be neglected. Of the variable gaseous constituents, only water vapor significantly affects the refraction, and that mainly in
the troposphere when the specific humidity is high enough.

We may regard the free air as an ideal gas, whose refractive index is a
linear function of density, in the first approximation. The height variation of
the refraction therefore depends entirely on the behavior of the pressure, temperature, and also humidity with height. As a result, the value of the refraction is subject to very considerable variation; various authors have precalcu-
lated discordant values by using different model atmospheres. In the context
of twilight research, however, we may neglect these differences and indeed

all fluctuations in the refraction, if we are determining heights along the trajectory of a ray of sunlight. We may even neglect the wavelength dependence. The author has shown [235] that in the first approximation the dispersion $\Delta \nu$ in the angle ν of refraction can be expressed as

$$\Delta \nu = \frac{\Delta n}{n-1}\, \nu,$$

(III.1, 37)

so that for the atmosphere in the visible range $\Delta \nu$ is some two orders of magnitude smaller than ν, or less than 1-2'. The refraction effects may therefore be evaluated approximately from Fesenkov's calculations [233] using the model atmosphere of Allen's tables [236] at $\lambda = 0.5\ \mu$. Detailed calculations of refraction effects under different assumptions as to the structure of the atmosphere have been performed by Link and his colleagues [225-228, 230, 292].

To begin, we note that the angle ν_ζ of astronomical refraction at the perigee of the ray decreases rapidly with increasing perigee height y because of the drop in the air density. Fesenkov's results [233] indicate that ν_ζ follows the approximate law

$$\nu_\zeta = 0.013 e^{-0.16 y},$$

(III.1, 38)

with y in kilometers and ν_ζ in radians. The error of the approximation is about 5%, which is no more than the uncertainty from the variable atmospheric conditions, and at any rate is insignificant for purposes of twilight analysis.

The relation (III.1, 38) implies that when $y \gtrsim 15$ km the refraction effects have become so weak that sunlight is scarcely deflected at all by the atmosphere, although refraction is still very appreciable, for example, when the earth's shadow forms on the lunar surface during eclipse, or even when the illumination of a high-orbit artificial satellite such as Echo is to be calculated [230, 292]. When $y \lesssim 15$ km, solar rays will be deflected noticeably toward the earth's surface, but their curvature will be perceptible only near perigee, and here the height of the trajectory will hardly be altered. We therefore only need correct the ray heights for refraction beyond perigee, and even then only if $y \lesssim 15$ km. Figure 87 illustrates the path of several rays with (broken) and without (solid lines) the refraction effect, as computed from a representative model atmosphere [124]. Also shown are the lines of sight $z = 0°$ and $z = 70°$, $A = 0°$, for observers at points having several different solar zenith distances (the vertical scale is exaggerated by a factor of five).

Arc distance from true terminator, km

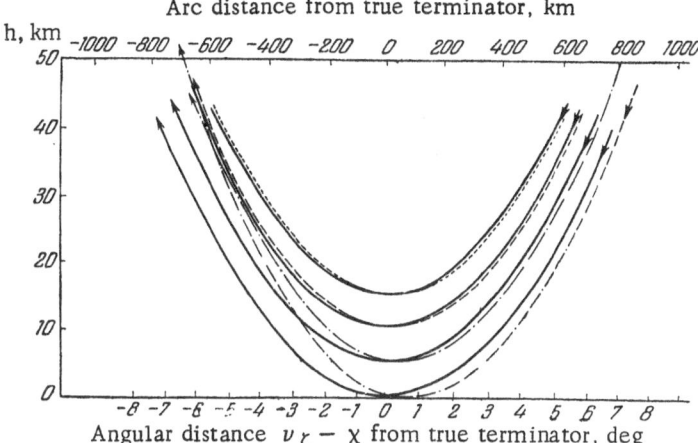

Angular distance $\nu_\zeta - \chi$ from true terminator, deg

Fig. 90. The trajectory of true (solid curves) and unrefracted solar rays of the same perigee height.

One result of this deflection of sunlight is to displace the true termina-tor from its unrefracted position (the "geometric terminator") by an angle $\triangle\chi$ (see Figs. 82) equal to the angle ν_ζ of astronomical refraction at perigee(Fig. 88). Since ν_ζ depends on y, the displacement of the terminator also does so, as shown by Fig. 89. Once again we see that the displacement may be ne-glected when $y \gtrsim 15$ km.

In Fig. 90, the solid curves represent the true path of solar rays having various perigee heights as a function of the angular distance $\nu_\zeta - \chi$ from the true terminator, according to Fesenkov's calculations [233]. The other curves show the path of the same rays with no allowance for refraction. Note espe-cially that although marked curvature sets in as a ray penetrates the denser layers of the atmosphere, there is hardly any effect on the height of the ray above the earth's surface. Before reaching the geometric terminator the true ray lies somewhat above the unrefracted ray, but inside the atmosphere the dif-ference in height remains practically constant, no more than 2-2.5 km for y = 0 or about 1.5 km for y = 5 km. The difference would have been much greater, however, if we had neglected the displacement of the terminator. In the immediate vicinity of the terminator, the trajectories of the true and un-refracted rays differ considerably, yet the heights are virtually the same, as before. Beyond the true terminator, the two rays begin to diverge sharply, and we can no longer neglect the differing heights of intersection with the line of sight. But even here we can replace the true ray by a straight line with the same perigee height y, but now with the zenith distance $\zeta - 2\nu_\zeta$ (see Fig. 88).

Since v_ζ is small, we can determine the refraction displacement of the height h at the intersection of the line of sight with a ray of perigee height y by expanding Eq. (III.1, 14) in powers of $2v_\zeta$ and retaining only the first two terms. With the notation (III.1, 26), we have for these terms

$$w = v\gamma + \frac{v^2}{2}\gamma^2 \operatorname{tg}^2 z (1 - \gamma \sin^2 A).$$

(III.1, 39)

Since

$$\frac{dv}{d\zeta} = -\cos\zeta, \quad \frac{d\gamma}{d\zeta} = \gamma \frac{\cos\varphi}{\cos\psi},$$

(III.1, 40)

we find, with Eqs. (III.1, 36) and some minor transformations,

$$\frac{dw}{d\zeta} = \gamma[1 + v\gamma \operatorname{tg}^2 z (1 - \gamma \sin^2 A)] \cos\zeta' + \frac{1}{2}v^2\gamma^3 \operatorname{tg}^2 z \sin^2 A \frac{\cos\varphi}{\cos\psi}.$$

(III.1, 41)

Since the correction terms containing v are important only at large z, where our cylindrical approximation is applicable only near the solar meridian, we may neglect the terms containing $\sin^2 A$. Then for the segment of the ray lying beyond the terminator, Eq. (III.1, 39) becomes

$$w = v\gamma + \frac{v^2}{2}\gamma^2 \operatorname{tg}^2 z (1 - \gamma \sin^2 z) + 2v_\zeta \gamma (1 + v\gamma \operatorname{tg}^2 z) \cos\zeta'.$$

(III.1, 42)

In the vicinity of the geometric terminator, $\cos\zeta' \ll 1$, and the refraction correction is insignificant, but it increases rapidly with the angle χ (Figs. 82). We also note that the correction is to be applied only when the observed point lies in the night hemisphere.

For the zenith region of the sky, in particular, the change in the height of the observed point due to refraction will be

$$\Delta h = 2v_\zeta \gamma [R \cos\zeta' - L \operatorname{tg} z \sin\zeta \cos A].$$

(III.1, 43)

As the line of sight approaches the horizon (in our cylindrical approximation, this is permissible only near the solar meridian), we must further allow for the curvature of the line of sight by refraction. Since the curvature arises primarily in the troposphere, there are three separate cases to consider.

a) If the observer is located on or near the earth's surface, so that L is small, then in geometrical work we may replace z by $z + v_z$, where v_z is the astronomical refraction at the point of observation, a function of L and z (see, for example, [236]), and may regard the line of sight as a straight line. The slight underestimate of the height of the observed point is of no consequence, and is even advantageous because it partially compensates for the underestimated height of the ray in the rectified trajectory. The corresponding correction for the height of the observed point may be taken as

$$\Delta w = \frac{dw}{dz} \, v_z.$$

(III.1, 44)

Reverting to Eq. (III.1, 39), noting that

$$\frac{d\gamma}{dz} = \frac{\cos \zeta \cos A}{\cos^2 \psi},$$

(III.1, 45)

and noting further that a correction is necessary only when z differs little from 90° and sin A ≪ 1, we obtain

$$\Delta w = \frac{v}{\sin^2 (\zeta - z)} \left[\cos \zeta + v \frac{\sin \zeta}{\sin (\zeta - z)} \right] v_z.$$

(III.1, 46)

b) If the observer is located high above the earth, with L ≫ 15 km, and if the line of sight either abuts against the earth's surface or passes far above it (y ≳ 15 km), then the refraction curvature of the line of sight may in general be neglected.

c) If the observer is located above L ≈ 15 km, and if the line of sight glides into the troposphere along the earth's surface, then it will have the same curvature as for a solar ray. Thus the refraction effect may be neglected before perigee, while beyond perigee a straight line may again be used, but with the zenith distance z replaced by $z + 2v_{\zeta}(y)$, where y is the perigee height of the line of sight:

$$y = (R + L) \sin z - R.$$

(III.1, 47)

Since the distance l from the observer to the perigee of the line of sight is

$$l = -(R + L) \cos z$$

or, by (III.1, 8),

$$u = - \cos z,$$

(III.1, 48)

we find, with the approximation (III.1, 35), that for

$$v \lessgtr - \cos z \cos \psi$$

(III.1, 49)

the zenith distance of the line of sight should be retained at the same value, while for

$$v \gtrless - \cos z \cos \psi$$

(III.1, 50)

it should be replaced by $z + 2 v_\zeta(y)$ and the observer should simultaneously be displaced in height by the amount

$$\Delta L = 2 v_\zeta (y)(R + L) \operatorname{ctg} z.$$

(III.1, 51)

One other refraction effect remains to be noted. When the solar rays pass the terminator they cease to be parallel, because of the y-dependence of the refraction angle v_ζ, and instead diverge in fanlike fashion. As a result the illumination of the atmosphere changes (apart from the effects of atmospheric extinction). Since this refraction divergence becomes important only at points sufficiently far from the terminator, we may as before regard the rays as straight lines, diverging from the terminator at angles of $\zeta - 2v_\zeta(y)$. The angular distance between two rays with perigee heights of y and $y + \Delta y$ will then be

$$\Delta v = - \left(\frac{d(2v_\zeta)}{dy} \right)_y \Delta y,$$

(III.1, 52)

with a corresponding linear distance of

$$\delta = \Delta y + \Delta v r,$$

(III.1, 53)

where r is the distance from the terminator, whence

$$\delta = \left[1 - 2 \left(\frac{dv_\zeta}{dy} \right)_y r \right] \Delta y.$$

(III.1, 54)

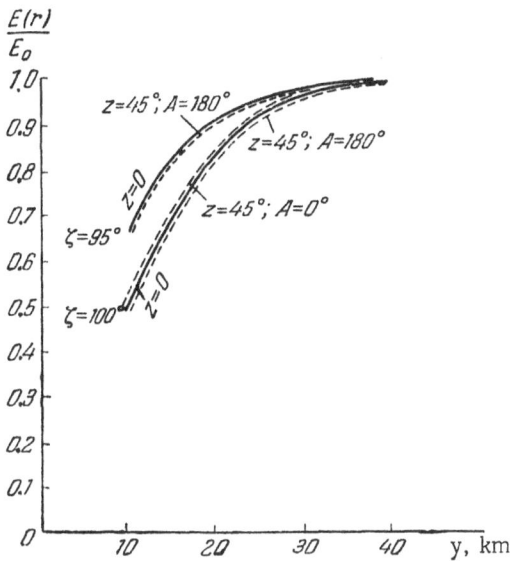

Fig. 91. Relative illuminance $E(r)/E_0$ as a function of perigee height y for selected ζ, z, and A.

The law of conservation of energy implies that in our cylindrical approximation the illuminance $E(r)$ in the lane bounded by rays of perigee height y and $y + \Delta y$ will be given by

$$E(r) = E_0 \frac{\Delta y}{\delta} \, ,$$

(III.1, 55)

where E_0 is the illuminance in the neighborhood of the terminator, or equivalently the illuminance outside the atmosphere. Hence

$$E(r) = \frac{E_0}{1 - 2 \left(\dfrac{dv_\zeta}{dy} \right)_y r} \, .$$

(III.1, 56)

For an approximate evaluation of r we return to Fig. 82a, which yields

$$r = (R + h) \cos(\pi - \zeta') \approx -R \cos \zeta' ,$$

(III.1, 57)

so that by the first of Eqs. (III.1, 36), with allowance for the refractive rotation of the ray,

$$\cos \zeta' = \cos \zeta + \frac{v \cos \varphi}{\cos \psi} - 2v_\zeta \sin \zeta.$$

(III.1, 58)

Finally, if we use Eq. (III.1, 38) and take $R \approx 6370$ km, we obtain

$$E(r) \approx \frac{E_0}{1 - 27e^{-0.16y} \left[\cos \zeta + \dfrac{v \cos \varphi}{\cos \psi} - 2v_\zeta \sin \zeta \right]},$$

(III.1, 59)

or in the solar meridian

$$E(r) \approx \frac{E_0}{1 - 27e^{-0.16y} [\cos \zeta + v \operatorname{ctg} (\zeta - z) - 2v_\zeta \sin \zeta]}.$$

(III.1, 60)

It is very important to recognize that this reduction in the illuminance because of refraction remains significant (for sufficiently large $\cos \zeta'$, that is, for large distances r) until y considerably exceeds 15 km, even as high as y = 30-40 km. Figure 91 illustrates how $E(r)/E_0$ depends on y for selected ζ, z, and A. The refraction-divergence effect depends quite strongly on ζ, but comparatively weakly on z.

So far as we know, Link [230] was the first to consider the refraction-divergence effect, in 1941. From a general theory of refraction in a molecular atmosphere, Link derived an expression free of the rough approximating assumptions upon which Eq. (III.1, 56) rests:

$$E = \frac{E_0}{1 + (R+h)\dfrac{dv}{dy} \cos \zeta'}.$$

(III.1, 61)

In view of Eq. (III.1, 57), we see that Eq. (III.1, 56) differs from Link's expression only in that v is replaced by $2v_\zeta$, which is perfectly acceptable in the region where refraction divergence is important.

We remark in conclusion that the distance from the observer to the geometric terminator, as measured along the earth's surface, is approximately equal to

$$b = 111 (\zeta - 90°) \text{ km}.$$

(III.1, 62)

Figure 89 may be used to correct b for the displacement of the terminator.

§2. Single Scattering of Sunlight by the Atmosphere

Having considered the geometry for the illumination of the atmosphere by sunlight, we are now prepared to develop a theory for the brightness and polarization of the light from the daytime and twilight sky. For the present we shall admit only single scattering of direct sunlight, and neglect the intrinsic glow of the atmosphere. Gruner [224] was the first to formulate the problem, and to solve it in a rough, idealized approximation. We shall not stop to review his reasoning, but shall give instead a generalized form of the theory as incorporated in the work of Fesenkov [5, 77, 83, 232, 237], Shtaude [85, 98, 99, 128, 238-240], Link [226, 227, 266, 294, 295], and Rozenberg [96, 130, 171].

Let $I_0(\lambda)$ denote the intensity of the direct solar radiation outside the atmosphere; this radiation is responsible for the illuminance

$$E_0 = I_0(\lambda)\,\omega_0,$$

(III.2, 1)

where ω_0 is the area of the solar disk in radian measure. Because of the complete depolarization of sunlight, $S_1^0 = I_0$, $S_2^0 = S_3^0 = S_4^0 = 0$. On its way to the point P under observation, a beam of sunlight is attenuated both by extinction and by refraction divergence. According to Eq. (III.1, 56), the corresponding illuminance at the observed point will be

$$E_P = \frac{T_P(\zeta, z, A, h, \lambda)\,I_0(\lambda)\,\omega_0}{1 - 2\left(\dfrac{dv_\zeta}{dy}\right)_y r},$$

(III.2, 2)

where T_P is the atmospheric transparency along the trajectory of the solar ray up to the irradiated point P, and y is the perigee height of the ray.

Light will be scattered within a volume element dV containing the point P, in particular in the observer's direction at the scattering angle φ. If we assume that the atmosphere is isotropic and that its optical properties vary with height only, in other words that the scattering matrix is simply

$$D_{ik} = D_{ik}(h, \varphi, \lambda),$$

(III.2, 3)

then by Eq. (II.2, 16) the components of the vector-parameter for the scattered light at the observer N will be

$$dS_i(\zeta, z, A, h, \lambda) = \frac{D_{i1}(h, \varphi, \lambda)\,E_P\,dV}{l^2\,d\omega},$$

(III.2, 4)

where l , as before, represents the distance of the observed point from the observer, and

$$d\omega = \frac{dV}{l^2 \, dl}$$

(III.2, 5)

is the solid angle subtended at the observer by the volume element dV. Along the path from the scattering volume to the observer, the light is once again attenuated because of extinction; thus, with Eq. (III.2, 2), we obtain

$$dS_i \, (\zeta, \, z, \, A, \, h, \, \lambda) = \frac{D_{i1} \, (h, \, \varphi, \, \lambda) \, T_N \, (z, \, A, \, h, \, L, \, \lambda) \, T_P \, (\zeta, \, z, \, A, \, h, \, \lambda) \, I_0 \, (\lambda) \, \omega_0 \, dl}{1 - 2 \left(\dfrac{dv_\zeta}{dy} \right)_v r}$$

(III.2, 6)

where $T_N(z, \, A, \, h, \, L, \, \lambda)$ is the atmospheric transparency along the trajectory of the ray from the point P to the point N. Finally, by integrating this expression over the full extent of the line of sight, we arrive at a formula for the components of the Stokes vector-parameter for the singly scattered light reaching the observer from an arbitrary direction (z, A):

$$S_i \, (z, \, A, \, \zeta, \, \lambda) = \int\limits_0^{l_{max}} \frac{D_{i1} \, (h, \, \varphi, \, \lambda) \, T_N \, (z, \, A, \, h, \, L, \, \lambda) \, T_P \, (\zeta, \, z, \, A, \, h, \, \lambda) \, I_0 \, (\lambda) \, \omega_0 \, dl}{1 - 2 \left(\dfrac{dv_\zeta}{dy} \right)_v r}$$

(III.2, 7)

where l_{max} is the greatest distance from the observer along the line of sight — either infinity or the distance to the intersection of the line of sight with the earth's surface.

We are now faced with the problem of explaining how the functions in the integrand of Eqs. (III.2, 7) depend on the geometric variables.

To begin with, it is clear that

$$dl = \frac{dh}{\cos z'} \, ,$$

(III.2, 8)

where cos z' is given by (III.1, 36). Moreover, we may use the approximations (III.1, 38) and (III.1, 56) in the denominator.

Outside the selective-absorption bands having rotational structure, the function $T_N(z, A, h, L, \lambda)$ may be expressed in accordance with Eqs. (II.2, 5), (II.2, 6), (II.3, 4), and (II.3, 7) in the form

$$T_N = \exp \left[- \int_0^l k(h, \lambda)\, dl \right] = \exp \left[- \int_L^h k(h, \lambda)\frac{dh}{\cos z'} \right] =$$
$$= \exp \{ - m(z)[\tau(h, \lambda) - \tau(L, \lambda)] \},$$

$$\text{(III.2, 9)}$$

where the air mass $m(z)$ in the line of sight is, in general, a function of the position of the observed point. This dependence, however, is usually weaker than the dependence of $m(z)$ on the variations in the optical structure of the atmosphere, and in most cases it may be neglected. In just the same way we may express the function $T_P(\zeta, z, A, h, \lambda)$ in the form

$$T_P = \exp \left[- m(\zeta')\tau'(h, \lambda) \right]$$

$$\text{(III.2, 10)}$$

if the observed point occurs ahead of perigee of the solar ray [$m(\zeta')$ is the air mass corresponding to a solar zenith distance ζ' at the observed point], or, as recommended by Shtaude [239, 240], in the form

$$T_P = \exp \left[- 2m\left(\frac{\pi}{2}\right)\tau'(y, \lambda) + m(\pi - \zeta')\tau'(h, \lambda) \right]$$

$$\text{(III.2, 11)}$$

if the observed point lies beyond perigee of the ray. Fesenkov [231, 233] has shown by special calculations that $m(\pi/2)$, as well as $m(\zeta \neq \pi/2)$, is practically independent of both the height y and the wavelength. One exception is the case where the extinction is caused by atmospheric constituents such as ozone or sodium, in a layer located at a certain height above the earth. We shall consider this case later, first when treating the problem of evaluating the air mass, and again in §5 of this chapter.

Since the atmospheric extinction coefficient shows a relatively gentle spectral dependence outside the selective-absorption bands, those bands having rotational structure merely entail a supplementary factor T_{NP_s} to allow for selective absorption along the path of the ray up to the scattering volume and beyond to the observer, a factor depending in a complicated way, as we have shown, on the total amount of absorbing matter along the trajectory.

We have, then, as the final form of the expression (III.2, 7),

$$S_i(z, A, \zeta, \lambda) = I_0(\lambda) \omega_0 \int_L^{0,\infty} \frac{D_{i1}(h, \varphi, \lambda) T_{NPS}}{1 - 27e^{-0.16y} \cos \zeta'} \times$$

$$\times \exp \left\{ -m(z)[\tau(h, \lambda) - \tau(L, \lambda)] - \right.$$

$$\left. - \left[\begin{array}{c} m(\zeta') \tau'(h, \lambda) \\ 2m\left(\dfrac{\pi}{2}\right) r'(y, \lambda) - m(\pi - \zeta') \tau'(h - \lambda) \end{array} \right] \right\} \frac{dh}{\cos z'}.$$

$$(III.2, 12)$$

The upper line in the large brackets refers to the region where the observed point lies ahead of perigee of the solar ray, the lower line to the region beyond perigee. The upper limit of integration is ∞ if the line of sight is unbounded, and 0 if it strikes the earth. We recall that the denominator enters only in the region beyond perigee of the solar ray, and that the change in ζ' due to refraction also has to be considered in this region (see § 1).

As an example, let us take the simplest case of calculating the brightness of the daytime sky for an observer located on the ground (L = 0), with absorption absent (k = o). Suppose that the scattering function $f_{11}(\varphi)$ does not depend on height, and that the angles z are not too large, so that cos z' ≈ cos z ≈ 1/m. Then, in view of Eq. (II.2, 18), the relation $\tau'(h) = \tau^* - \tau(h)$, and the fact that under daytime conditions $\zeta' \approx \zeta < 90°$, we obtain

$$I \equiv S_1 = I_0 \omega_0 \frac{f_{11}(\varphi)}{4\pi} m(z) e^{-m(\zeta) \tau^*} \int_0^\infty e^{[m(\zeta) - m(z)] \tau(h)} \sigma(h) \, dh,$$

$$(III.2, 13)$$

or, since σ(h)dh = dτ,

$$I = I_0 \omega_0 \frac{f_{11}(\varphi)}{4\pi} m(z) e^{-m(\zeta) \tau^*} \int_0^\infty e^{[m(\zeta) - m(z)] \tau} \, d\tau,$$

$$(III.2, 14)$$

which yields, upon integration,

$$I = I_0 \omega_0 \frac{f_{11}(\varphi)}{4\pi} m(z) \frac{e^{-m(z) \tau^*} - e^{-m(\zeta) \tau^*}}{m(\zeta) - m(z)}.$$

$$(III.2, 15)$$

We arrive, then, at a well-known formula describing the brightness distribution of the daytime sky as a function of ζ, allowing only for single scattering
of direct sunlight (see, for example, [231, 240]).

§3. The Air-Mass Problem

According to the formula (III.2, 12) for the components of the Stokes
vector-parameter, we can only evaluate the extinction if we know how the
value of the air mass m depends on the zenith distance z and, in general, on
the height h. Under twilight conditions, the air masses for z close to 90° are
of special interest. To the extent that m, as we have seen in Chap. II, § 3,
represents the integrated mean secant of the zenith distance, m(z, h) will show
a substantial dependence on the structure of the atmosphere. As the optical
state of the atmosphere changes, not only will its optical thickness τ vary,
but the form of the function m(z, h) will also, affecting the behavior of the
so-called Bouguer lines [9, 19]. Moreover, because of the horizontal inhomogeneity that usually obtains in the atmosphere [19], an effect particularly striking
ing in the case of twilight phenomena at a characteristic scale of hundreds or

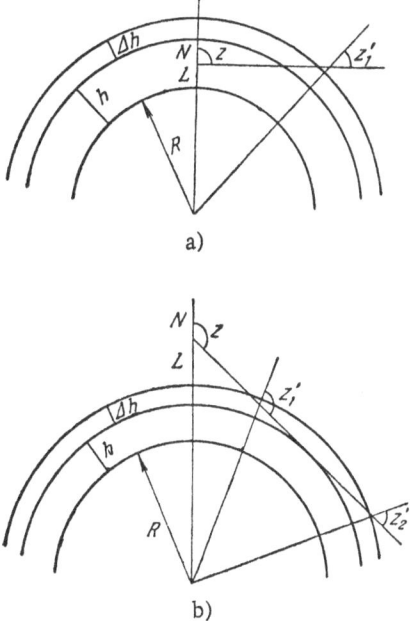

Fig. 92. Establishing the function m(z) for
a thin layer.

even thousands of kilometers, the quantity m will not only vary with time but, in general, with the azimuth as well. Finally, refraction alters the path of a ray through the atmosphere, thereby changing the optical thickness that is traversed, and hence the corresponding air mass.

All these factors make it rather futile to try to calculate m reliably, especially for values of z close to 90°, where each of the effects becomes strongest. We clearly need an experimental determination of m for every particular case (an average value would not suffice since m is so variable). But this situation still does not remove the need for a theoretical study of how m depends on the various factors, and for establishing the limits within which m can vary.

It is the curvature of the atmosphere that has the greatest effect on the behavior of m(z, h). To open the discussion, then, let us suppose that at height h above the earth's surface there is a thin spherical layer of thickness Δh, with the volume extinction coefficient k (Fig. 92). We shall neglect the effect of the rest of the atmosphere for the present. Furthermore, let the observer be stationed at the point N at height L, and let the line of sight have the zenith distance z at the point N. Returning to the first of Eqs. (III.1, 9), we find upon differentiation that

$$(\cos z + u)\, du = (1 + w)\, dw$$

$$(III.3, 1)$$

or, by Eqs. (III.1, 8),

$$dl = \frac{1+w}{\cos z + u}\, dh,$$

$$(III.3, 2)$$

where dl represents an elementary length along the ray (or the line of sight) in the layer. Appealing again to the first of Eqs. (III.1, 9), expand with respect to u to obtain

$$dl = \frac{1+w}{\sqrt{(1+w)^2 - \sin^2 z}}\, dh \approx \frac{dh}{\sqrt{\cos^2 z + 2w}}\,,$$

$$(III.3, 3)$$

where w^2 has been neglected in comparison with unity. If L < h with the observer located below the layer (Fig. 92a), the ray will intersect the layer just once, and the path length within the layer will be

$$\Delta l = \frac{\Delta h}{\sqrt{\cos^2 z + 2w}}\,,$$

with w > 0 by (III.1, 8). But if L > h, with the observer located above the layer (Fig. 92b), the ray will intersect the layer twice (or not at all), so that the total path length within the layer becomes

$$\Delta l = \frac{2\Delta h}{\sqrt{\cos^2 z + 2w}} \, ,$$

with both w and cos z negative. Comparing now Eqs. (III.3, 2) and (III.1, 33), we see that

$$\pm \sqrt{\cos^2 z + 2w} = \cos z'$$

(III.3, 4)

at the point where the ray intersects the layer; for L < h, cos z' is positive, while for L > h both signs will occur (Fig. 92b). For z close to 90°, we must also allow for the refraction variation in the angle z'. If the layer is at a high enough level the correction will evidently be equal to the astronomical refraction ν_z, so that the cos z' determined by the relation (III.3, 4) should be replaced by cos $(z' + \nu_z)$.

For z = 0 the optical thickness of the layer is $\tau = k\Delta h$, but in the direction z, $m\tau = k\Delta l$, whence

$$m = \begin{cases} \sec z' & \text{for} \quad L < h, \\ 2 \sec z' & \text{for} \quad L > h. \end{cases}$$

(III.3, 5)

If the layer is not spherical, the expressions for L < h will remain valid only upon allowing for the z-dependence of the height of the layer; for L > h, the doubled secant will be replaced by the sum of two secants corresponding to different values of h, that is, w.

If cos z ≫ w, we may neglect the difference between z' and z, obtaining the simple expression

$$m = \sec z,$$

(III.3, 6)

which naturally holds strictly for a plane layer. In the real atmosphere it holds well enough when z ≤ 60°, but if z ≥ 60° a correction must be made for the curvature of the layer.

To do this when z is not too large, we expand Eq. (III.3, 4) in powers of w sec² z and retain only the first two terms of the expansion:

$$\cos z' \approx \cos z (1 + w \sec^2 z),$$

(III.3, 7)

so that if $L < h$,

$$\sec z' \approx \sec z \, (1 - w \sec^2 z) = \sec z - w \sec^3 z; \tag{III.3, 8}$$

subsequent terms may be neglected whenever $z \lessgtr 75°$.

We proceed now to evaluate the air mass for the whole thickness of the atmosphere above the level L. By definition,

$$m(z, L) = \frac{\int\limits_{L}^{\infty} k(h) \sec z'(h) \, dh}{\int\limits_{L}^{\infty} k(h) \, dh} = \overline{\sec z'}. \tag{III.3, 9}$$

Thus, from the viewpoint of the m(z) relationship the atmosphere is entirely equivalent to a thin layer whose height and form satisfy the requirement $\sec z'_{layer} = \overline{\sec z'}_{atm}$. Shtaude [241] was the first to point out this fact, on the basis of her analysis of the general properties of the function m(z).

As we have seen, when $z \lessgtr 60°$ there is hardly any difference between z and z', the relation (III.3, 6) holding independently of the structure of the atmosphere and the height L of the observer. At greater angles z we shall use Eq. (III.3, 8). Thus

$$m(z, L) = \sec z - \overline{w} \sec^3 z, \tag{III.3, 10}$$

where

$$\overline{w} = \frac{\int\limits_{L}^{\infty} k(h) \, w \, dh}{\int\limits_{L}^{\infty} k(h) \, dh}, \tag{III.3, 11}$$

or, by (III.1, 8),

$$\overline{w} = \frac{\int\limits_{L}^{\infty} k(h) \, h \, dh}{(R+L) \int\limits_{L}^{\infty} k(h) \, dh} - \frac{L}{R+L} = \frac{\overline{h} - L}{R+L}, \tag{III.3, 12}$$

where $\overline{h}(L)$, the mean optical height of the atmosphere for an observer located at the level L, depends on the atmospheric structure. In particular, if

$$k(h) = k_0 e^{-h/H}, \tag{III.3, 13}$$

where H is the optical height of a homogeneous atmosphere, then by direct integration we readily find that

$$\bar{h} = L + H,$$

(III.3, 14)

whence

$$\bar{w} = \frac{H}{R+L} \approx \frac{H}{R} .$$

(III.3, 15)

Returning to Eq. (III.3, 10), we see that the correction is proportional to the optical height of the homogeneous atmosphere at the level L. If the atmosphere contains no aerosols or selectively absorbing materials, H will correspond to the ordinary height of the homogeneous atmosphere; this quantity remains relatively constant [185], which accounts for Fesenkov's finding [231, 233] that m is practically independent of the height of the observer.

Fesenkov's calculations referred, however, to an idealized mean atmosphere. The actual aerosol distribution with height, particularly in layers near the ground, can be represented neither by the model adopted in [233] nor by the exponential law used in [231] and suggested above by us. In the real atmosphere, we would therefore expect to find, for z close to 90°, a marked dependence of m on height, and substantial departures of the m(z) law from that corresponding to any model. In this regard we may mention that experiments [19] have revealed no correlation between the horizontal and oblique transparencies of the atmosphere, as would not have been the case if there were an m(z) law universally valid up to z = 90°. One may adopt such a law only when making estimates of the contribution of some particular effect, not for purposes of measurement. On the other hand, if z is not too large the influence of the optical structure of the atmosphere will remain relatively weak, and the z-dependence of m will be a perfectly regular one. We shall estimate below the scale of the fluctuations in m for z close to 90°. But we must first examine the effects of refraction, as they become important at the angles z of interest to us.

This topic was first studied by Laplace, who assumed a purely gaseous isothermal atmosphere. For L = 0 we then have, as Laplace showed,

$$m \approx \frac{\nu_z}{(n_0 - 1) \sin z} ,$$

(III.3, 16)

where ν_z is the angle of astronomical refraction and n_0 is the refractive index at sea level. Under normal conditions, $n_0 - 1 \approx 0.000293$, so that for z = 90° we obtain m ≈ 44. Subsequently, Bemporad discarded the isothermal-atmos-

TABLE 10. Air Mass m as a Function of Zenith Distance

z, deg	sec z	Laplace	Bemporad	Muller	$(\cos z + 0.025\, e^{-11\cos z})^{-1}$
0	1.00	1.00	1.00	1.00	1.00
30	1.15	1.15	1.15	1.14	1.15
45	1.41	1.41	1.41	1.45	1.41
60	2.00	1.99	2.00	2.18	2.00
70	2.92	2.90	2.90	3.30	2.92
75	3.86	3.81	3.82	4.32	3.85
80	5.76	5.56	5.60	6.03	5.65
83	8.20	7.68	7.77	7.81	7.60
85	11.5	10.2	10.4	10.0	10.4
86	14.3	12.1	12.4	11.8	12.3
87	19.1	14.8	15.4	14.4	15.1
88	28.65	18.84	19.79	18.02	19.4
89	57.29	—	26.95	—	26.3
90	∞	44	(35—40)	—	40

phere hypothesis, introducing height variations of the temperature and pressure into the calculation. It was found that the closer z is to 90°, the stronger is the influence of variations in atmospheric conditions on the value of the air mass. In particular, at z = 87° different model atmospheres lead to values of m between 14.84 and 15.36, while at z = 87° a change in ground temperature from −60 to +60°C implies a decrease of 5 in m [242]. In recent years several authors, particularly Link [230, 292, 293, 296] and Fesenkov [231-233], have performed calculations of the function m(z) for various model atmospheres; at large z, their results for m differ by about 1-2, and at z = 90° they range from 35 to 42. However, Muller has reduced photometric data for stars at various zenith distances, obtaining a mean m(z) law differing considerably from all the theoretical calculations. Table 10 compares the functions m(z) as given by Eq. (III.3, 6), by the Laplace and Bemporad formulations, and by Muller's data. The table also includes values for the air mass as computed from an approximate empirical formula suggested by Rozenberg:

$$m = (\cos z + 0.025\, e^{-11\cos z})^{-1}.$$

(III.3, 17)

We see that this formula reproduces the function m(z) accurately enough (see also Table 11), and its simplicity makes it very convenient for considering the qualitative aspect of the twilight phenomenon. It will be clear that we can hardly expect to go further and find a single universal formula, for the variability of atmospheric conditions prevents such a representation from covering every particular case.

TABLE 11. Air Mass m(z) for Selected Heights L of the Observer [292]

L, km	z, deg 90°	89°40'	89°20'	89°	88°	87°
0	38.35	33.6	29.4	26.4	19.4	15.3 summer
	40.77	35.6	29.4	27.4	20.0	15.6 winter
2	37.51	33.5	29.6	26.5	19.5	15.3 summer
	40.19	35.0	29.6	27.2	20.0	15.5 winter
4	38.43	33.8	29.8	26.8	18.7	15.3 summer
	40.23	35.1	28.8	27.5	20.0	15.5 winter
6	39.26	34.4	29.4	27.1	19.9	15.5 summer
	40.88	35.7	30.2	27.8	20.2	15.7 winter
$\cos z + 0.025\, e^{-11\cos z}$	40.0	34.0	30.0	26.3	19.4	15.1

In Table 11 we give values of m(z) for z close to 90° at various heights L of the observer, as computed by Link and Neužil [292] for the summer and winter seasons. Note that the L-dependence of m is much weaker than the seasonal variation, or for that matter the day-to-day variation of m. This behavior agrees fully with the results of Fesenkov's calculations [233], which indicate that m is practically independent of L.

We come now to one further effect which so far has not received proper attention in connection with the transparency of the atmosphere and which, in our opinion, makes it pointless even to formulate the air-mass problem for large z with an observer located at or near the earth's surface, so far as the short-wave region of the spectrum is concerned. This effect — the multiple scattering of light — violates the conditions for the Bouguer law, and makes it necessary to operate with the instrumental rather than the idealized transparency of the atmosphere (Chap. II, §2).

To estimate the magnitude of this effect, we return to Eq. (II.2, 59), and inquire as to what air masses are necessary for the effect to become appreciable.

Assume that the field of view of the measuring device coincides with the angular size of the sun, namely $\omega = \omega_0 \approx 9 \cdot 10^{-5}$ rad^2 (we note that because of refraction the value of ω_0 depends slightly on the zenith distance). Then Eq. (II.2, 59) will take the form

$$\frac{I_{\text{scat}}}{I_{\text{dir}}} = 3 \cdot 10^{-5} \frac{g^2(z)\, l}{e^{-\tau \sec z} \sec z\,[(1-A)\,\tau + l]}. \qquad (III.3, 18)$$

TABLE 12. Values of sec z Corresponding to Selected
Values of τ, A, and I_{scat}/I_{dir}

τ	A	I_{scat}/I_{dir}				
		0.01	0.1	1	10	100
0.1	arbitrary	>80	>80	>80	>80	>80
0.2	arbitrary	49	62	75	>80	>80
0.3	0	31.5	40	49	57	64
	0.8	30.5	39	47	56	63
0.4	0	23	30	35.5	42	48
	0.8	21	28.5	35.0	41	47
1.0	0	8.7	11.2	13.7	18.2	18.6
	0.8	8.0	10.6	13.2	15.6	18.0

If we take the rough estimates $g(z) = 1$, $l = 1$, and solve this equation graphically for several values of τ, we will find the values of sec z corresponding to a given value of I_{scat}/I_{dir}. Table 12 presents the results of the calculation for A = 0 (typical of conditions above the ocean) and A = 0.8 (the conditions above snow cover).

In view of the sphericity of the earth's atmosphere, the sec z in the expression (III.3, 18) and in Table 12 should be interpreted as the air mass m(z). However, since the expression (III.3, 18) is itself valid only for a plane layer, the estimates given in Table 12 should be regarded as provisional only.

If we utilize the tables of [16], which were computed by machine solution of the equation of transfer for a purely gaseous atmosphere (again without allowance for the sphericity), we obtain the following values for I_{scat}/I_{dir}: for sec z = 10 and albedo A = 0, 3% for $\tau = 1$ and 0.05% for $\tau = 0.5$; for sec z = 10 and A = 0.8, 7% for $\tau = 1$ and 0.07% for $\tau = 0.5$; for sec z = 5, 0.08% for $\tau = 1$ if A = 0 and 0.14% if A = 0.8, in good accord with the data of Table 12. We note that for $\tau = 1$ and $\tau = 0.5$, Eq. (II.2, 59) gives satisfactory agreement with the tables of [16] for the value $l = \frac{4}{3}$ corresponding to molecular scattering; the departures do not exceed a factor of 2, and this can be explained entirely by the uncertain angular dependence of g(z) and the maintenance of a weak azimuthal dependence. In fact a comparison shows that g(z) does not greatly differ from unity.

Returning now to the real atmosphere, let us take the data given in Allen's tables [236] for the mean vertical optical thickness, representing settled, fair weather with the observer located at sea level (Table 13).

TABLE 13. Mean Spectral Dependence of τ^*

λ, mμ	0.34	0.36	0.38	0.40	0.45	0.50	0.55	0.60	0.65	0.70	0.80	0.90	1.0
τ^*	0.84	0.67	0.55	0.46	0.31	0.23	0.19	0.17	0.127	0.093	0.063	0.049	0.040

A comparison of Tables 13 and 12 demonstrates at once that in the long-wave region of the spectrum, $\lambda \gtrsim 0.65$ μ, the Bouguer law applies unconditionally to all the cases covered by Eq. (III.2, 12). In the interval 0.50 $\mu \lesssim \lambda \lesssim$ 0.65 μ, it remains valid at all zenith distances for an observer located on the earth's surface, but fails for rays with perigee y = 0, whose attenuation must be allowed for over the entire path through the atmosphere. For $\lambda \lesssim 0.5$ μ, the Bouguer law begins seriously to break down even for an observer on the ground with a line of sight inclined to the horizon; the failure occurs most readily for small λ. In particular, at $\lambda \lesssim 0.5$ μ the sun, in effect, no longer sets below the horizon but rather is lost in the haze of multiply scattered light. Incidentally, since the ratio I_{scat}/I_{dir} at the time of sunset, when m = 40, is some 100 times as great in the blue spectral region as in the green, a green flash is observed at sunset [243]. Similarly, the effect is a striking one in color photographs of Venus when the planet is at the horizon.

We shall find it useful to estimate the perigee height y at which we may begin to use the Bouguer law to compute the attenuation of direct sunlight penetrating the entire thickness of the atmosphere, that is, to use the factor exp $[-2m\,(\pi/2)\,\tau'(y)]$ in Eq. (III.2, 12). Table 12 shows that we may do so only if $\tau'(y) \lesssim 0.1$. Returning to Fig. 62 (Chap. II, §3), and observing that the presence of aerosols will as a rule approximately double the true value of τ' at $\lambda \approx 0.5$ μ compared to the value computed for a purely gaseous atmosphere, we find the perigee heights entered in Table 14.

We have assumed here that the spectral dependence of τ for the aerosol component is proportional to λ^{-1}. It will be shown presently that the restrictions on which Table 14 is based have hardly any effect on the region of applicability of Eq. (III.2, 12).

We conclude, then, that the Bouguer law may be used with confidence when $m\tau^* \leq 6$-7. This value refers to the case where the aperture of the instrument coincides in angular size with the source of light. If some of the source remains outside the field of view, the estimates will not change. But if the field of view of the instrument also includes some of the field around the source, with $\omega > \omega_0$, the contribution of scattered light will increase, and the limit quoted above will shift toward smaller $m\tau^*$.

TABLE 14. Limit of Applicability of the Bouguer Law
for Tangential Solar Rays

λ, μ	0.3	0.4	0.5	0.6	0.7	0.8	0.9
y_{min}, km	18	13	7.5	4	1	0	0

As a special case, we have the measurement of the brightness of an extended source of light, such as when the upper layers of the atmosphere are
illuminated, during twilight hours, by the direct rays of the setting sun. In this
event our estimates will remain in force, but with the reservation that the instrument will also receive scattered light from the rest of the sky. From now
on we shall regard the terms "secondary scattering" or "multiply scattered
light" as including this correction. Estimates for the correction and procedures
for experimental evaluation will be discussed in Chaps. IV and V.

If the atmosphere is illuminated from the outside by a diffuse rather
than a directed light beam, with an approximately uniform angular intensity
distribution, then the Bouguer law definitely will not be applicable, whatever
the optical thickness of the atmosphere. To calculate the radiant flux received by the detector, Eq. (II.2, 58) should be used directly, setting $g(\vartheta) = 1$.

Finally, the unavoidable presence of aerosols in the air produces, as we
have seen in Chap. II, a strong forward elongation of the scattering diagram.
If the optical thickness $m\tau^*$ is not too great, this aureole effect (with allowance for forward multiple scattering also) will result in an increased brightness of the background of scattered light when the instrument is directed toward
the source of light. The increase in the background level will amount to about
two orders of magnitude under atmospheric conditions, once again displacing
the limit of applicability of the Bouguer law toward smaller $m\tau^*$. However,
one should recognize that for the aureole (even with forward multiple scattering) the optical path length traversed by the light will not differ appreciably
from $m\tau^*$. Hence the aureole will be attenuated in the same proportion as
the direct sunlight. We infer that the Bouguer curves for log I as a function of
m will remain valid, but their slope will change slightly to reflect the effective decrease in optical thickness (that is, in the extinction coefficient), and
the effect will depend substantially on the size of the field of view of the instrument. When $m\tau^* \lesssim 6$-7, that is, when we first encounter signs of a departure from the Bouguer law, the contribution of the aureole against the general background of scattered light will begin to decline rapidly, and the aureole
will cease to appear significantly against the background shortly before the
light source itself vanishes in the background.

§4. Anomalous Transparency and the Reversal Effect

The ozone layer in the atmosphere is responsible for characteristic twilight effects that illustrate clearly the behavior described at the end of the
preceding section. We refer here to the reversal effect (Umkehreffekt) discovered by Götz [244] in 1929, and to the anomalous-transparency effect discovered by Rodionov, Pavlova, and Stupnikov [245] in 1936. Both phenomena
concern the dependence on solar zenith distance of the color index $CE_{\lambda_2}^{\lambda_1}$ for
a pair of wavelengths on the long-wave side of the Hartley absorption band
of ozone (at $\lambda \approx 0.3\ \mu$). The intensity of sunlight that has crossed the ozone
layer on the way to the ground should evidently decline monotonically as the
air mass m_ζ (or the solar zenith distance) increases; the decline should be
most rapid when the ozone absorption coefficient ($k^{ozone} = \alpha^{ozone}$), or the
optical thickness τ_0 of the ozone layer, is large. If $\lambda_1 > \lambda_2$, the ozone absorption will be smaller at λ_1 than at λ_2, that is, $\tau_0(\lambda_1) < \tau_0(\lambda_2)$, and the color index defined by Eq. (I.3, 1),

$$CE_{\lambda_2}^{\lambda_1} \approx 2.5\, m\, [\tau(\lambda_1) - \tau(\lambda_2)],$$

where the Bouguer law is used, should also decrease monotonically with increasing ζ.

Yet if we point a detector at the zenith and measure the ζ-dependent
sky brightness at λ_1 and λ_2, as Götz did, we find that $CE_{\lambda_2}^{\lambda_1}(\zeta)$ indeed diminishes monotonically at small ζ; but at a certain value of ζ, varying from
day to day, the function $CE_{\lambda_2}^{\lambda_1}(\zeta)$ passes through a minimum and, reversing
direction, begins to increase (Fig. 93). This is the reversal effect, and
we may call its graph the reversion curve. The position of the minimum depends on the choice of the wavelength pair; in particular, if one holds λ_1 constant and displaces λ_2 shortward, the minimum will regularly move toward
smaller ζ. Moreover, the position of the minimum, and indeed the behavior
of the entire reversion curve, depends decisively on the total ozone content
in the atmosphere: an increase in the ozone content will again displace the
curve toward smaller ζ. Finally, the shape of the reversion curve is highly
variable, often differing strongly for different wavelength pairs, and cases have
been observed where the reversal effect is missing altogether. As an example,
Fig. 94 illustrates reversion curves derived from observations by Rodionov and
Pavlova [246]; they represent means over 12 days.

Rodionov, Pavlova, and Stupnikov [245] discovered a related effect when
the detector is pointed directly at the sun ($z = \zeta$, $A = 0$), as illustrated in Fig. 95.

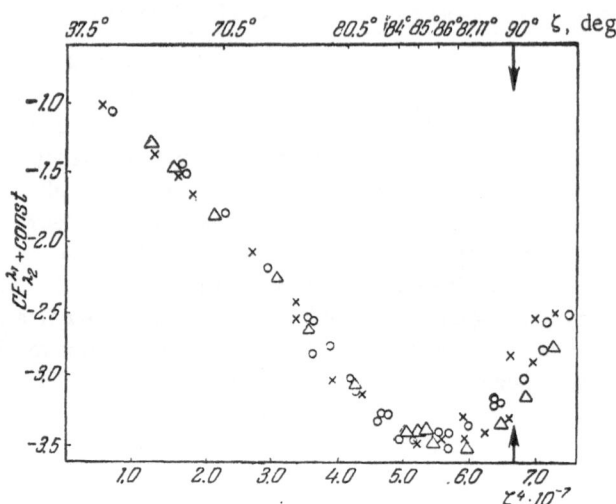

Fig. 93. Sample $CE_{\lambda_2}^{\lambda_1}(\zeta)$ curves observed at the
zenith, showing the reversal effect. Following estab-
lished convention, the abscissa is the quantity $\zeta^4 \cdot 10^{-7}$.
The various symbols represent different wavelength
pairs. All the curves have been adjusted to agree in
color index at small ζ.

They called it the a n o m a l o u s - t r a n s p a r e n c y e f f e c t, because if the
Bouguer law holds over the entire range of ζ, the effect can arise only from a
relative increase in the transparency at λ_2 compared with the transparency at
λ_1 ($\lambda_2 < \lambda_1$).

   Ever since they were discovered, these two effects have been described
in all treatises on atmospheric physics, and their explanation has been the subject of incessant controversy. Götz offered the following explanation for the
reversal effect. When the instrument is pointed at the zenith it will receive
only scattered sunlight, with scattering occurring both above and below the
ozone layer (Fig. 96). However, there is an important difference between the
two cases. When light is scattered above the layer, it reaches the scattering
volume without attenuation and crosses the ozone layer in the vertical direction, along the trajectory from the scattering region to the observer. The
transparency of the ozone layer for these rays will be $e^{-\tau_0}$. But when light is
scattered below the layer, it will have penetrated the ozone layer prior to the
scattering event, so that the corresponding transparency of the layer will be

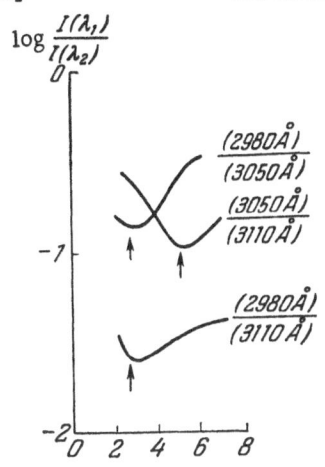

Fig. 94. The reversal effect as observed at the zenith, for selected wavelength pairs. The arrows indicate the computed positions of minimum. The abscissa is the air mass m_0 of the ozone layer.

$e^{-m_0(\zeta)\tau_0}$, where $m_0(\zeta)$ is the air mass for the ozone layer. Since most of the air lies below the ozone layer, scattering in the lower layers will predominate at small ζ. But at some value of ζ these regions will be subordinated to the air layers lying above the ozone layer, with a corresponding change in the behavior of the color index $CE^{\lambda_1}_{\lambda_2}(\zeta)$. It was on these grounds that Götz suggested using the reversal effect to derive the structure of the ozone layer with respect to height. The theory for this procedure has since been further developed in papers by several authors (see, for example, [247]), and has now become a fundamental tool for ozonometric studies. It is noteworthy that the height distributions of ozone obtained in this way accurately reflect the true state of affairs, at least in general outline.

On the other hand, Rodionov has shown that if one adopts the Bouguer law, no considerations of the Götz type can account for the anomalous-transparency effect. In Rodionov's view, the effect can only be explained on the assumption that the attenuation coefficient for the aerosol component of the atmosphere falls off with decreasing wavelength in the ultraviolet region. This assumption rests on a selective-transparency effect discovered by Rodionov, Pavlova, Rdultovskaya, and Reinov [248]. Because of the latter effect, direct measurements of the atmospheric transparency imply that the aerosol component of atmospheric extinction declines rapidly with penetration into the ultraviolet (Fig. 97). It follows at once that the anomalous-transparency effect arises from two concurrent factors: attenuation of light by ozone, increasing toward shorter wavelengths, and attenuation by aerosols, increasing in the opposite direction. Since the ozone layer lies far above the aerosol layer, the relative role of the two factors depends on ζ, as the $m(\zeta)$ relation differs for the upper and lower atmosphere (Fig. 98). Ozone attenuation prevails at small ζ, and aerosol attenuation at large ζ.

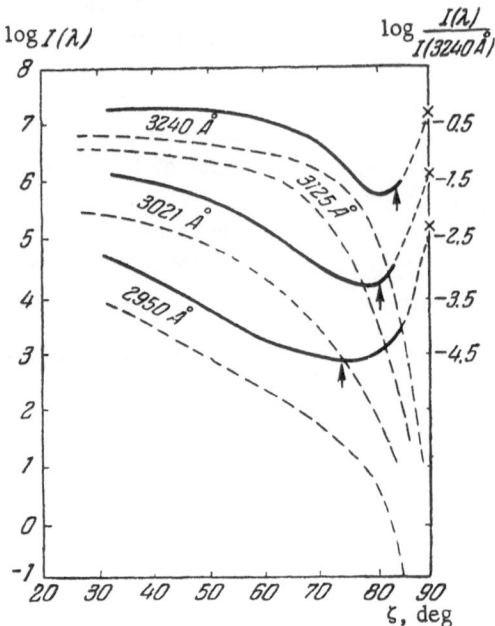

Fig. 95. Broken curves, log I(λ) as a function of
ζ. Solid curves, the function $CE_{\lambda_2}^{\lambda_1}(\zeta)$ – the anomalous-transparency effect. Instrument pointed at the
sun. The brightness is expressed in arbitrary units.

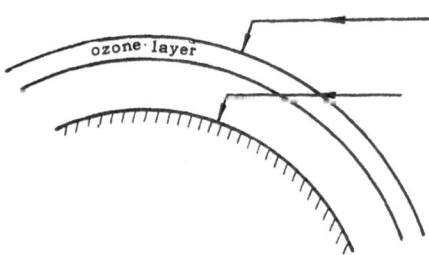

Fig. 96. Götz's explanation of the reversal effect.

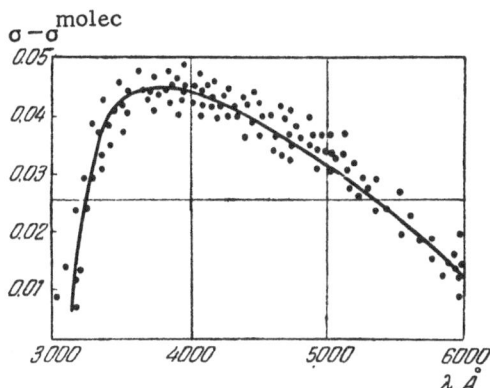

Fig. 97. The selective-transparency effect.

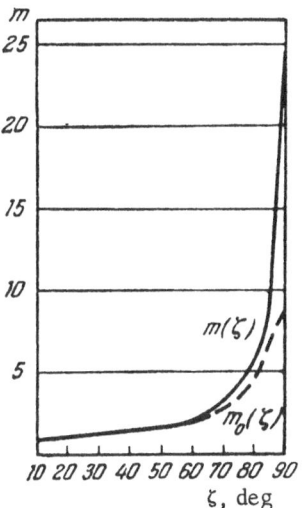

Fig. 98. Air mass m(ζ) for the entire atmosphere, according to Bemporad (solid curve), and for the ozone layer, according to Rodionov (dashed).

Rodionov [249] has also called attention to the similarity of the reversal and anomalous-transparency effects, and to the relatively close agreement between the positions of minimum for observations of direct and scattered light. We cannot but concur in Rodionov's opinion that the similarity of the two effects indicates that they have a common origin. Thus in Rodionov's view the reversal effect is also to be explained through the action of aerosols; we would then have to regard Götz's ideas as untenable, and his method inapplicable for investigating the height structure of the ozone layer.

Strong arguments can be advanced against any such rejection. First, it is hard to believe that the data on the structure of the ozone layer as obtained by the reversal method and by direct probing of the atmosphere could be in chance agreement. Secondly, it is well

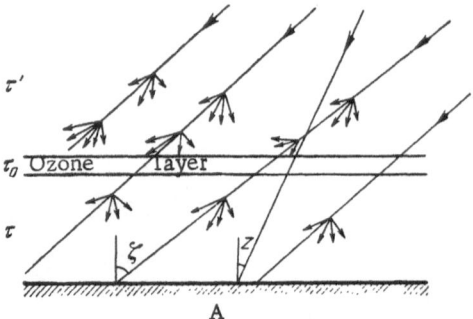

Fig. 99. Illustration of the reversal and anomalous-transparency effects.

known that the aerosol component of the air is far from constant in character,
and that the aerosol attenuation of light exhibits a highly variable spectral behavior [19]. Yet the selective-absorption effect can only be explained through
a very stable particle distribution with size, at any rate for the large particles
that are most efficient optically. Thirdly, the strongly variable aerosol concentration that is so characteristic of the real atmosphere can in no way be
reconciled with the relatively high stability of the anomalous-transparency
and reversal phenomena. Finally, the anomalous-transparency and reversal
effects set in at the very solar zenith distances where, as we have seen, multiple scattering breaks down the Bouguer law on which the explanation of the
effects had been based.

This last circumstance enables us to assert that all three phenomena —
the reversal, anomalous-, and selective-transparency effects — actually do
have a common nature; however, the explanation does not lie in the presence
of atmospheric aerosols, but in multiple scattering, which until now has been
excluded from the analysis. (An attempt by Pekeris [250] to incorporate secondary scattering must be considered unconvincing from the standpoint of
modern transfer theory.) We shall see below that it is possible to preserve
Götz's general idea, together with the application of his method to a determination of the height structure of the ozone layer, although it becomes
necessary to make some important revisions in the theory for the interpretation of the reversion curves.

To allow fully for the effects of multiple scattering at large solar zenith
distances, that is, under the twilight conditions where the reversal and anomalous-transparency effects are observed, one must solve an equation of transfer
that incorporates the sphericity of the atmosphere. So far this has been im-

practicable even for the simplest models. Nevertheless, even with a plane model one can consider the problem semiquantitatively and develop the essential aspects. We shall find that so long as attention is confined to very strong effects, where the quantities vary by several orders of magnitude, a plane treatment will yield sufficiently reliable quantitative estimates.

Consider, then, the idealized arrangement of Fig. 99. Suppose that the ozone is concentrated in a thin layer of optical thickness τ_0. Since $\alpha \gg \sigma$ ($\beta \gg 1$) within this layer, the Bouguer law is valid there for arbitrary ζ, since multiple-scattering effects become appreciable only in the event that $\beta \lesssim 1$. [We recall that Eqs. (II.2, 58) and (II.2, 59) hold only if $\beta \ll 1$.] The bulk of the air lies beneath the ozone layer; it has an optical thickness of τ, with $\beta \ll 1$ because ozone is absent. There is also air above the ozone layer, with $\beta \ll 1$ and an optical thickness $\tau' \ll \tau$. At the point A, we place a measuring device with an angular field of ω, and point it at zenith distance z in the solar vertical (the sun's zenith distance is ζ).

In the spectral region where the reversal and anomalous-transparency effects are observed ($\lambda \approx 0.3 \, \mu$), the optical thickness of the lower layers of the atmosphere is $\tau \approx 1.2$, allowing for the aerosol component. Thus at large ζ we should use the concept of instrumental rather than idealized transparency for the lower layer of the atmosphere [Eq. (II.2, 59)]. If we now introduce the attenuation of direct sunlight in the ozone layer and in the upper layer of the atmosphere, the expression for the instrumental transparency of the atmosphere will take the form

$$t = \frac{g(\zeta) g(z)}{\pi} \cos \zeta \, \frac{\omega l}{(1-A)\tau + l} \exp\left[-(\tau_0 + \tau)\sec z\right] +$$
$$+ \exp\left[-(\tau_0 + \tau + \tau')\sec \zeta\right]\delta_{z,\zeta}, \tag{III.4, 1}$$

where $\delta_{z,\zeta}$ is the Kronecker symbol, and it is assumed that $\omega \geq \omega_0$.

However, light is also scattered in the upper layer of the atmosphere, and light scattered above the ozone layer and subsequently rescattered in the lower atmospheric layer will also enter the measuring device. The intensity of the light scattered by the upper layer is given by the relation (III.2, 15), which may, since τ' is considered small, be written as

$$I_p \approx I_0 \omega_0 \frac{f_{11}}{4\pi} \sec z \, \tau' \left(1 - \frac{\sec \zeta + \sec z}{2} \tau'\right). \tag{III.4, 2}$$

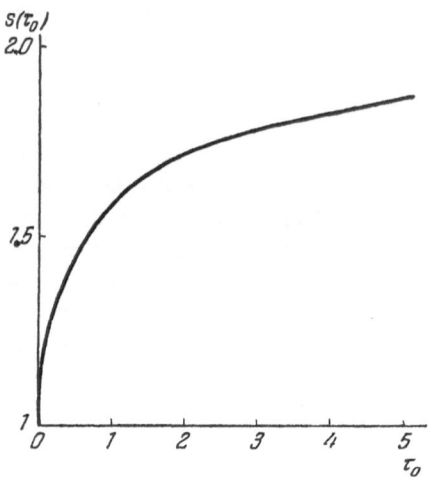

Fig. 100. The function $s(\tau_0)$.

Since the light scattered in the upper layer forms a comparatively uniform radiation field, the Bouguer law by no means need fail for a given optical thickness of the lower layer. We shall use here the expression (II.2, 58) for the instrumental transmissivity, taking $g(\vartheta) = 1$. To this end we must find the flux density Φ_0 of the descending scattered radiation directly beneath the ozone layer. It is readily shown that

$$\Phi_0 \approx \frac{I_0\omega_0\tau'}{2}\left(1 - \frac{\sec\zeta}{2}\tau'\right)\int_0^1 e^{-\tau_0\sec z}\,d\cos z =$$

$$= \frac{I_0\omega_0}{2}\tau'e^{-\tau_0}\,s\,(\tau_0)\left(1 - \frac{\sec\zeta}{2}\tau'\right), \tag{III.4, 3}$$

where we have taken $f_{11}(\varphi) = 1$, a spherical scattering diagram, and where

$$s\,(\tau_0) = 1 - \tau_0\,e^{-\tau_0}\,\text{Ei}\,(-\tau_0). \tag{III.4, 4}$$

Figure 100 shows the form of the function $s(\tau_0)$. In the expression for the instrumental transparency, we must correspondingly introduce an additional term to cover the diffuse radiation flux (III.4, 3), thus obtaining, instead of Eq. (III.4, 1),

$$t = \frac{g(z)}{\pi} \frac{l\omega}{(1-A)\,\tau + l} \left[g(\zeta) \cos\zeta\, e^{-(\tau_0 + \tau')\sec\zeta} + \right.$$
$$\left. + \frac{\tau'}{2} e^{-\tau_0} s(\tau_0) \left(1 - \frac{\sec\zeta}{2}\tau'\right) \right] + e^{-(\tau_0 + \tau + \tau')\sec\zeta} \delta_{z,\zeta}. \quad \text{(III.4, 5)}$$

In the spectral region of interest to us, at the edge of the Hartley band ($\lambda \approx$ 0.3 μ), the optical thicknesses vary approximately over the ranges

$$10^{-2} \lessgtr \tau_0 \lessgtr 5, \quad 1 \lessgtr \tau \lessgtr 1.4,$$

and $\tau' = c\tau$, where the constant $c \ll 1$ is independent of λ and is governed only by the height of the ozone layer. Here the constant c is approximately equal to the ratio of the atmospheric pressure at the level of the ozone layer to the pressure at sea level. If we now take into account the sphericity of the earth's atmosphere, then we should replace the sec ζ by the air masses m(ζ) for the atmosphere and $m_0(\zeta)$ for the ozone layer (Fig. 98). Note, however, that for very large m(ζ) we may expect marked departures due to the spherical atmosphere. Moreover, if we set $l = \frac{4}{3}$, corresponding to molecular scattering, adopt g(ζ) = 1 as a rough estimate, and neglect the attenuation of direct sunlight in the atmospheric region above the ozone, then we obtain

$$t \approx \frac{0.4\,\omega}{(1-A)\,\tau + \frac{4}{3}} \left[\frac{e^{-\tau_0 m_0(\zeta)}}{m(\zeta)} + \frac{c\tau}{2} e^{-\tau_0} s(\tau_0) \left(1 - \frac{m}{2} c\tau\right) \right] +$$
$$+ e^{-\tau_0 m_0(\zeta) - \tau m(\zeta)} \delta_{z,\zeta}. \quad \text{(III.4, 6)}$$

Suppose that the instrument is first pointed in a direction away from the sun (toward the zenith, for example); thus we take $\delta_{z,\zeta} = 0$, corresponding to the conditions for observation of the reversal effect. Since c is small, the first term in brackets will be considerably greater than the second at small ζ, and will decline almost exponentially with increasing $m_0(\zeta)$. Because of the ζ-dependence of m, the second term will also decrease with increasing ζ, but this drop will only become appreciable at very large ζ (since c is small), namely at $\zeta \approx 85°$. Thus at a certain value of ζ corresponding to equality between the two terms in brackets, that is, when

$$\frac{1}{s(\tau_0)} e^{-\tau_0 [m_0(\zeta) - 1]} = \frac{c}{2} m\tau \left(1 - \frac{c}{2} m\tau\right), \quad \text{(III.4, 7)}$$

a radical change will occur in the rate with which the radiation flux arrives at the recording device. The effect appears distinctly in Fig. 101, which shows

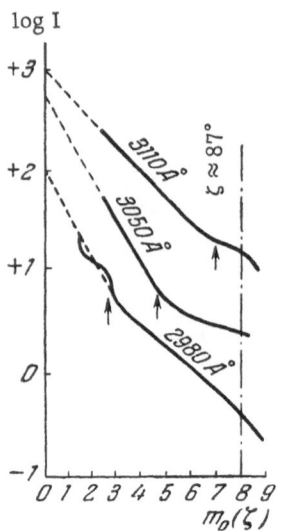

Fig. 101. Zenith observations of log I as a function of $m_0(\zeta)$, according to the data of Fig. 94. Brightness expressed in arbitrary units.

the somewhat smoothed experimental curves on which Fig. 94 was based, drawn as a function of $m_0(\zeta)$. The accelerating drop of the curves for $\zeta \gtrsim 86°$ is due qualitatively to the presence of the factor $(1 - cm\tau/2)$ in Eq. (III.4, 6), as well as to the breakdown of Eq. (II.2, 59) at large ζ because of the sphericity of the atmosphere. The latter effect should also explain the mild slope of the log $I(m_0)$ curves after their turning point.

Figure 101 indicates that a decrease in wavelength, or an increase in τ_0, will displace the turning point m_{0t} of the log $I(m_0)$ curves toward smaller m_0. This behavior follows at once from Eq. (III.4, 7); differentiating the condition, we obtain

$$\left.\begin{array}{c} \dfrac{\partial m_{0t}}{\partial \tau_0} \approx -\dfrac{m_0 + \dfrac{d \ln s(\tau_0)}{d\tau_0} - 1}{\tau_0 + \dfrac{1}{m}\dfrac{dm}{dm_0}}, \\[4mm] \dfrac{\partial m_{0t}}{\partial (\ln c)} \approx \dfrac{1}{\tau_0}, \qquad \dfrac{\partial m_{0t}}{\partial (\ln \tau)} \approx \dfrac{1}{\tau_0}, \end{array}\right\} \qquad (\text{III.4, 8})$$

so that m_{0t} depends relatively weakly on τ and c but strongly on τ_0, and an increase in any of these quantities will displace the turning point toward smaller values of m_0.

If we adopt an ozone content of 0.3 cm and use the tables given in ⌊247⌋ for the ozone absorption coefficient, then with a height $h \approx 30$ km for the ozone layer (see Fig. 42), that is, with $c \approx 1 \cdot 10^{-2}$, we may use Eq. (III.4, 7) to find the positions of the turning points for the wavelengths indicated in Fig. 101. The resulting values are marked by arrows in the figure. Since the assumptions adopted here are so rough, we would not expect to find better agreement either in the absolute values or in their λ-dependence. Note that we are considering an especially idealized model, with its thin, uniform ozone layer.

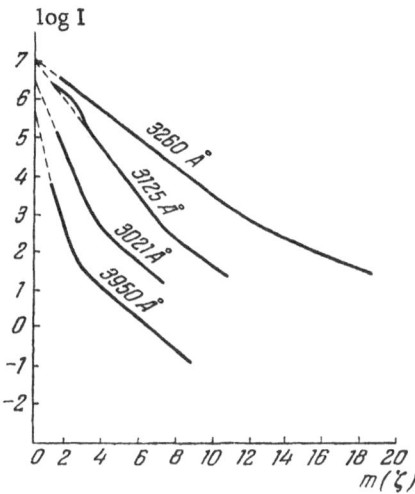

Fig. 102. Observations at the sun of log I as a
function of m(ζ). Brightness expressed in arbitrary units, according to the data of Fig. 95.
The abscissa is $\overline{m}(\zeta) = (m_0 \tau_0 + m \tau)/(\tau_0 + \tau)$.

To the extent that the log $I(m_0)$ curves run more steeply at small ζ and
bend sooner for large values of τ_0, the differences log $I(\lambda_1)$ − log $I(\lambda_2)$ ($\lambda_1 >$
λ_2), which are proportional to $CE_{\lambda_2}^{\lambda_1}$, will first decrease (increase in absolute value), then reach a minimum near the turning point m_{0t} of the log $I(\lambda_2)$
curve [but not the log $I(\lambda_1)$ curve] where the two curves have the same slope
[d log $I(\lambda_1)$/dm$_0$ = d log $I(\lambda_2)$/dm$_0$], and finally increase again because of the
log $I(\lambda_1)$ and log $I(\lambda_2)$ curves. In other words, the position of the minimum of
the reversion curve will approximately coincide with the turning point of the
log $I(\lambda_2)$ curve. This situation is illustrated by Fig. 94, in which the position
of the arrows is the same as in Fig. 101. We note that the position of minimum
on the $CE_{\lambda_2}^{\lambda_1}(\zeta)$ curves is far more sharply defined than the turning point on
the log $I(m_0)$ curves, in particular because of the uncertainty we have mentioned in the computed value of m_0. It is very important to observe that the
position of the minimum on the reversion curve depends very weakly on τ,
that is, on atmospheric conditions; it is almost independent of the direction of
observation, and entirely independent of the size of the instrumental field
of view.

In principle, then, Götz's explanation of the reversal effect retains its force, but with the reservation that it is not single, but multiple scattering that makes the chief contribution in the lower layer of the atmosphere. As a result, the techniques for quantitative analysis of the reversion curves have to be improved.

In particular, we can allow for the height structure of the ozone layer in the first approximation by replacing the expression (III.4, 3) with

$$\Phi_0 = \frac{I_0 \omega_0}{2} \int\limits_A^\infty e^{-[\tau_0^* - \tau_0'(h)]} s\left[\tau^* - \tau_0'(h)\right] e^{-\tau_0'(h)\, m_0(h)} \frac{d\tau'}{dh}\, dh,$$

$$(III.4, 9)$$

where $\tau'_0(h)$ and $\tau'(h)$ are the vertical optical thicknesses of the ozone and the air above level h, and the lower limit A of the integral corresponds to the bottom of the ozone layer. We must also make the corresponding changes in the second term in brackets in Eq. (III.4, 5).

Let us now reconstruct the conditions corresponding to an observation of the anomalous transparency of the atmosphere; that is, we point our instrument at the sun ($z = \zeta$, δ_z, $\zeta = 1$), and regard the field of view ω as not much greater than the angular size $\omega_0 \approx 9 \cdot 10^{-5}$ of the sun.

Returning to Eq. (III.4, 6),let us first suppose that the measurements are being conducted in the spectral region where τ_0 is negligibly small, so that the second term in brackets becomes insignificant. We then revert to the expressions (III.4, 1), for $\tau_0 = 0$, and (III.3, 18) (see end of §3). At a certain value of ζ, the rapidly falling exponential term representing the direct sunlight will become smaller than the second term, due to multiple scattering, and far less dependent on ζ. Hence the steep drop in the function log I(m) will become a comparatively mild one, and the larger τ is, the sooner this will happen. As a result, we should observe a reversal effect here too, starting when the Bouguer law breaks down and the condition $I_{scat} \approx I_{dir}$ is satisfied. From Table 12 (§3), we see that we can only observe this effect in the ultraviolet spectral region, or in the event that the lower layers of the atmosphere are highly turbid ($\tau^* \gtrsim 0.5-1$).

If we pass to the region of the spectrum where ozone absorption is substantial, then as we have seen, the second term in brackets will begin to predominate over the first term long before the Bouguer law fails in the absence of ozone. Hence the failure of the Bouguer law will here be governed by light which has first been scattered above the ozone layer and then multiply scattered in the lower atmospheric layers before reaching the instrument. Once again we find that the steep drop in the function log I(m) is replaced by a more gentle behavior (Fig. 102), so that a reversal effect in the $CE_{\lambda_2}^{\lambda_1}(m)$ relation will again occur, as is in fact observed (Fig. 95).

The transition from one log I(m) relation to the other now takes place when

$$e^{-\tau_0 m_0 - \tau m} = \frac{0.4\omega}{(1-A)\,\tau + \frac{4}{3}} \frac{c}{2}\,\tau e^{-\tau_0}\,s\,(\tau_0),$$

$$(III.4, 10)$$

with $m_0(\zeta)$ depending significantly on the height of the ozone layer. Differentiating this equation, we obtain

$$\left. \begin{array}{ll} \dfrac{dm_t}{d\tau_0} = -\dfrac{m_0 - 1 + \dfrac{d\ln s}{d\tau_0}}{\tau + \tau_0\,\dfrac{dm_0}{dm}}, & \dfrac{dm_t}{d\tau} = -\dfrac{m + \dfrac{\frac{4}{3}}{(1-A)\,\tau + \frac{4}{3}}}{\tau + \tau_0\,\dfrac{dm_0}{dm}}, \\[4ex] \dfrac{dm_t}{dc} = -\dfrac{1}{c\left(\tau + \tau_0\,\dfrac{dm_0}{dm}\right)}, & \dfrac{dm_t}{d\omega} = -\dfrac{1}{\omega\left(\tau + \tau_0\,\dfrac{dm_0}{dm}\right)}, \\[4ex] \dfrac{dm_t}{dA} = -\dfrac{\tau}{\left(\tau + \tau_0\,\dfrac{dm_0}{dm}\right)\left[(1-A)\,\tau + \frac{4}{3}\right]}, \end{array} \right\}$$

$$(III.4, 11)$$

where m_t, as before, is the value of m at the turning point of the log I(m) curve.

We remark first of all that unlike the situation for the reversal effect, the position of the turning point on the log I(m) curve now depends on the instrumental field of view ω and the albedo A of the earth's surface as well as on the scattering diagram in the lower layers of the atmosphere (since the quantity l depends on the form of the diagram). An increase in ω or the albedo will displace the turning point toward smaller values of ζ. Moreover, the position of the turning point, and hence also the minimum on the anomalous-transparency curves, depends not only on τ_0, but also on τ, that is, on atmospheric conditions; and in both cases an increase in the optical thickness will displace the turning point toward smaller ζ. Finally, just as for the reversal effect, the position of the turning point depends on c, that is, on the height of the ozone layer. This wealth of auxiliary factors influencing the anomalous-transparency effect makes it considerably less suitable for ozonometric measurements than the reversal effect.

If we again take $c = 1 \cdot 10^{-2}$, $\tau = 1.2$, and an ozone content of 0.3 cm, also assume that $\omega \leq \omega_0$ and $A = 0.8$ (from measurements by Rodionov and his colleagues, conducted mostly on the snowcap of Mt. Elbrus), and use Rodionov's computations for the values of $m_0(\zeta)$, then we can obtain without difficulty the positions of the turning points on the log $I(\overline{m})$ curves, where $\overline{m} = (m_0 \tau_0 + m\tau)/(\tau_0 + \tau)$, or the corresponding positions of minimum on the anomalous-transparency curves, as marked by arrows in Fig. 95. We at once find good qualitative accord between theory and experiment. Both the positions of the precomputed points and especially the τ_0-dependence of the positions are transmitted essentially in a regular manner. However, it is no less striking to observe that the precomputed positions of the turning points (and minima) are displaced systematically, and far beyond the limits of the admissible experimental error or the uncertainty in the quantities used in the calculation, toward larger values of ζ. One should furthermore note the relatively steep slope of the log $I(\overline{m})$ curves in Fig. 102 for $\overline{m} > \overline{m}_t$, which also fails to agree with the theoretical expectations. The cause of the two effects is not hard to find. It lies in our neglect of the aureole. The fact is that the anomalous-transparency effect sets in well before the source of radiation is submerged in the haze produced by multiple scattering of the direct rays in the lower atmospheric layers. This means that the aureole effect is still very strongly in evidence, and that it exceeds by two orders of magnitude the brightness of the field around the source [more precisely, only the first bracketed term in Eq. (III.4, 6)]. Actually, then, we cannot neglect this term, and its inclusion will displace the resulting values of m_t approximately in the direction of the observed positions (an exact computation has not yet been performed). Moreover, as m increases the aureole will gradually disappear, as will the source itself, so that the contribution of the first bracketed term will gradually diminish, and the log $I(\overline{m})$ curves will drop relatively steeply for $\overline{m} > \overline{m}_t$. Thus a fully adequate calculation of the anomalous-transparency effect cannot, in all probability, be carried out without a more detailed solution of the equation of transfer, with allowance for the sphericity of the earth's atmosphere and the shape of the scattering diagram of the free air. This situation only serves to emphasize how little the effect is suited for ozonometric purposes.

We turn now to the third effect, the appearance of "selective" atmospheric transparency in the ultraviolet spectral region, an effect discovered, as we have mentioned, by Rodionov and his colleagues [248], and subsequently observed by Polyakova [251]. Unfortunately, the descriptions of the experiments, particularly in [251], do not yield sufficient information for a detailed quantitative analysis. However, the authors have obviously neglected the background of multiply scattered light, which as we have seen is very strong in the ultraviolet region. Those quantitative estimates that can be made for the case

[248] show that the background must play a substantial role, increasing rapidly with decreasing wavelength, and can lead to effects which in character and in order of magnitude are close to the "selective" transparency effect detected by these authors. Further support for this conclusion comes from the fact that the effect was never reported in [248] at small ζ, and always increased smoothly with ζ both in the morning and in the evening. This circumstance casts some doubt on the reality of the "selective" transparency effect itself, and it may be attributable to a careless interpretation of the observational data. In any event, a thorough experimental investigation is needed, with the use of a point source of light.

§ 5. Attenuation of the Sun at Twilight

Let us now consider how the atmospheric transparency T_P for direct sunlight (see Chap. III, §2) depends on the perigee height y of a solar ray. Because of the rapid decline in atmospheric density with height, the form of the function T_P (y, λ) outside the selective-absorption bands of atmospheric gases is governed essentially by the optical structure of the atmosphere in the perigee region of the solar ray — that is, by the lower strata, the troposphere and stratosphere. Since these layers of the atmosphere are subject to strong weather effects, a strict formulation, of the problem cannot prescribe a "typical" behavior of any kind for the function T_P (y, λ); empirical data have to be taken directly from the observations, together with concurrent measurements of the brightness of the twilight sky. In principle, it would be desirable for this information to be acquired independently, such as by measurements of the brightness of artificial earth satellites as they emerge from the earth's shadow, by measuring the brightness of the sun during twilight hours at various altitudes, by employing balloons or rockets, or by probing the atmosphere with searchlight beams [19]. However, as we shall see presently, with certain reservations the information can also be obtained by analyzing data for the brightness in different parts of the twilight sky, at different depressions of the sun below the horizon. We are assisted here by the admissibility of a particular function T_P(y, λ) which is valid over a very wide range of optical conditions in the atmosphere.

If the atmosphere is spherically symmetric, with a negligible horizontal inhomogeneity, then outside the selective-absorption bands of atmospheric gases we may write the function T_P (y, λ) in the form (III.2, 10) or (III.2, 11); that is, we may investigate its behavior by using the expression

$$T_P = e^{-m_{\text{eff}} \tau'(y)},$$

$$(III.5, 1)$$

where m_{eff} is a suitable effective air mass depending on the position of the point under observation, P, along the beam of sunlight illuminating it.

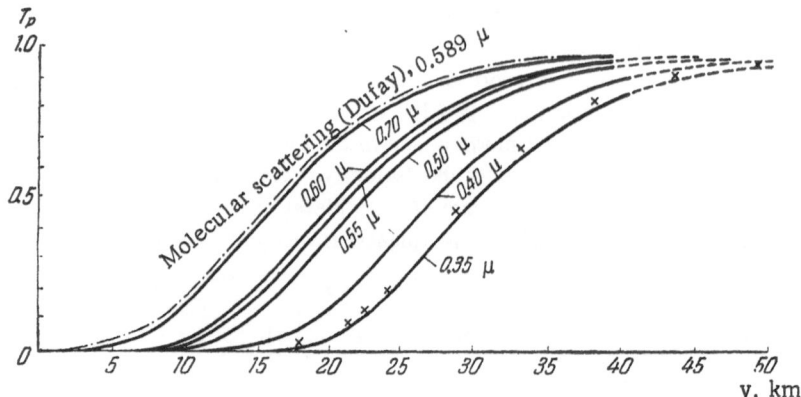

Fig. 103. T_p as a function of y for selected λ, after Fesenkov [233].

Several authors have calculated functions T_p (y) for various λ, with differing assumptions for the behavior of the scattering and absorption coefficients σ and α of the free air with respect to wavelength and height. In particular, many calculations have been made for $\lambda = 589$ mμ in connection with
the twilight flash of atmospheric sodium [145, 151, 152, 252, and elsewhere];
pure molecular scattering has been assumed. At various times, Fesenkov [5,
233], Link [225-228, 230, 292], Hulburt [91], Polyakova [253], Shtaude [128],
Volz and Goody [282], and others have undertaken calculations for various
wavelengths and model atmospheres, for application to the theory of twilight
phenomena, and with a few exceptions [233, 253, 282] the aerosol contribution to atmospheric attenuation of light has always been neglected. Moreover,
the change in the path of a ray due to refraction has been incorporated [233],
as well as the refraction-divergence effect [230, 282, 292] (see end of § 1).
Figure 103 presents the functions $T_p = e^{-2m(\pi/2)\tau'(y)}$ for selected wavelengths,
corresponding to the transparency of the entire thickness of the atmosphere
along a ray of perigee height y, that is, for $m_{eff} = 2m(\pi/2)$. These values
were obtained by Fesenkov [233], who assumed 1) proportionality of σ to the
air density, 2) a standard density distribution with height, and 3) a standard
wavelength dependence of the vertical transparency of the atmosphere, in accordance with [236]. Curves for T_p (y) derived under other assumptions about
the optical structure of the atmosphere may differ considerably from those
shown in Fig. 103, particularly in their wavelength dependence. The situation is illustrated by the same figure, which for comparison includes a function T_p (y) for $\lambda = 589$ mμ, as computed by J. Dufay for pure molecular scattering. Significantly, the function T_p (y) is always monotonic, similar in

shape to the current-voltage characteristic of a radio tube, with a well-defined linear portion and a point of inflection near $T_p = 0.4$. The shape of the curves depends only weakly on wavelength; this parameter mainly displaces the curve in y, with a displacement toward shorter λ corresponding (in the absence of selective absorption) to a shift in the curve toward larger y. In fact, we obtain a similar result under any reasonable assumption for the optical structure of the cloudless atmosphere. Even if the lower layers of the atmosphere are highly turbid, there will be little change in the trend of the curves, with distortion (a drop in the already very small values of T_p) only at small y, where the atmosphere is practically opaque.

The position y_i of the point of inflection on the $T_p(y)$ curve is specified, in accordance with Eq. (III.5, 1), by the condition

$$m_{\text{eff}} \left(\frac{d\tau'}{dh} \right)^2_{h=y_i} = \left(\frac{d^2\tau'}{dh^2} \right)_{h=y_i}$$

(III.5, 2)

or, from Eqs. (II.2, 6) and (II.3, 7), by

$$m_{\text{eff}} \, k_{y_i} = K_{y_i},$$

(III.5, 3)

where

$$K = - \frac{d \ln k}{dh} .$$

(III.5, 4)

If the function k(h) is available, Eq. (III.5, 3) will yield the height y_i. In particular, if we have

$$k = k(h_0) \, e^{-K(h-h_0)}$$

(III.5, 5)

near the point of inflection, then

$$y_i = h_0 - \frac{1}{K} \ln \frac{K}{m_{\text{eff}} \, k(h_0)} .$$

(III.5, 6)

If, moreover, the exponential relation (III.5, 5) remains invariant over a sufficiently great range in height above y_i and if there are no layers of cloud for $h > y_i$, we shall have

$$\tau'(y_i) \approx \frac{k(h_0) \, e^{-K(h-h_0)}}{K} = \frac{1}{m_{\text{eff}}}$$

(III.5, 7)

in view of Eqs. (III.5, 3) and (III.5, 5). Substituting Eq. (III.5, 7) into Eq. (III.5, 1), we obtain

$$T_P(y_i) \approx e^{-1},$$

(III.5, 8)

in good agreement with Fig. 103.

Since

$$\left(\frac{dT_P}{dy}\right)_{y_i} = -m_{\text{eff}}\left(\frac{d\tau'}{dh}T_P\right)_{y_i} = m_{\text{eff}}\,k_{\nu_i}\,T_P(y_i),$$

(III.5, 9)

we find, by Eq. (III.5, 3), that in the neighborhood of $y = y_i$

$$T_P(y) \approx T_P(y_i)[1 + K(y - y_i)]$$

(III.5, 10)

or, on the assumption (III.5, 5), and using Eq. (III.5, 8),

$$T_P(y) \approx e^{-1}[1 + K(y - y_i)].$$

(III.5, 11)

We do not obtain a significant improvement by including the next term of the expansion, which is proportional to $(y - y_i)^3$, nor is it expedient to include the term, both because of the instability of the optical state of the atmosphere, and from the standpoint of interpreting twilight phenomena (see Chaps. IV and V). If selective absorption is absent, we may therefore approximate the function $T_P(y)$ by the expression

$$T_P(y) = \begin{cases} 0 & \text{for} \quad y \leqslant y_i - \frac{1}{K}, \\ e^{-1}[1 + K(y - y_i)] & \text{for} \quad y_i - \frac{1}{K} \leqslant y \leqslant y_i + \frac{e-1}{K}, \\ 1 & \text{for} \quad y \geqslant y_i + \frac{e-1}{K}. \end{cases}$$

(III.5, 12)

Figure 104 illustrates an approximation for the case $\lambda = 0.7\ \mu$, one of those considered by Fesenkov (Fig. 103).

In this figure, a cross marks the point of inflection on the curve, as given by Eq. (III.5, 8). The dotted line corresponds to $K = 0.13$ km^{-1}, the dot-dash line to $K = 0.15$ km^{-1}. The first line forms a better approximation in the region of high transparency, and the second is better for low transparency. The assumption (III.5, 5) agrees well with Fig. 103 for $K = 0.15$ km^{-1} and

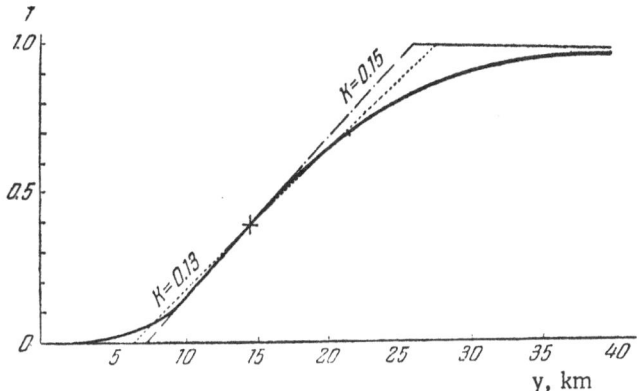

Fig. 104. An approximate function $T_P(y)$.

$m_{eff} = 2m\,(\pi/2) = 80$, as shown by the crosses near the $\lambda = 0.35\,\mu$ curve in Fig. 103, which is computed from Eq. (III.5, 1) under these conditions.

Figure 103 shows that if there is no selective absorption, the point of inflection on the $T_P(y)$ curves ranges in position from 10 to 30 km, depending on the wavelength. If there is no stratospheric cloud cover within this range, we may take the value of K as approximately $-d(\ln \rho)/dh$, following Fesenkov; here K varies only slightly with height and time, remaining between 0.13 and 0.15 km^{-1} (see Fig. 52). On the other hand, Fig. 104 indicates that these small variations in K will have little effect on the function $T_P(y)$, and that the approximations (III.5, 12), first suggested by Rozenberg [96, 171] and recently applied by Volz and Goody [282], will represent the general behavior of the function quite well.

We may add that the troposphere develops a sharply defined optical instability (in particular, because of the abundant cloud structure) only at small y, that is, only along the lower portion of the curve, and only in the infrared spectral region ($\lambda \gtrsim 0.7\,\mu$) is the effect of any importance.

If the form of the function $T_P(y)$ remains comparatively insensitive to atmospheric conditions and changes in wavelength, the position y_i of the point of inflection will, on the contrary, show a strong dependence, as we have seen.

To estimate this effect, let us return to Eq. (III.5, 6), writing it in the form

$$y_i = h_0 - \frac{1}{K}\,\ln\frac{K}{m_{eff}} + \frac{1}{K}\,\ln k\,(h_0) = y_{i0} + \frac{1}{K}\,\ln k\,(h_0). \tag{III.5, 13}$$

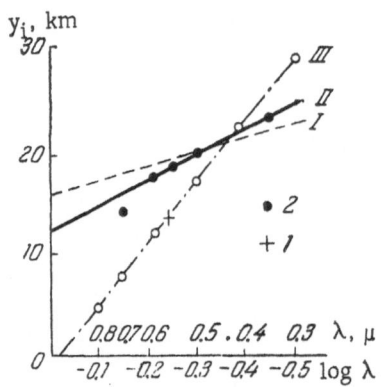

Fig. 105. The height y_i of the point of inflection along the T_p (y) curve as a function of log λ, for various optical conditions in the atmosphere. I) Aerosols, Junge distribution; II) n = 1.6; III) molecular atmosphere; 1) Dufay's data; 2) Fesenkov's data.

Since m_{eff} and K are practically independent of λ, the λ-dependence of y_i is governed wholly by that of the attenuation coefficient k, that is, by the atmospheric conditions. In particular, if $k \sim \lambda^{-n}$, then $\ln k \sim - n \ln \lambda$. For molecular scattering, n = 4, while for scattering by aerosols with the particle size following the Junge distribution (II.2, 36), n = 1.

Figure 105 presents y_i as a function of λ for molecular scattering (y_{i0} is assigned from Dufay's data [252] for $\lambda = 0.589$ μ; see Fig. 103), for pure aerosol scattering (y_{i0} assigned arbitrarily), for the case of Fesenkov's calculations [233] (dots), and for n = 1.6 (y_{i0} assigned from Fesenkov's data for $\lambda = 0.55$ μ; Fig. 103). We see that all of Fesenkov's values, except for the case $\lambda = 0.70$ μ, follow the straight line for n = 1.6.

To determine the height y_i of the point of inflection, we may again use Eq. (III.5, 7), if we know the relation $\tau'(h)$. In particular, by taking $m_{eff} = 80$ and using the data of Fig. 62 for the relation $\tau'(h)$ in the case of molecular scattering, we obtain the values y_i shown by circles in Fig. 105; these are in good agreement with the calculations of Dufay and with Eq. (III.5, 13).

One should, however, note especially in Fig. 105 the strong dependence of the slope of the lines y_i (log λ) upon n, that is, on the optical state of the atmosphere. It follows at once that in particular cases we must establish the values of $y_i(\lambda)$ experimentally, with the behavior of the T_p (y) curve in the vicinity of y_i being regarded as linear. Nevertheless, for purposes of orientation it is useful to adopt some definite optical structure of the atmosphere in the computations. We may, for example, take the model used by Fesenkov [233], but with the choice n = 1.6. We then obtain, for $m_{eff} = 80$,

$$y_i \approx 12.5 (1 - 2\log\lambda) \text{ km,} \qquad \text{(III.5, 14)}$$

where λ is expressed in microns; thus, setting $K = 0.14$ km^{-1}(Fig. 104), we have

$$T_P(y,\ \lambda) \approx \begin{cases} 0 \quad \text{for} \quad y \leqslant 5 - 25\log\lambda \text{ km,} \\ 0.05y - 0.25 + 1.25\log\lambda \text{ for } \quad 5 - 25\log y \leqslant \\ \qquad\qquad \leqslant y \leqslant 25(1 - \log\lambda) \text{ km,} \\ 1 \quad \text{for} \quad y \geqslant 25(1 - \log\lambda)\text{km.} \end{cases}$$

$$(III.5,\ 15)$$

One should keep in mind, however, that substantial departures from the approximations (III.5, 15) may be expected under different conditions.

The expressions we have given refer to the case $m_{\text{eff}} = 2m(\pi/2)$. Let us now consider the case where $m_{\text{eff}} = m(\zeta' < \pi/2)$ or $m_{\text{eff}} = m(\pi - \zeta' < \pi/2)$ for $|\zeta' - \pi/2| \ll 1$. We have seen above (§3, Tables 10-11) that the relation $m(\zeta)$ provides a satisfactory approximation to the expression (III.3, 17), which for $\cos\zeta = \alpha \ll 1$ may be written in the form

$$m\left(\frac{\pi}{2} - \alpha\right) \approx (\alpha + 0.025\,e^{-11\alpha})^{-1},$$

$$(III.5,\ 16)$$

where $\alpha = \pi/2 - \zeta$ is expressed in radian measure.

On the other hand, allowing for the refraction curvature of solar rays (§1), we should take, by (III.1, 36),

$$\alpha_1 \approx \frac{\pi}{2} - \zeta' + \nu_\zeta \approx \cos\zeta + \frac{v\cos\varphi}{\cos\psi} + \nu_\zeta$$

$$(III.5,\ 17)$$

before the true perigee of the ray is reached, and

$$\alpha_2 \approx \zeta' - \frac{\pi}{2} + \nu_\zeta \approx -\cos(\zeta - 2\nu_\zeta) - \frac{v\cos\varphi}{\cos\psi} + \nu_\zeta \approx -\cos\zeta - \frac{v\cos\varphi}{\cos\psi} - \nu_\zeta$$

$$(III.5,\ 18)$$

beyond true perigee, where ν_ζ is given by the relation (III.1, 38).

The region ahead of true perigee is specified by the condition

$$\frac{v\cos\varphi}{\cos\psi} > -\cos\zeta - \nu_\zeta$$

$$(III.5,\ 19)$$

and the region beyond true perigee by the condition

$$\frac{v \cos \varphi}{\cos \psi} < - \cos \zeta - v_\zeta.$$

$$\text{(III.5, 20)}$$

If we now take $| \zeta - \pi/2 = \xi | \ll 1$, then

$$\alpha_1 \approx - \xi + \frac{v \cos \varphi}{\cos \psi} + v_\zeta$$

$$\text{(III.5, 21)}$$

and

$$\alpha_2 \approx \xi - \frac{v \cos \varphi}{\cos \psi} - v_\xi.$$

$$\text{(III.5, 22)}$$

By substituting Eq. (III.5, 21) or (III.5, 22) into Eq. (III.5, 16), we obtain approximate expressions for $m(\zeta' < \pi/2 + v_\zeta)$ in the region ahead of true peri-gee and $m(\pi - \zeta' < \pi/2 + v_\zeta)$ in the region beyond true perigee.

One can easily see that by neglecting refraction, the m_{eff}-dependence of $T_P(y, \lambda)$ will reduce simply to a displacement in y_i, depending on m_{eff}; by Eq. (III.5, 6) we have

$$y_i = y_i \ (m_{eff} = 80) - \frac{1}{K} \ln \frac{80}{m_{eff}} \, ,$$

$$\text{(III.5, 23)}$$

so that a drop in m_{eff} will diminish the height of the point of inflection, dis-placing the $T_P(y)$ curves toward smaller values of y. The existence of the opaque surface of the earth will merely lead here, at small m_{eff}, to a cutoff in the lower part of the $T_P(y)$ curve, with hardly any change in the upper part. Allowing for refraction complicates the calculation somewhat but does not change the situation qualitatively.

Let us first assume that the point under observation lies ahead of the true perigee of the solar ray. We recall that it is not T_P which enters the expres-sion (III.2, 12) for the twilight-sky brightness, but the product $T_{NP} = T_N T_P$, where T_N is given by the expression (III.2, 9); T_N and T_P now depend on h, that is, on y, and m_{eff}, which enters the expression for T_P, also depends on y. Hence the height y_i of perigee of the ray passing through the inflection point h_i on the $T_{NP}(h)$ curve is given, rather than by Eq. (III.5, 2), by the condition

$$2 \left[(m_\zeta - m_z) \tau' - 1 \right] \frac{dm_\zeta}{dh} \frac{d\tau'}{dh} + \tau'^2 \left(\frac{dm_\zeta}{dh} \right)^2 - \tau' \frac{d^2 m_\zeta}{dh^2} -$$

$$- (m_\zeta - m_z) \frac{d^2\tau'}{dh^2} + (m_\zeta - m_z)^2 \left(\frac{d\tau'}{dh} \right)^2 = 0, \quad \text{(III.5, 24)}$$

and according to Eqs. (III.1, 28) and (III.1, 29),

$$\frac{dm_\zeta}{dh} = \gamma \frac{dm_\zeta}{dy} = -\gamma m^2 (1 - 0.275 \, e^{-11a}) \frac{d\alpha}{dy} \approx \gamma m^2 \frac{d\alpha}{dy},$$

(III.5, 25)

while by Eqs. (III.5, 21) and (III.1, 38),

$$\frac{d\alpha}{dy} \approx \frac{1}{R} \frac{\cos \varphi}{\cos \psi} - 0.002 \, e^{-0.16 y}.$$

(III.5, 26)

By estimating the terms in Eq. (III.5, 24), allowing for the small value of y and the fact that $m_\zeta < 40$, and rejecting terms of a higher order of smallness, we may replace Eq. (III.5, 24) by the approximation

$$\tau' m_\zeta (m_\zeta \tau' - 2) p^2 - 2 m_\zeta [(m_\zeta - m_z) \tau' - 1] p \frac{d\tau'}{dn} +$$
$$+ (m_\zeta - m_z)^2 \left(\frac{d\tau'}{dh}\right)^2 - (m_\zeta - m_z) \frac{d^2\tau'}{dh^2} = 0,$$

(III.5, 27)

where

$$p = \gamma m \frac{d\alpha}{dy}.$$

In the event that an exponential relation k(h) of the form (III.5, 5) obtains near and above the point of inflection, we obtain from Eq. (III.5, 27):

$$\tau'(h_i) \approx \frac{1}{m_\zeta - m_z} \left[1 + \frac{1}{\left(\dfrac{K}{p} + \dfrac{m_\zeta}{m_\zeta - m_z}\right)^2} \right],$$

(III.5, 28)

where we have used the smallness of m_z/m_ζ, or

$$\tau'(h_i) \approx \frac{1}{m_\zeta - m_z} \left[1 + \frac{1}{\left[1 + \dfrac{1}{m_\zeta}\left(\dfrac{K}{\gamma \dfrac{d\alpha}{dy}} + m_z\right)\right]^2} \right],$$

(III.5, 29)

whence, with Eqs. (III.5, 16), (III.5, 22), and (III.5, 26), and a known relation $\tau'(h)$, it is not difficult to find h_i.

If the observed point P lies beyond true perigee of the solar ray, then according to Eq. (III.2, 11),

$$T_{NP} = T_N \, e^{-80 \, \tau'(y) + m(\pi - \zeta') \, \tau'(h)}$$

(III.5, 30)

and the condition for an inflection in the $T_{NP}(h)$ curve takes the form

$$2\left[(m_\zeta - m_z)\,\tau'(h) - 1\right]\left(\frac{dm_\zeta}{dh}\right)_h\left(\frac{d\tau'}{dh}\right)_h + \tau'^2(h)\left(\frac{dm_\zeta}{dh}\right)_h -$$

$$-\tau'(h)\left(\frac{d^2m_\zeta}{dh^2}\right)_h - (m_\zeta - m_z)\left(\frac{d^2\tau'}{dh^2}\right)_h + (m - m_z)^2\left(\frac{d\tau'}{dh}\right)_h^2 +$$

$$+\frac{160}{\gamma}\left(\frac{d\tau'}{dh}\right)_y\left[\left(\frac{dm_\zeta}{dh}\right)_h\tau'(h) + (m_\zeta - m_z)\left(\frac{d\tau'}{dh}\right)_h\right] +$$

$$+\frac{80^2}{\gamma^2}\left(\frac{d\tau'}{dh}\right)_y^2 - \frac{80}{\gamma}\left(\frac{d^2\tau'}{dh^2}\right)_y = 0, \qquad\qquad \text{(III.5, 31)}$$

with Eq. (III.5, 26) being replaced, according to Eq. (III.5, 18), by

$$\frac{d\alpha}{dy} = -\frac{1}{R}\frac{\cos\varphi}{\cos\psi} + 0.002\,e^{-0.16\,v}. \qquad\qquad \text{(III.5, 32)}$$

Again, disregarding terms of a higher order of smallness and adopting an exponential relation of the type (III.5, 5), we obtain, after some minor reductions (in view of the smallness of m_z/m_ζ):

$$m_\zeta^2\left(\frac{p}{K} + \frac{m_\zeta - m_z}{m_\zeta}\right)^2 (\tau')^v\, e^{-KH} -$$

$$-\frac{1}{m_\zeta - m_z}\left[1 + \frac{1}{\left(\frac{K}{p} + \frac{m_\zeta}{m_\zeta - m_z}\right)^2}\right]e^{-KH} -$$

$$-\frac{160}{\gamma}\left[\frac{p}{K} + \frac{m_\zeta - m_z}{m_\zeta}\,\tau'\,e^{-KH} + (80\,\tau' - 1)\frac{80\tau^{1-\gamma}}{\gamma^2}\right] = 0, $$

$$\text{(III.5, 33)}$$

where

$$H(y) = h - \gamma y \qquad\qquad \text{(III.5, 34)}$$

represents the height of the boundary of the earth's shadow under the condition that the refraction $v_\zeta = v_\zeta/(y_i)$ is independent of y. For $KH \gg 1$, this expression enables us to recover Eq. (III.5, 7) for $m_{eff} = 80$, as would have been expected. On the other hand, in the special case $\gamma = 1$ we arrive at the condition

$$\tau'(y_i) = \cfrac{\cfrac{m_\zeta^2}{m_\zeta - m_z}\left(\cfrac{p}{K} + \cfrac{m_\zeta - m_z}{m_\zeta}\right)^2 \left[1 + \cfrac{1}{\left(\cfrac{K}{p} + \cfrac{m_\zeta}{m_\zeta - m_z}\right)^2}\right] e^{-2KH} + 80}{\left[m_\zeta\left(\cfrac{p}{\zeta} + \cfrac{m_z - m_\zeta}{m_\zeta}\right)e^{-KH} - 80\right]^2}$$

(III.5, 35)

or, since $m_z \ll m_\zeta$,

$$\tau'(y_i) \approx \frac{m_\zeta\left(2\dfrac{p^2}{K^2} + 2\dfrac{p}{K} + 1\right)e^{-2KH} + 80}{\left[m_\zeta\left(\dfrac{p}{K} + 1\right)e^{-KH} - 80\right]^2}.$$

(III.5, 36)

This last expression permits us to follow the displacement in the perigee y_i of the solar ray passing through the inflection point h_i on the curve $T_{NP}(h) = T_N(h)T_P(h)$ as ζ varies, for the case where the point P lies beyond true perigee of the ray.

If we use Fig. 62 for the relation $\tau'(h)$, corresponding to a pure molecular atmosphere, then Eqs. (III.5, 29) and (III.5, 36) will reduce to the relations $y_i(\xi = \zeta - \pi/2)$ illustrated in Fig. 106. We see at once that the perigee of the solar ray illuminating the inflection point P_i recedes abruptly from the earth's surface even before sunset, that the shorter the wavelength the sooner this happens, and that by the time $\xi \approx 0.07$ ($\zeta \approx 94°$), y_i has reached its maximum value, and experiences no further change for $\zeta > 94°$. Here $y_{i, max}$ increases with decreasing λ.

Figure 107 shows how ξ depends on the true height h_i of the inflection point P_i on the transparency curve $T(h)$, for a constant zenith distance z of the line of sight, and $\gamma \approx 1$. Note the characteristic bends in the $h_i(\xi)$ curve due to the changing height $y_i(\xi)$. At first, when the sun is still above the horizon, h_i increases only because of the increase in y_i. Near the true terminator, the increase in y_i begins to be accompanied by a rapid rise in the earth's shadow, and the $h_i(\xi)$ curve bends sharply upward. Near $\xi \approx 0.05$, the rise in $y_i(\xi)$ decelerates considerably, and the subsequent course of $h_i(\xi)$ reflects only the continually accelerating motion of the earth's geometric shadow. It is for this reason that we find a sharp decline in the slope of the $h_i(\xi)$ curve near $\xi \approx 0.05$. Thus the steepest part of the curve occurs at a height of about 15 km (for the wavelength region 0.6-0.7 μ).

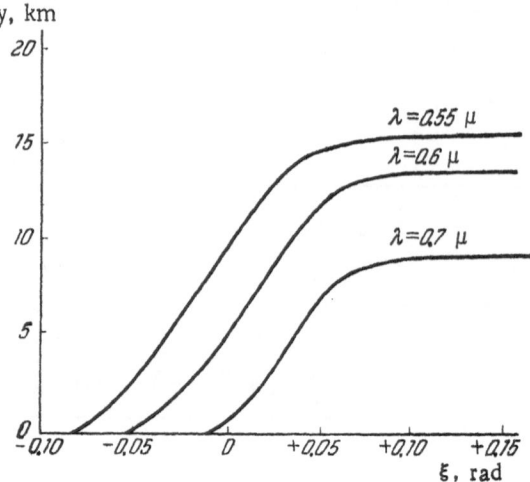

Fig. 106. The relation $y_i(\xi)$ for selected λ in a molecular atmosphere.

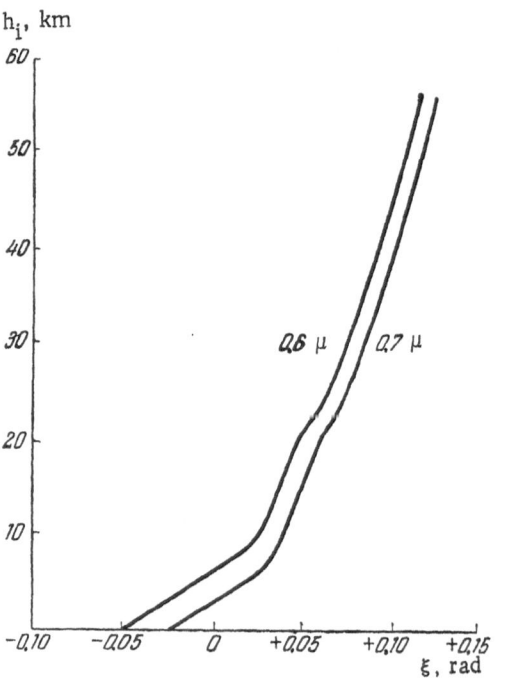

Fig. 107. The relation $h_i(\xi)$ for selected λ in a molecular atmosphere.

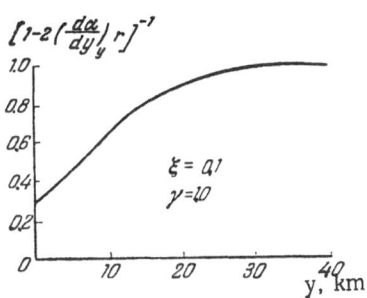

$$\left[1-2\left(\frac{d\alpha}{dy}\right)_y r\right]^{-1}$$

Fig. 108. A typical curve for the refraction-divergence correction as a function of y.

These considerations apply only to a molecular atmosphere. In the case of a turbid atmosphere, the behavior of the $h_i(\xi)$ curves will be governed wholly by the relation $r'(h)$, that is, by the optical structure of the atmosphere. In this situation, the preceding discussion indicates that the most important factor in the relation $h_i(\xi)$ at small ξ will be the state of the troposphere and the overlying layer of the stratosphere, the regions particularly subject to weather influences. Hence the region of small ξ on the $h_i(\xi)$ curves will be extremely variable. On the other hand, when $\zeta \gtrsim 94°$ the perigee of the ray illuminating the inflection point h_i will fall (if λ is not too large) above the troposphere, and the behavior of the $h_i(\xi)$ curve at large ξ will depend comparatively weakly on the weather conditions (and not at all on the conditions prevailing in the troposphere).

We see from the relations (III.5, 12) and Figs. 103 and 107 that if the height of the observed point satisfies $h \gtrsim h_i + [(e-1)/K]\gamma$, then the transparency $T_{NP} = T_N T_p \approx e^{-m} z^{r'(h_N)}$, the atmospheric transparency along the line of sight. In practice this will occur, as we see from Figs. 103 and 107, for $h \gtrsim 25$ km in the red spectral region, and for $h \gtrsim 35$-40 km in the violet.

It remains to consider the factor $[1 - 2(d\alpha/dy)_y r]^{-1}$ due to the refraction divergence of solar rays (§§1 and 2). The value of this factor depends heavily on z and ζ, and it rapidly approaches unity as y increases. In particular, for $\gamma = 1.0$ and $\xi = 0.1$ its y-dependence takes the form shown in Fig. 108. Figure 109 illustrates the corresponding changes in $T_p(y)$. We see that an allowance for this factor will displace the $T_p(y)$ curves toward larger y with hardly any change in their shape; the larger y_i is, the smaller the displacement will be, and it will not exceed 1-2 km in the visible region of the spectrum.

For given z and ξ, the brightness of the twilight sky is determined by the dependence of the quantity $T \equiv T_N T_p [1 - 2(d\alpha/dy)_y r]^{-1}$ upon h along the line of sight [see Eq. (III.2, 12)]. We can derive this relation either by projecting on the line of sight the corresponding functions of y, with allowance for the relation h(y) (see §1), or by a direct calculation. Several authors have made such computations, under various assumptions for the structure of the atmosphere. Figure 110 presents T(h) curves for selected ξ (labels along curves), as obtained by Volz and Goody [124] for the case z = 70° (a) for a molecular

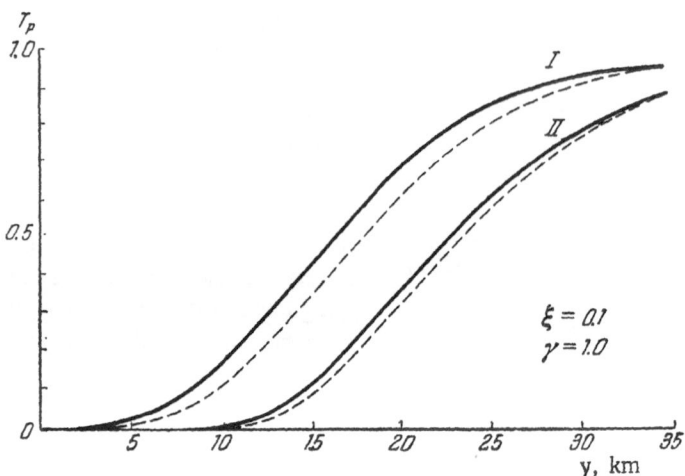

Fig. 109. Typical changes in the function Tp(y) when a correction is made for refraction divergence. I) Molecular scattering, $\lambda = 0.589\ \mu$; II) Fesenkov's analysis, $\lambda = 0.5\ \mu$.

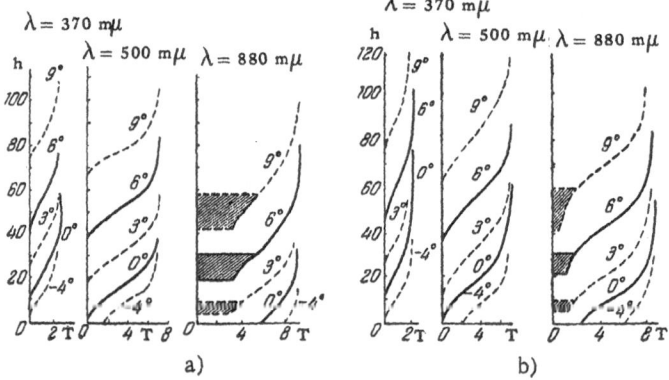

Fig. 110. The function T(h) for $z = 70°$ and selected ξ, for (a) a pure atmosphere and (b) an aerosol atmosphere, at selected λ. The shaded areas indicate the regions of the earth's geometric shadow where light penetrates because of refraction.

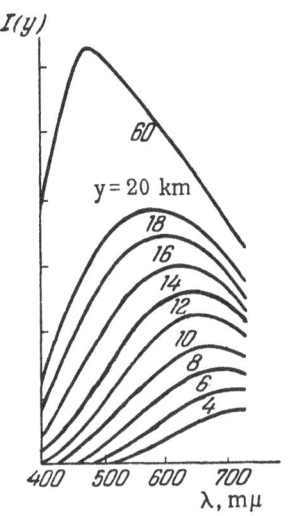

$I(y)$

y = 20 km

400   500   600   700
λ, mμ

Fig. 111. The spectrum of
direct sunlight as a function
of y.

atmosphere, and (b) for an atmosphere with
aerosols, under certain assumptions, apparently fairly realistic, about the aerosol distribution with respect to height and particle
size (the effect of ozone is negligible at
these wavelengths). We see clearly that refraction effects only become important at
long wavelengths, where h_i falls in the vicinity of the earth's geometric shadow; they are
of no importance in the short-wave region,
where h_i is considerably greater than the
height of the geometric shadow.

The strong T(y) and $y_i(\lambda)$ dependences
both result in a substantial change in T(λ)
for given h, that is, in the spectral composition of the light penetrating the atmosphere. Figure 111 shows an example of this
relation for pure molecular scattering (with
no correction for the influence of ozone)
[91]. Note especially the strong reddening
of the sunlight as y decreases.

Let us now consider the changes produced in the situation we have described by the selective absorption of light by atmospheric gases — ozone and
sodium. The characteristic concentration of these constituents in a more or
less well-defined layer at a certain height above the earth's surface results in
a substantial change in the relation between the air mass m, the height of the
observer, and the direction of the ray. Figure 112 illustrates how the total number N of air and ozone molecules along the path of a light beam of 1-cm^2 cross
section through the entire thickness of the earth's atmosphere depends on
the height y of perigee of the beam (corrected for refraction), according to
computations by Venkateswaran, Moore, and Krueger [184]. Whereas for air
the curve is a very smooth one, differing little from a $-\log N \sim y$ relation and
ensuring an even behavior for the $T_p(y)$ curves, in the case of ozone a prominent maximum occurs at $y \approx 25$ km, the region of greatest ozone concentration, where the ray crosses the ozone layer at an angle. There is a corresponding effect on the height variation of the atmospheric transparency in the vicinity of the Chappuis bands. We note that the rotational structure of these bands
is not resolved, so that even if instruments of modest spectral resolution are
employed one can use the Bouguer law, taking $T_{P_s} = \exp - \int k_{oz} d\ell$ in Eq. (III.2,

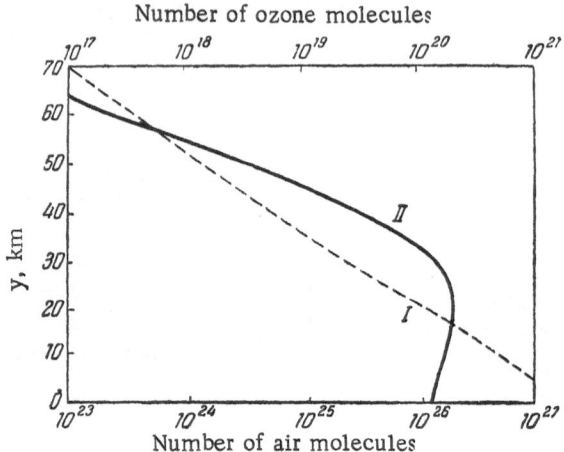

Fig. 112. The total abundance of (I) air and (II) ozone along the trajectory of a light beam through the atmosphere, as a function of the perigee height of the beam.

12), with the integration extending along the entire trajectory of the ray. Several such calculations have been made for the $\lambda = 589$ mμ region, in connection with the problem of the twilight flash of sodium.

Figures 113 present the results of calculations by Blamont, Donahue, and Stull [151], based on the assumptions of pure molecular scattering in the atmosphere, and a Gaussian distribution of ozone with height (maximum at height h_0, halfwidth t_0, total ozone content c_0). We notice immediately the sharp difference between the y-dependence of T_pTp_s for an atmosphere containing ozone and the function $T_p(y)$ for pure air. The curves depart from their smooth behavior and shift toward larger y, with an accompanying reduction in their slope. Roughly speaking, however, they retain their Z-shape, so that in the first approximation all the considerations developed above remain applicable. It should further be noted that the inflection point y_i is displaced in this case approximately toward $T(y_i) = 0.5$.

A change in the total ozone content (Fig. 113a), and hence in the wavelength (in the ozone absorption coefficient), will affect the slope of the curves, y_i, and the value of $T(y_i)$ in a relatively smooth way. The diffusivity of the ozone layer (Fig. 113b) has a comparatively mild effect on the $T(y)$ curve. These curves are most sensitive to changes in the height of the ozone layer (Fig. 113c). Yet even these changes will, for an approximation of the type (III.5, 10), be responsible for an error of no more than 1-2 km in y for constant T, or $\approx 10\%$ in T for constant y.

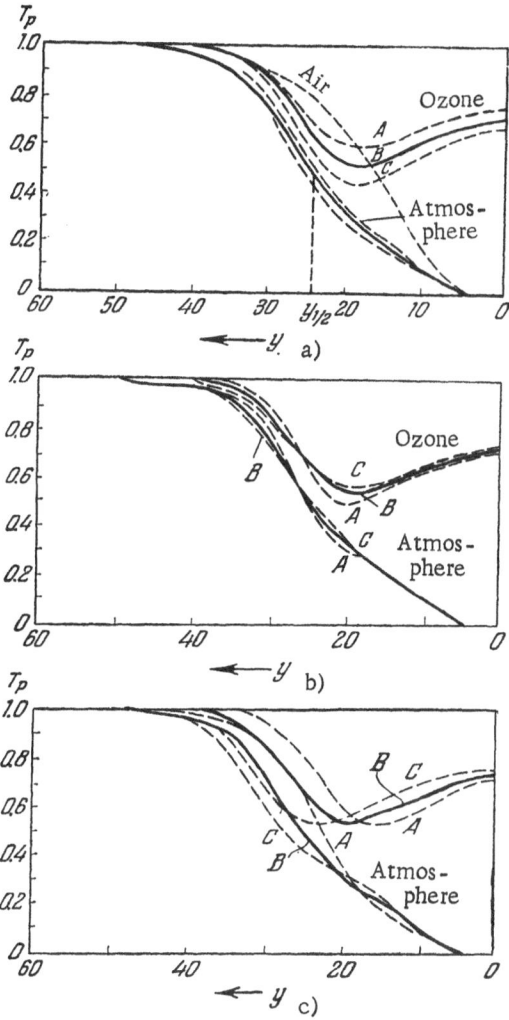

Fig. 113. The transparency Tp_s of ozone and $T_pT_{p_s}$ of pure air with ozone for a ray traversing the entire thickness of the atmosphere, as a function of perigee height y. a) Effect of total ozone content (A, $c_0 = 2$ mm; B, $c_0 = 2.5$ mm; C, $c_0 = 3$ mm); b) effect of half-width of ozone layer (A, $t_0 = 12.4$ km; B, $t_0 = 15.6$ km; C, $t_0 = 18.8$ km); c) effect of height of ozone layer (A, $h_0 = 19$ km; B, $h_0 = 23$ km; C, $h_0 = 27$ km).

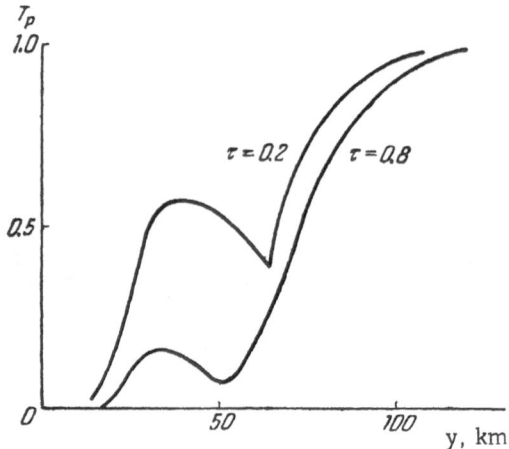

Fig. 114. Sample functions $T_P(y)$ for the region of
the atmospheric-sodium absorption, for selected
values of τ_{Na}.

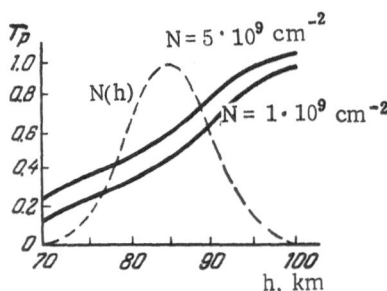

Fig. 115. Sample functions $T_P(h)$ for
the region of the atmospheric-sodium absorption, with the observed
point located inside the sodium layer.

Affairs are much more complicated for the atmospheric transparency in the neighborhood of the
absorption by atmospheric sodium
vapor. On the one hand, we are confronted with a sharp spectral dependence for the optical thickness τ_{Na}
of the sodium layer, and on the other
with the large value of τ_{Na} (for the
central portion of the absorption
lines, τ_{Na} is close to 1). Figure 114
shows some examples of the y-dependence of the atmospheric transparency $T_P T_{P_S}$ in the neighborhood
of the absorption by atmospheric sodium (corrected for molecular scat-
tering and ozone absorption), for given z and ζ and selected τ_{Na}, as computed
under the assumption that the center of gravity of the Na layer lies at the level
$h_{Na} = 85$ km. So long as the height h of the observed point P is less than h_{Na},
the direct sunlight along the trajectory to P will cross the sodium layer only
once. If $h > h_{Na}$, it will cross the layer twice, and as h increases the optical

thickness of the layer along the trajectory of the ray will first increase and then decrease; the greater τ_{Na} is, the lower the sharp minimum in T will be located, for $y \approx 60$-70 km.

We should remember, however, that in general we will be interested in computing Tp mainly in cases where the point under observation falls inside the sodium layer, and hence on the ascending branch of Tp for sufficiently great y. Figure 115 presents the corresponding functions $T_p(h)$ for two values of the total sodium content in the layer, as computed under certain assumptions for the height structure of the layer [152]. We readily see that an approximation of the type (III.5, 10) will be applicable in this case also. But in addition the large value of τ_{Na} compels us to take multiple scattering into account in the illuminated layer itself, that is, the departure of the illuminance of the layer from $Tp I_0 \omega_0$. In the first approximation, these effects have been treated by Gal'perin [254] and by Chamberlain et al. [153-157, 255]. We may add that the functions $T_p(h)$ will differ for the two components of the Na doublet because of the difference in the corresponding values of $\tau_{Na}(\lambda)$.

CHAPTER IV

THE ANATOMY OF
TWILIGHT PHENOMENA

§ 1. A First Approximation

A great many factors, themselves subject to strong fluctuations, combine to produce the wealth of twilight phenomena and their variations in space and time. As a result, an exceptionally diverse range of behavior can occur, but by the same token the twilight period becomes very difficult to analyze adequately. We have already shown that a rigorous analysis of twilight phenomena would require solving the equation of transfer, and that a solution still remains unattainable, or at least none has been achieved. But even at best we can expect to do nothing more than use computing machines to arrive at a limited set of solutions for special and highly idealized cases. This procedure would be of no avail as a basis for developing methods of solving the inverse problem — using twilight phenomena for studying the upper layers of the atmosphere. Thus the only course open to us has been to formulate an approximate theory for twilight with the roles of the various factors clearly expressed. We are faced with the need to explain the anatomy of the several twilight phenomena, that is, to learn what the most important relationships are between the character of the phenomena and the structure of the atmosphere.

We can obtain a fairly clear understanding of the nature of the chief twilight phenomena caused by the single scattering of sunlight in the atmosphere by means of our analysis in Chap. III, §§ 1 and 5, of the geometric configurations for solar rays, and of how their extinction by the atmosphere depends on various circumstances. We are in a position to adopt here several simplifications that will help to distinguish the most important factors, and will render the pattern of twilight phenomena more graphic and accessible for both qualitative and approximate quantitative analysis, without demanding especially laborious computations.

To this end, let us return to Eq. (III.2, 12), which enables us to calculate the components S_i of the Stokes vector-parameter for the singly scattered
light received from the twilight sky, and write it in the form

$$S_i(z, A, \zeta, \lambda) = I_0(\lambda)\omega_0 P^m(\lambda) m \int_0^\infty D_{i1}[h, \varphi(z, A, \zeta), \lambda] \times$$

$$\times T[y(h), \zeta, z, A, \lambda] \, dh =$$

$$= I_0(\lambda)\omega_0 P^m(\lambda) m \int_0^\infty D_{i1}[h(y), \varphi(z, A, \zeta), \lambda] \; T[y, \zeta, z, A, \lambda]\frac{dh}{dy} dy,$$

$$\text{(IV.1, 1)}$$

where for definiteness we have taken $L = 0$, $P(\lambda)$ is the vertical transparency of
the atmosphere, $m(z)$ is abbreviated as m, and we have considered m as not
too large and with a negligible dependence on h; moreover,

$$T(y, \zeta, z, A) = \frac{T_{NP_s} T p e^{m[\tau'(0)-\tau'(h)]}}{1 - 27e^{-0.16y} \cos\zeta'} \qquad \text{(IV.1,2)}$$

The functions $h(y)$ or $y(h)$ depend parametrically on ζ, z, A, and the
structure of the atmosphere both in the vicinity of the observed point P, that
is, at the level h, and in the region of the perigee y of the corresponding ray.

We shall further regard the function $T(y)$ as sufficiently well approxi-
mated by the characteristic Z-shaped curve, with the value $T(y_i)$ and the slope
of the linear branch sensitive essentially only to the presence of selectively
absorbing constituents (ozone, sodium), but with the position of the inflection
point y_i depending significantly on wavelength and (for small h and large λ)
on ζ, z, and A. In other words, the results of Chap. III, §5, imply that the
function $T(y, \zeta, z, A, \lambda)$ may be written with sufficient accuracy in the form

$$T(y, \zeta, z, A, \lambda) = T[y - y_i(\zeta, z, A, \lambda)] \qquad \text{(IV.1. 3)}$$

and, if we use the linear approximation (III.5, 10),

$$T(y, \zeta, z, A, \lambda) = \begin{cases} 0 & \text{for} \quad y \leqslant a(\zeta, z, A, \lambda), \\ \dfrac{y-a}{b} & \text{for} \quad a(\zeta, z, A, \lambda) \leqslant y \leqslant a+b, \quad \text{(IV.1, 4)} \\ 1 & \text{for} \quad y \geqslant a+b, \end{cases}$$

where in the event that selective absorption is absent we have, by the approximation (III.5, 11),

$$b \approx \frac{e}{K}, \quad y_i \approx a + \frac{1}{K} \approx a + \frac{b}{e}, \qquad (IV.1, 5)$$

while in the more general case

$$y_i = a + cb. \qquad (IV.1, 6)$$

The results of Chap. III, §5, imply that b differs appreciably from e/K only if selective absorption occurs, and that an increase in b usually entails a comparatively slight increase in c. But the principal dependence of T on ζ, z, A, and λ is centered in the quantity $a(\zeta, z, A, \lambda)$, with the dependence of a on ζ, z, and A being important only for small h and large λ, where refraction makes a significant contribution.

We have, then, an acceptable approximation if we write Eq. (IV.1, 1) in the form

$$S_i(z, A, \zeta, \lambda) = I_0(\lambda)\,\omega_0 P^m(\lambda)\,m \int_0^\infty D_{i1}[h(y), \varphi(z, \zeta, A), \lambda] \times$$

$$\times T[y - y_i(\zeta, z, A, \lambda)]\frac{dh}{dy}\,dy \qquad (IV.1, 7)$$

or, in the approximation (IV.1, 4),

$$S_i(z, A, \zeta, \lambda) = I_0(\lambda)\,\omega_0 P^m(\lambda)\,m \times$$

$$\times \left\{ \frac{1}{b} \int_a^{a+b} D_{i1}[h(y), \varphi(\zeta, z, A), \lambda](y - a)\frac{dh}{dy}\,dy + \right.$$

$$\left. + \int_{h(a+b)}^\infty D_{i1}[h, \varphi(\zeta, z, A), \lambda]\,dh \right\}. \qquad (IV.1, 8)$$

Recalling now that by Eq. (II.2, 18),

$$D_{ik}(h, \varphi) = \frac{1}{4\pi} f_{ik}(h, \varphi)\sigma(h) \qquad (IV.1, 9)$$

and that by Eqs. (II.2, 6), (II.2, 1), and (II.3, 7),

$$\int_h^\infty \sigma(h)\, dh = \tau_s'(h),$$
(IV.1, 10)

where the subscript s means that we are concerned with the optical thickness for scattering only, we may by analogy introduce the value of the optical thickness for scattering in a given direction,

$$\tau_{ik}'(h,\ \varphi) = \int_h^\infty D_{ik}(h,\ \varphi)\, dh,$$
(IV.1, 11)

where the subscript s may now be dropped: the subscript ik itself specifies that a scattering diagram is involved, that we are dealing with the scattering rather than the extinction of light. The relation (IV.1, 8) may then be written in the form

$$S_i(z,\ A,\ \zeta,\ \lambda) = I_0(\lambda)\, \omega_0\, P^m(\lambda)\, m \times$$

$$\times \left\{ \frac{1}{b} \int_a^{a+b} D_{i1}[h(y),\ \varphi(\zeta,\ z,\ A),\ \lambda] \frac{dh}{dy}(y-a)\, dy + \right.$$

$$\left. + \tau_{i1}'[h(a+b)] \right\}.$$
(IV.1, 12)

§ 2.   The Twilight Layer and Twilight Ray

Suppose first that the geometric boundary of the shadow rises to so great a height H that we may neglect the effects of selective absorption and refraction; thus y_i (and hence a) are to be independent of ζ, z, and A. This situation will certainly prevail for $H \gtrless 15$ km, in the range $0.35\ \mu \lessgtr \lambda \lessgtr 0.5\ \mu$ (for large y, the departures from the Bouguer law due to multiple-scattering effects will be negligible; see Chap. III, § 3). By Eqs. (III.1, 28) and (III.1, 29) we will then have, if z is not too large,

$$h(y) = H + \eta; \quad \eta = \gamma y,$$
(IV.2, 1)

so that the expression (IV.1, 7) becomes

$$S_i(z,\ A,\ \zeta,\ \lambda) = I_0(\lambda)\, \omega_0 P^m(\lambda)\, m \int_0^\infty D_{i1}(H+\eta, \lambda)\, T\left(\frac{\eta}{\gamma},\ \lambda\right) d\eta.$$

(IV.2, 2)

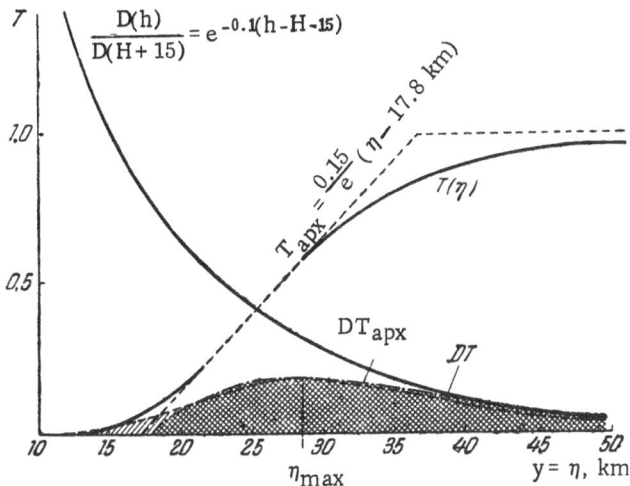

Fig. 116. $D_{11}T$ as a function of η for $\gamma = 1$, $K(y) = 0.15$, $K(H + \eta) = 0.10$, and $\lambda = 0.4 \mu$.

Note that the integrand here is the product of two factors: the factor D_{i1} depends only on the structure of the relatively high atmospheric layers at the level $H + \eta$, while the factor T depends on the structure of the lower layers at the level y_i.

If we exclude the case where well-defined strata of cloud occur in the upper atmosphere, above the level of the earth's shadow (in our case, tropospheric clouds will have hardly any effect on the behavior of the functions D_{i1} and T), then the integrand will contain the product of a rapidly, monotonically decreasing function $D_{i1}(h)$ and a monotonically increasing function $T(y)$. Since both functions are essentially positive, the integrand should reach a maximum at some value $\eta = \eta_{max}$. The position and prominence of this maximum in the product $D_{i1}T$ will depend, first, on the value of the logarithmic gradient in D_{i1} at the level of the point under observation, that is, on the value of $K(H + \eta)$; and secondly, on the curvature in the $T(y)$ relation, that is, on the value of $K(\eta/\gamma)$ at the perigee level of the solar ray illuminating the observed point. Figure 116 illustrates the case where $\gamma = 1$, the scattering coefficient declines exponentially with height, $K(h/\gamma) = K_1 = 0.15$ in the perigee region, $K(H + \eta) = K_2 = 0.10$ near the observed point, and the $T(y)$ relation follows Fesenkov's data for $\lambda = 0.40 \mu$ (Fig. 103). The area of the shaded region corresponding to the product $D_{11}T$, is proportional to the brightness of the twilight sky for $\gamma = 1$, at given H. The figure also contains the linear approximation to the function $T(y)$, with the corresponding area $D_{11}T_{apx}$ (heavy shading). We see that the approximation actually introduces only minor distortions into the value of the integral, or the sky brightness.

Figure 116 indicates that the overwhelming majority of the singly scattered light received by an observer on the ground comes from direct sunlight that has been scattered in a relatively thin layer of the atmosphere. The dense underlying layers of air are largely in shadow and scatter very little light, while the strongly illuminated upper layers have too small a scattering coefficient. Fesenkov [77] was the first to point out this circumstance, which he [83, 237] and Shtaude [85, 98, 99, 128, 238, 240] have used to develop an approximate theory for the twilight phenomenon (see also [91, 96, 124]). Figure 117 shows some examples of the height structure of this effective t w i - l i g h t   l a y e r for z = 70°, A = 0, various $\xi = \zeta - 90°$, and selected wavelengths, as computed by Volz and Goody [124] for a molecular atmosphere (Fig. 117a), an atmosphere with aerosol turbidity (Fig. 117b), and a molecular atmosphere in the region of selective ozone absorption (Fig. 117c). As the sun sinks below the horizon and the height H of the earth's geometric shadow increases, the twilight layer progressively rises, but its structure changes comparatively little (because of the varying γ), maintaining an effective thickness of ≈ 20 km in height. At short wavelengths, the layer has a very distinct shape, practically independent of refraction effects. In the long-wave spectral region, however, part of its profile is severely distorted by refraction, as would be expected, since at these wavelengths refraction sharply alters the $T(y)$ relation, or more accurately the value of y_i. The position and shape of the twilight layer also depend relatively little on the aerosol content, provided it is valid to assume [124] that the height distribution of the aerosols differs little from the march of air density. On the other hand, selective ozone absorption markedly affects the profile of the layer, especially at great heights (Fig. 117c). Yet even in this case there is a clearly defined twilight layer for effective scattering of sunlight.

A displacement of the line of sight along the solar vertical results in analogous effects, since $H = H_0 \gamma$, where H_0 is the height of the earth's shadow at $\gamma = 1$, and $\gamma = \gamma(z, A, \zeta)$. But along with the variation in H the profile of the layer also changes somewhat, since an increase in γ corresponds to a decrease in the slope of the $T(\eta/\gamma)$ curve because of the varying scale along the axis of abscissas in Fig. 116.

Moreover, Fig. 117 graphically demonstrates that the effective scattering layer drops toward the ground considerably later than sunrise, the more so the shorter is the wavelength, that is, the greater the optical thickness of the atmosphere. [We shall find presently that this follows directly from the behavior of the function $T(y)$ for ζ near 90°; see also Chap. III, §5.] Shtaude [240] was the first to point out that the rise of the scattering layer at sunset defines the beginning of twilight at a time before the sun has fallen to the horizon; one should regard the onset of twilight as the time when the maximum of the effective scattering layer breaks away from the earth's surface.

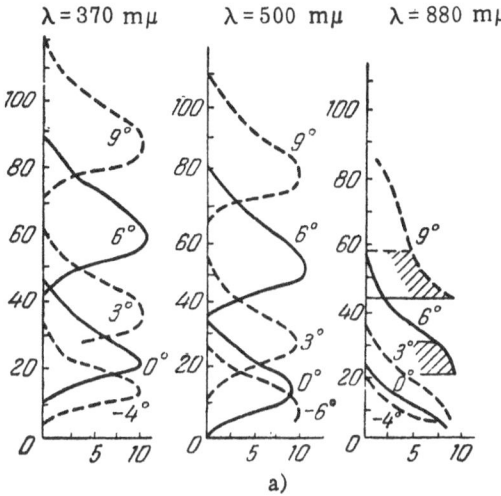

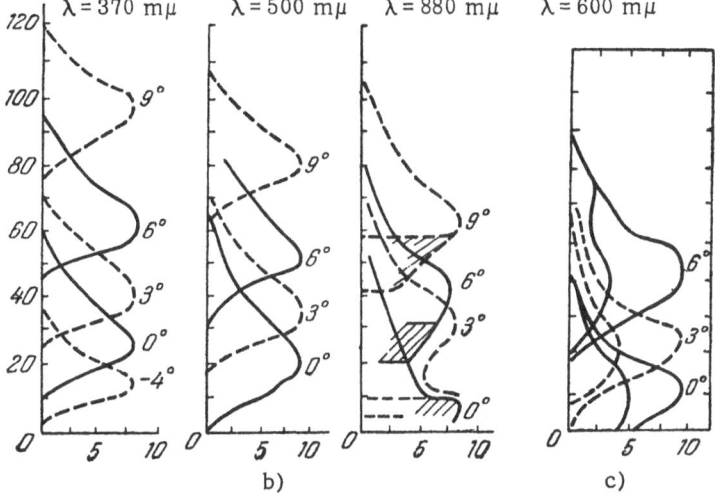

Fig. 117. Height profiles of the twilight layer for z = 70°, A = 0, selected ξ (labels on curves), and selected spectral regions. The portions
of the geometric shadow illuminated because of refraction are indicated by shading.

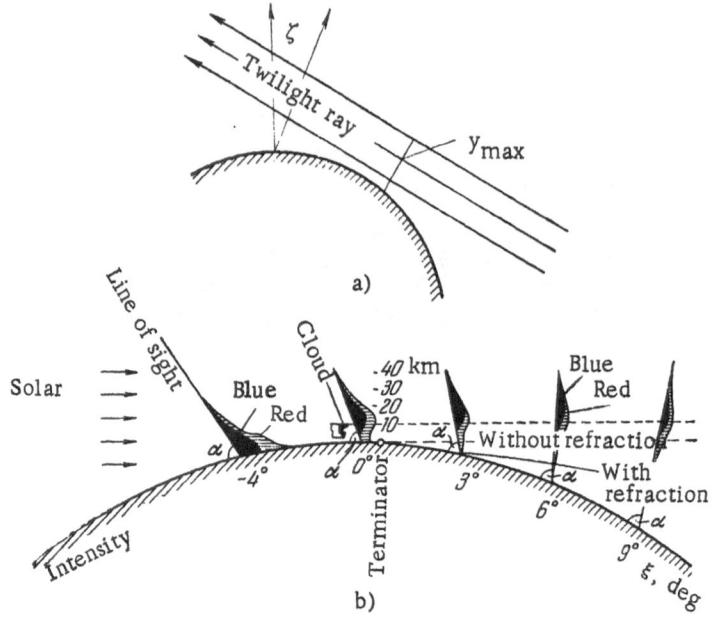

Fig. 118. The twilight ray.

In a first and rather crude approximation, then, we may regard the globe as "blanketed" near its terminator by a comparatively thin (about 20 km thick), luminous layer, the "twilight layer," responsible for the brightness of the twilight sky and for the twilight illumination of the earth's surface. The brightness of the glow depends, in this scheme, exclusively on the scattering power of the atmosphere at the level of the twilight layer. The height of the layer varies both with ζ and with the position of the observed point. In the night hemisphere, the height rises rapidly with increasing H, while with approach to the terminator the layer sinks continually lower, reaching the ground only at a certain elevation of the sun above the horizon. Properly speaking, it is not until this moment that day begins on the earth's surface. The height of the layer also depends materially on the wavelength, especially when the scattering coefficient of the free air is high.

Shtaude [128, 240, and elsewhere] has made a thorough study of the behavior of the twilight layer. She has called particular attention [85] to the fact that when the earth's shadow rises far enough above the surface, the relation between h and y tends to stabilize (near the zenith, $h = H + \gamma y$), and the twilight layer is illuminated by rays having the same perigee y, a value practically independent of H, that is, of ζ. The corresponding pencil of rays which, upon being scattered, makes the main contribution to the twilight-sky brightness, has been called the t w i l i g h t   r a y (Fig. 118a). However, at small H and large λ, where refraction phenomena and the dependence of y_i on ζ, z, and A are important, the concept of a twilight ray loses its validity, although the twilight layer remains fully intact.

In Fig. 118b, after Volz and Goody [282], the theoretically computed height distributions of the brightness in the atmospheric regions illuminated by the sun are shown for several depressions of the sun below the horizon, for the blue and red portions of the spectrum, and for $\alpha = 90° - z = 20°$ (as in Fig. 87, the vertical scale is exaggerated by a factor of five). We clearly see that the twilight ray has already formed at $\xi = \zeta - 90° \approx 3$-$4°$.

The concepts of the twilight layer or twilight ray have been used by Fesenkov [77, 83, 237] and Shtaude [128, 240, and elsewhere] in developing their method for twilight probing of the atmosphere. We shall find below that this formulation is a relatively crude one and is capable of substantial refinement. First, however, we wish to examine the effects of atmospheric structure in the twilight-layer region [that is, the height distribution $D_{i1}(h)$] upon the shape and position of the maximum in the brightness of the layer. This question is of special interest because the observational data on the twilight-sky brightness, as first demonstrated convincingly by Mironov [92], yield a logarithmic gradient K of $D_{i1}(h)$ that declines significantly with increasing h.

Shtaude [128, 240] and Rozenberg [96] have investigated in some detail the behavior of the maximum in the twilight-layer brightness with respect to K. The situation becomes particularly simple if we resort to the approximation (IV.1, 4) and take $K = K(H) = $ const in the twilight-layer region. The first integral in braces in Eq. (IV.1, 8) then becomes

$$\frac{1}{b} \int_a^{a+b} D_{i1}(h) \frac{dh}{dy} (y-a) \, dy =$$

$$= \frac{D_{i1}[h(a)]}{b} \int_a^{a+b} e^{-K\gamma(y-a)} \gamma (y-a) \, dy =$$

$$= \frac{D[h(a)]}{\gamma b} \int_0^{\gamma b} e^{-Kx} \, x \, dx, \qquad \text{(IV.2, 3)}$$

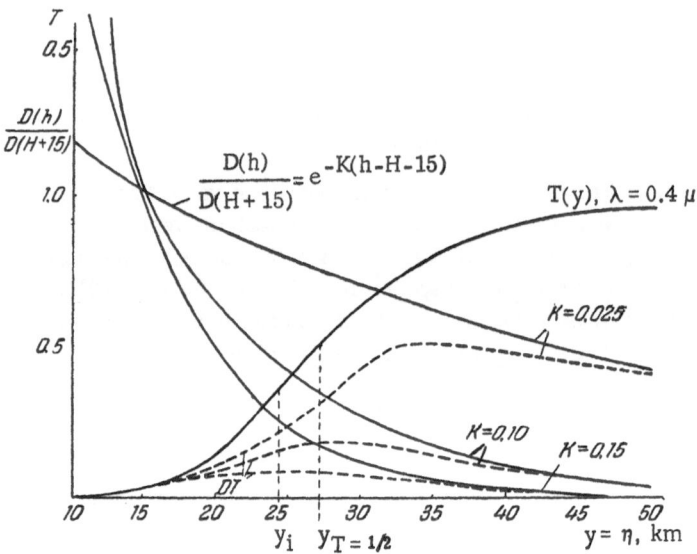

Fig. 119. Displacement of the maximum and deformation of
the profiles of the twilight layer with respect to K(H).

where $x = \gamma(y - a)$. The maximum of the expression in the integrand occurs
for $x = 1/K(H)$ or, in view of Eq. (IV.1, 5),

$$\eta_{max} = \gamma y_i + \begin{cases} \dfrac{1}{K(H)} - \dfrac{1}{K(y)}, & \text{if} \quad K(H) \geqslant e^{-1} K(y), \\[2ex] \dfrac{e-1}{K(y)}, & \text{if} \quad K(H) \leqslant e^{-1} K(y). \end{cases}$$

$$(IV.2, 4)$$

Since y_i increases rapidly with decreasing λ, the height of the twilight-layer
maximum rises toward shorter wavelengths, in accordance with our previous
discussion (Figs. 117 and 118b). While η_{max} is only a few kilometers in the
long-wave region, it reaches 25-35 km in the violet.

But if K(H) decreases, η_{max} will increase rapidly, as shown by Fig. 119,
which presents the functions $D_{11}T$ for K(H) = 0.15, 0.10, and 0.025, with $\lambda =$
0.4 μ. So strong a dependence of η_{max} on K(H) leads to considerable uncertainty in the value of η_{max} under actual conditions, since the quantity K may
in general vary strongly and unpredictably with height.

Moreover, a decrease in $K(H)$ corresponds to a rapid thickening of the twilight layer. While for $K(H) = 0.15$ the layer occupies an interval of about 20 km in height (Figs. 116, 117), for $K(H) = 0.25$ it already extends over a thickness of many tens of kilometers, so that the very concept of a twilight layer becomes less meaningful [96]. [In Figs. 117, the curves are drawn for a very weak height dependence of $K(H)$.] It is these two circumstances, the strong K-dependence of η_{max} and the spreading of the layer with decreasing K, that are responsible for the difficulties in applying the twilight-layer concept.

§3. Twilight Phases and the Effective Shadow Boundary

To some extent, the difficulties with the twilight layer can be circumvented by adopting an effective boundary for the earth's shadow, a concept introduced by Rozenberg [96, 130, 171] and in an indirect form by Shtaude [128].

Let us return to the expression (IV.1, 7). Since the function $T(y)$ is monotonic and D_{i1} and T are essentially positive, we may write the integral in Eq. (IV.1, 1) as

$$\int_0^\infty D_{i1}[h,\ \varphi(z, A, \zeta),\ \lambda]\, T\, [y(h),\ \zeta,\ z,\ A,\ \lambda]\, dh =$$

$$= \int_{\overline{H}(\zeta,\, z,\, A,\, \lambda)} D_{i1}[h,\ \varphi(\zeta,\ z,\ A),\ \lambda]\, dh, \qquad \text{(IV.3, 1)}$$

where the height $\overline{H}(\zeta, z, A, \lambda)$ has a value to be determined. Physically, this operation, which is perfectly rigorous from the mathematical standpoint, amounts to replacing the true height H of the earth's geometric shadow by some effective height $\overline{H}(\zeta, z, A, \lambda)$, together with the compensating assumption that $T = 0$ when $h < \overline{H}$, while $T = 1$ when $h > \overline{H}$. Then with Eq. (IV.1, 11) we obtain, in place of Eq. (IV.1, 1),

$$S_i(z,\ A,\ \zeta,\ \lambda) = I_0(\lambda)\, \omega_0 P^m(\lambda)\, m\tau'_{i1}(\overline{H}). \qquad \text{(IV.3, 2)}$$

All the complexities of the problem are now transferred to determining the value of $\overline{H}$, with its dependence on ζ, z, A, λ, and the state of the atmosphere.

Soon after Rozenberg [96, 171] derived the formula (IV.3, 2), Shtaude [128] arrived at the same expression from different considerations, involving only the brightness of the twilight sky (that is, S_1); in her treatment, Shtaude obtained the formula as a refinement of the twilight-layer concept, with $\tau'_{i1}(h_{max})$ appearing in place of $\tau'_{i1}(\overline{H})$.

Consider first the case where refraction phenomena can be neglected and where z is not too great. Then from Eqs. (IV.3, 2), (IV.1, 12), and (IV.2, 1), we obtain

$$\tau'_{i1}\,(H+\overline{\eta}) = \tau'_{i1}\,[H + \gamma\,(a+b)] +$$

$$+\frac{1}{\gamma b}\,'\!\!\int_{\gamma a}^{\gamma\,(a+b)} D_{i1}\,[H+\eta,\;\varphi(\zeta,\;z,\;A),\;\lambda]\,[\eta-\gamma a\,(\lambda)]\,d\eta$$

$$\text{(IV.3, 3)}$$

or

$$\int_{\eta}^{\gamma\,(a+b)} D_{i1}\,[H+\eta,\;\varphi(\zeta,\;z,\;A),\;\lambda]\,d\eta =$$

$$= \frac{1}{\gamma b}\int_{\gamma a}^{\gamma\,(a+b)} D_{i1}\,[H+\eta,\;\varphi(\zeta,\;z,\;A),\;\lambda]\,[\eta-\gamma a\,(\lambda)]\,d\eta.$$

$$\text{(IV.3, 4)}$$

Suppose now that in the height range from $H+\gamma a$ to $H+\gamma(a+b)$, D_{i1} varies exponentially with h; that is,

$$D_{i1}\,(h) = D_{i1}\,(H+\gamma a)\,e^{-K\,(H)\,(\eta-\gamma a)} \qquad \text{(IV.3, 5)}$$

Again taking

$$x = \eta - \gamma a \qquad \text{(IV.3, 6)}$$

and performing the integration, we have

$$e^{-K\,(H)\,(\overline{\eta}-\gamma a)} = \frac{1-e^{-K\,(H)\,\gamma b}}{K\,(H)\,\gamma b} \qquad \text{(IV.3, 7)}$$

or, since $y_{1/2} = a + b/2$, where $T(y_{1/2}) = 1/2$,

$$e^{-K\,(H)\,(\overline{\eta}-\gamma y_{1/2})} = \frac{2\,\mathrm{sh}\left[\dfrac{K\,(H)\,\gamma b}{2}\right]}{K\,(H)\,\gamma b} \approx 1 + \frac{[K\,(H)\,\gamma b]^2}{24},$$

$$\text{(IV.3, 8)}$$

whence

$$\overline{\eta} \approx \gamma y_{1/2} - K\,(H)\,\frac{\gamma^2 b^2}{24}. \qquad \text{(IV.3, 9)}$$

In particular, if the relations (IV.1, 5) hold, and if $K(H) = K(y)$, $\bar{\eta} \approx \gamma y_i$; if $K(H) \ll K(y)$, then $\bar{\eta} \approx \gamma y_{1/2}$. Thus over a wide range of variation in $K(H)$, $\bar{\eta}$ will remain in the neighborhood of γy_i, differing from this quantity by no more than 2-5 km. Since we usually have $b = 20$-25 km, $d\bar{\eta}/dK \approx -20\gamma^2$, and the expression (IV.3, 9) may be written approximately in the form

$$\bar{\eta} \approx \gamma y_i + [K(y) - K(H)] \, 20\gamma^2. \qquad (IV.3, 10)$$

The value of $\bar{\eta}$ is also comparatively insensitive to variations in b. The relation (IV.3, 9) implies that

$$\frac{d\bar{\eta}}{db} \approx \frac{\gamma}{2} \left(1 - \frac{K(H)\,\gamma b}{6} \right) < \frac{\gamma}{2}, \qquad (IV.3, 11)$$

so that variations in b will affect η on only half the scale with the main result being a displacement in the inflection point y_i.

Variations in λ have the most important influence on $\bar{\eta}$. In fact, since $K(H)$, γ, and b are practically independent of λ,

$$\frac{d\bar{\eta}}{d\lambda} \approx \gamma \frac{dy_i}{d\lambda} \qquad (IV.3, 12)$$

and in such a case as the atmosphere considered by Fesenkov (Fig. 103), where the relation (III.5, 14) applies, we have

$$\frac{d\bar{\eta}}{d\lambda} \approx -\gamma \frac{25}{\lambda} \text{ km}/\mu. \qquad (IV.3, 13)$$

As an example, Fig. 120 introduces the effective boundary of the earth's shadow for the case $K(H) = 0.10$ and $\gamma = 1$, with $\lambda = 0.4\ \mu$ and $K(y) = 0.15$ [the $T(y)$ relation is taken from Fig. 103]; Eq. (IV.3, 9) is used to compute the height $\bar{\eta}$.

So far our discussion has touched only on the case where the effective boundary of the earth's shadow lies at a substantial height, and where there is no contribution from refraction or from a dependence of y_i on ζ, z, and A. But even if these restrictions are relaxed, the condition

$$\bar{H} = h_i + [K(y) - K(H)] \, 20 \left(\frac{dh}{dy} \right)^2_{h_i} \qquad (IV.3, 14)$$

should hold with adequate accuracy; that is, the effective shadow boundary will fall near h_i.

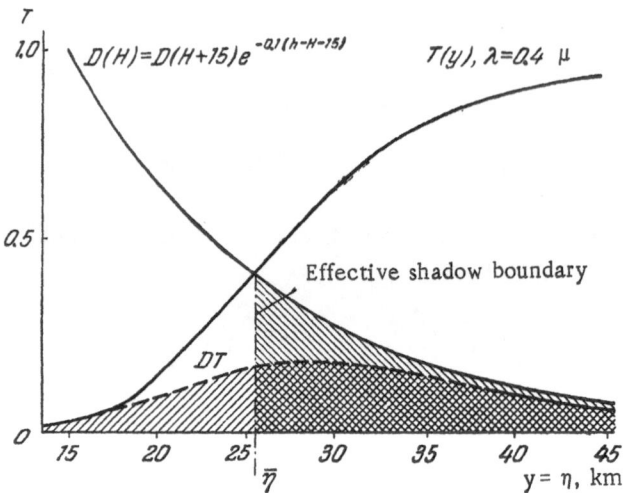

Fig. 120. The effective boundary of the earth's shadow
(a sample case).

Appreciable departures will occur only if a becomes negative $(a = -\Delta)$, that is, if $T(y) \neq 0$; or else if $T = 0$ up to perigee height $y = a + \Delta$ but with $T(y)$ reverting to Eqs. (IV.1, 4) for $y > a + \Delta$, as can happen if there is a high cloud cover near perigee. In both cases, the lower limit in the integral on the right-hand side of Eq. (IV.3, 4) is replaced by $a + \Delta$ ($\Delta \leq b$), so that, with the condition (IV.3, 5), we have instead of Eq. (IV.3, 7) the expression

$$e^{-K(H)\left[\bar{\eta}-\gamma\left(a+\frac{b}{2}+\frac{\Delta}{2}\right)\right]} =$$

$$= \frac{2\left[1+\dfrac{K(H)\,\gamma\Delta}{2}\right]}{K(H)\,\gamma b}\,\text{sh}\left[\frac{K(H)\,\gamma\,(b-\Delta)}{2}\right] +$$

$$+ \frac{\Delta}{b}\,\text{ch}\left[\frac{K(H)\,\gamma\,(b-\Delta)}{2}\right]$$

(IV.3, 15)

or, upon developing the hyperbolic functions in series and retaining only the first two terms of the expansion,

$$e^{-K(H)\left[\bar{\eta}+\gamma\left(a+\frac{b}{2}+\frac{\Delta}{2}\right)\right]} = \left[1+\frac{K(H)\,\gamma\Delta\left(1-\dfrac{\Delta}{b}\right)}{2}\right] \times$$

$$\times \left\{ 1 + \left[1 + \frac{2\frac{\Delta}{b}}{1 + \frac{K(H)\,\gamma\Delta}{2}\left(1 - \frac{\Delta}{b}\right)} \right] \frac{K^2(H)\,\gamma^2(b-\Delta)^2}{24} \right\},$$

(IV.3, 16)

whence

$$\bar{\eta} \approx \gamma y_{1/2} + \gamma \frac{\Delta}{2} - \frac{1}{K(H)} \ln\left[1 + \frac{K(H)\,\gamma\Delta\left(1 - \frac{\Delta}{b}\right)}{2} \right] -$$

$$- \left[1 + \frac{2\frac{\Delta}{b}}{1 + \frac{K(H)\,\gamma\Delta\left(1 - \frac{\Delta}{b}\right)}{2}} \right] \frac{K(H)\,\gamma^2(b-\Delta)^2}{24}.$$

(IV.3, 17)

By expanding the logarithm in series and neglecting the relatively small de-parture from unity of the denominator in the second brackets, we obtain approximately

$$\bar{\eta} \approx \gamma y_{1/2} + \frac{\gamma\Delta^2}{2b} - \left(1 + \frac{2\Delta}{b}\right) \frac{K(H)\,\gamma^2(b-\Delta)^2}{24}.$$

(IV.3, 18)

Now let us suppose that the earth's shadow falls at a fairly high level, and that at height $h_c = H + \eta_c$ there is a scattering layer of optical thickness τ_c and a scattering coefficient whose behavior with height is described by a δ-function: $\sigma_c = \tau_c \delta(h - h_c)$. If τ_c is small, we may repeat the operations leading to Eq. (IV.3, 9) to obtain

$$\bar{\eta} = \gamma y_{1/2} - K(H) \frac{\gamma^2 b^2}{24} - \frac{\tau_c}{K(H)\,\gamma b} \eta_c +$$

$$+ \begin{cases} \dfrac{a+b}{K(H)\,b}\tau_c & \text{for} \quad \eta_c > \bar{\eta}, \\[2ex] \dfrac{a}{K(H)\,b}\tau_c & \text{for} \quad \eta_c < \bar{\eta}. \end{cases}$$

(IV.3, 19)

Thus, from the time when $\eta_c = \gamma(a+b)$ onward, $\bar{\eta}$ will rise until it reaches the value

$$\bar{\eta}_1 = \frac{\gamma y_{1/2} - K(H)\dfrac{\gamma^2 b^2}{24} + \dfrac{a+b}{K(H)\,b}\tau_c}{1 + \dfrac{\tau_c}{K(H)\,\gamma b}} = \eta_{c1}.$$

(IV.3, 20)

Subsequently $\bar{\eta}$ will drop along with η_{C} until it reaches

$$\bar{\eta}_2 = \bar{\eta}_1 - \frac{\tau_c \gamma b}{K \gamma b + \tau_c} = \eta_{c2} < \gamma y_{1/2} - K\,(H)\frac{\gamma^2 b^2}{24}\,,$$

(IV.3, 21)

whereupon it will again rise, returning to its original position when $\bar{\eta}_\mathrm{C} = \gamma\,a$. The rates of increase of $\bar{\eta}$ in the intervals $\bar{\eta}_1 \leq \eta_\mathrm{C} \leq \gamma(a+b)$ and $\gamma\,a \leq \eta_\mathrm{C} \leq \bar{\eta}_2$ are the same, namely

$$\frac{d\bar{\eta}}{d\zeta} = -\tau_\mathrm{c}\,\frac{d\eta_\mathrm{c}}{d\zeta}\,.$$

(IV.3, 22)

It remains now to consider the position $\bar{H}$ of the effective boundary of the earth's shadow under daytime conditions. To this end, we may use Eq. (III.2, 15), which reduces to the form (IV.3, 2) if we take

$$\tau'\,(\bar{H}) = \frac{1 - e^{-\Delta m \tau^*}}{\Delta m}\,,$$

(IV.3, 23)

where $\Delta m = m_\zeta - m_\mathrm{z}$, or equivalently

$$\tau'\,(\bar{H}) = e^{-\frac{\Delta m \tau^*}{2}}\,\frac{2\,\mathrm{sh}\left(\dfrac{\Delta m \tau^*}{2}\right)}{\Delta m}\,.$$

(IV.3, 24)

For a provisional estimate, we may assume once again that the scattering coefficient in the layer of air near the ground varies exponentially with height, with a logarithmic gradient equal to K:

$$\tau'\,(\bar{H}) = \tau^* e^{-K\bar{H}}\,.$$

(IV.3, 25)

Taking logarithms in Eq. (III.2, 5), we then have

$$\bar{H} = \frac{\Delta m \tau^*}{2K} - \frac{1}{K}\ln\left[\frac{2\,\mathrm{sh}\left(\dfrac{\Delta m \tau^*}{2}\right)}{\Delta m \tau^*}\right]$$

(IV.3, 26)

Since $\Delta m \tau^*/2$ is small, we may develop the hyperbolic sine in series to obtain the formula

$$\bar{H} \approx \frac{\Delta m \tau^*}{2K} - \frac{1}{K}\ln\left[1 + \frac{(\Delta m)^2 \tau^{*2}}{24}\right]$$

(IV.3, 27)

or, upon expanding the logarithm,

$$\bar{H} \approx \frac{\Delta m \tau^*}{2K} \left(1 - \frac{\Delta m \tau^*}{12} \right).$$
(IV.3, 28)

We see at once from this last expression that, independently of m_ζ, if $\Delta m \leq 0$ the concept of the effective shadow boundary is not meaningful physically, since the shadow is actually bounded by the earth's surface. But if $\Delta m > 0$ ($m_\zeta > m_z$), the effective shadow boundary will rise above the earth's surface, and the larger Δm or τ^*, the higher it will rise.

We are able now to follow the behavior of $\bar{H}(\zeta)$ over the entire period from midday to night. For definiteness, let us take as an example the zenith region of the sky, with $m_z \approx 1$ and $\gamma \approx 1$.

When the sun itself is in the zenith, $\bar{H} = 0$, by Eq. (III.3, 28). As the sun moves toward the horizon, $\bar{H}$ will begin to rise slowly, the more so if τ^* is small. For a molecular atmosphere and $\zeta = 60°$, $\bar{H}$ will reach the approximate values $\bar{H} = 1.2$ km at $\lambda = 0.4$ μ, and $\bar{H} = 0.2$ km at $\lambda = 0.6$ μ. As ζ continues to increase, $\bar{H}$ will rise more rapidly, and the larger τ^* is, the sooner $\bar{H}$ will rise; by $\zeta = 80°$, $\bar{H}$ will already have reached the values 5 km at $\lambda = 0.4$ μ, and 1 km at $\lambda = 0.6$ μ (Fig. 121). Because of the rapid increase in Δm for $\zeta \gtrsim 80°$, the rise in $\bar{H}$ will continue to accelerate. Moreover, in this range of solar zenith distances the inflection point y_i on the transparency curve $T(y)$ will also begin to rise above the ground (see Fig. 106). The appropriate computations for $\bar{H}$ can be based not only on Eq. (IV.3, 28), which begins to break down at large τ^*, but also on Eq. (IV.3, 18), with Eq. (III.5, 29) used for the ζ-dependence of $h_i = y_i$ (see Fig. 106). In Fig. 121, the dashed curve shows the results of such a computation at $\lambda = 0.6$ μ, for a molecular atmosphere. They evidently agree well with the results of computations using Eq. (IV.3, 28), as would be expected since τ^* is not large at $\lambda = 0.6$ μ. Again, if the line of sight extends beyond the true perigee of a ray through height $\bar{H}$, Eq. (III.5, 29) will be replaced by Eq. (III.5, 36) (see Figs. 106, 107). But until the perigee height of the inflection point y_i has attained its limiting value, that is, until $\zeta \approx 94°$ (see Fig. 106), we must take Eq. (IV.3, 18) into account. Thus in the range of zenith distances from about $\zeta = 80°$ to true perigee, that is, to $\zeta = 90° + \nu_\zeta$, the chief effects governing the rise in $\bar{H}$ will be the increase in y_i (Fig. 106) and the corresponding variation in Δ [Eq. (IV.3, 18)]. From true perigee until about $\zeta = 94°$, these effects will be supplemented by the elevation of the earth's geometric shadow – the $H(\zeta)$ relation. Finally, for $\zeta \gtrsim 94°$ only this last effect will remain, since the value of y_i will now have stabilized.

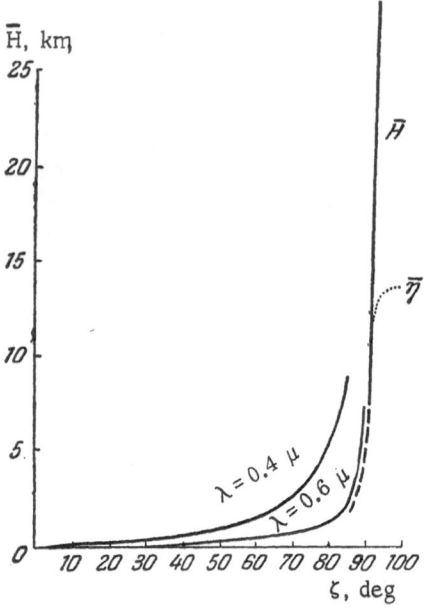

Fig. 121. Height $\overline{H}$ of the effective boundary of the earth's shadow at the zenith
as a function of ζ.

There are two items to be noted here. First, the specific limits of influence of any particular effect, as well as the magnitude of the influence,
will depend materially on the wavelength (apart from the effect of elevation
of the earth's geometric shadow). Secondly, the influence will also have a
substantial dependence on the direction of the line of sight, that is, on z and
A. This means that the phases of twilight at any moment will differ both for
different spectral regions and for different parts of the sky. In fact, the sky
brightness, as observed from the ground, is proportional, by Eq. (IV.3, 2), to
the optical thickness $\tau'_{ik}(\overline{H})$ for light scattered at height $\overline{H}$, and the brightness variations are associated directly with variations in $\overline{H}$. It therefore appears appropriate to subdivide the twilight into phases, according to the character of the $\overline{H}(\zeta)$ variations.

As we have seen, the value of $\overline{H}$ changes only very slightly during the day
as ζ increases. The transition to twilight — to the period of rapid fall in sky
brightness, and hence in $\tau'_{ik}(\overline{H})$ also — is characterized by a rapidly rising $\overline{H}$
as the sun approaches the horizon. Figure 121 (see also Figs. 2 and 15) shows

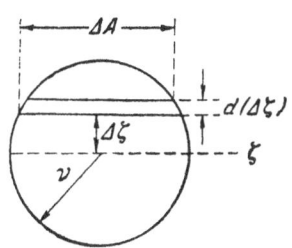

Fig. 122. A correction for the angular size of the sun.

how diffuse and incapable of precise definition is the transition from day to twilight. Indeed, the transition occurs at different times for different wavelengths and parts of the sky.

On the other hand, at some value of ζ, which differs with the part of sky but is approximately the same for all wavelengths outside the selective-absorption bands (for the zenith, at $\zeta \approx 94°$), the value of y_i stabilizes, and the increase in $\bar{H}$ with ζ no longer depends on refraction effects, being determined only by the progressive elevation of the earth's geometric shadow; thus

$$\bar{H} = H\left(\zeta\right) + \bar{\eta} \approx H\left(\zeta\right) + \gamma y_i. \qquad (IV.3, 29)$$

It is from this time onward that twilight can be regarded as fully established in a given part of the sky.

We arrive, then, at a fairly clear distinction between two phases of twilight:

1. semitwilight, the period between daytime and the establishment of twilight in a given part of the sky;

2. total twilight, beginning when twilight is established in a given part of the sky, at practically the same time for all wavelengths.

However, if we turn to the observational data (Chap. I, §§3 and 4), we are at once confronted with the fact that at a certain value of ζ, depending on the part of the sky (for the zenith, at $\zeta \approx 100°$), many twilight phenomena change radically in character. We have therefore to distinguish a third and last phase of twilight:

3. deep twilight, the period when, as we shall find, important contributions to the brightness come from multiple scattering in the atmosphere, the intrinsic airglow, and other sources of light in the night sky. Deep twilight concludes with the transition to night, a process which again takes place at different times in different parts of the sky (see §7 for further details).

Semitwilight is notable for rapid changes in the coloration of the sky, since the phenomenon proceeds quite differently in different spectral regions — that is, $\bar{H}$ is a strong and ζ-dependent function of λ during the semitwilight period. These are the changes characteristic of sunset (see §5 for details).

Thus s u n s e t is nothing but the assemblage of semitwilight processes in a given part of the sky for all wavelengths (we shall reserve the term "semitwilight" for relatively narrow spectral bands only). The times of onset and conclusion and the degree of development of sunset differ over the sky. In particular, while deep twilight may already have set in near the Gegenschein, sunset phe-nomena will still be under way along the solar horizon (see Chap. I, § 2). For the most brightly illuminated region above the solar horizon it is therefore convenient to retain the term "twilight-glow (sunset) segment," without imply-ing that sunset phenomena are associated with this segment only.

Equation (IV.3, 2) further enables us readily to incorporate the true angu-lar size of the sun. Consider Fig. 122, which shows the solar disk with its angu-lar radius ν. If we neglect the refractive distortion of the shape of the solar disk, and take ζ as the zenith distance of its center, then the horizontal angu-lar extent ΔA (A is the azimuth) of a cross section of the disk at distance $\Delta \zeta$ from its center will be given by the obvious relation

$$(\Delta \zeta)^2 + \frac{(\Delta A)^2}{2} = \nu^2. \qquad \text{(IV.3, 30)}$$

Select now a band of width $d(\Delta \zeta)$ at distance $\Delta \zeta$ from the center of the disk. If we regard the band as a source of light, we may replace the ω_0 in Eq. (IV.3, 2) by the quantity $\Delta A \, d(\Delta \zeta)$, and introducing the relation $\bar{H}(\zeta)$, obtain

$$dS_i(z, \, A, \, \zeta, \, \zeta + \Delta \zeta, \, \lambda) =$$
$$= I_0(\lambda) \, P^m(\lambda) \, m \, \Delta A \, d(\Delta \zeta) \, \tau'_{i_1} \left[\bar{H}(\zeta) + \frac{d\bar{H}}{d\zeta} \, \Delta \zeta \right] =$$
$$= I_0 P^m(\lambda) \, m \tau'_{i_1} \, [\bar{H}(\zeta)] \, \Delta A \, d(\Delta \zeta) +$$
$$+ I_0 P^m(\lambda) \, m \left(\frac{d\tau'_{i_1}}{dH} \, \frac{d\bar{H}}{d\zeta} \right)_\zeta \Delta \zeta \, \Delta A \, d(\Delta \zeta). \qquad \text{(IV.3, 31)}$$

After integration with respect to $d(\Delta \zeta)$ over the entire area of the disk, the first term returns us to Eq. (IV.3, 2), since

$$\int\limits_{-\nu}^{+\nu} \Delta A \, d(\Delta \zeta) = \omega_0.$$

Moreover, in view of Eq. (IV.3, 30) we have

$$\Delta \zeta \, \Delta A \, d(\Delta \zeta) = \frac{1}{2} \, \Delta A \, d(\Delta \zeta)^2 = \frac{1}{2} \, (\Delta A)^2 \, d(\Delta A), \qquad \text{(IV.3, 32)}$$

so that

$$\int_{-\nu}^{+\nu} \Delta\zeta\,\Delta A\,d(\Delta\zeta) = \int_{0}^{2\nu} (\Delta A)^2\,d(\Delta A) = \frac{2\nu^3}{3} = \frac{8}{3}\,\frac{\nu}{\pi}\,\omega_0,$$

(IV.3, 33)

since $\pi\nu^2 = \omega_0$. Thus an integration of Eq. (IV.3, 31) over the entire area of the disk leads to the expression

$$S_i(z, A, \zeta, \lambda) = I_0\omega_0 P^m(\lambda)\,m\left\{\tau'_{i1}\,[\overline{H}(\zeta)] + \frac{d\tau'_{i1}}{d\zeta}\,\frac{8\nu}{3\pi}\right\} =$$

$$= I_0\omega_0 P^m(\lambda)\,m\tau'_{i1}\left[\overline{H}\left(\zeta + \frac{8\nu}{3\pi}\right)\right].$$

(IV.3, 34)

In other words, allowing for the angular size of the solar disk amounts to re-placing the true zenith distance ζ of the sun by the quantity $\zeta + 8\nu/3\pi$, an effective displacement of the sun by about 13'.5 in the direction toward the horizon.

§4. Atmospheric Inhomogeneities and the Logarithmic Brightness Gradient

For the brightness $I = S_1$ of the twilight sky, Eq. (IV.3, 2) takes the form

$$I(z, A, \zeta, \lambda) = I_0(\lambda)\,\omega_0 P^m(\lambda)\,m\tau'_{11}\,[\overline{H}(z, A, \zeta, \lambda)].$$

(IV.4, 1)

The logarithmic gradient of this quantity,

$$q = -\frac{d\ln I}{dH} = -\frac{d\ln I}{d\zeta}\,\frac{d\zeta}{dH},$$

(IV.4, 2)

is evidently equal to

$$q = \frac{D_{11}(\overline{H})}{\tau'_{11}(H)}\,\frac{d\overline{H}}{dH}$$

(IV.4, 3)

and in the case of an exponentially varying $D_{11}(h)$ with the logarithmic gradient $K(H)$ we have

$$q = K(\overline{H})\,\frac{d\overline{H}}{dH}.$$

(IV.4, 4)

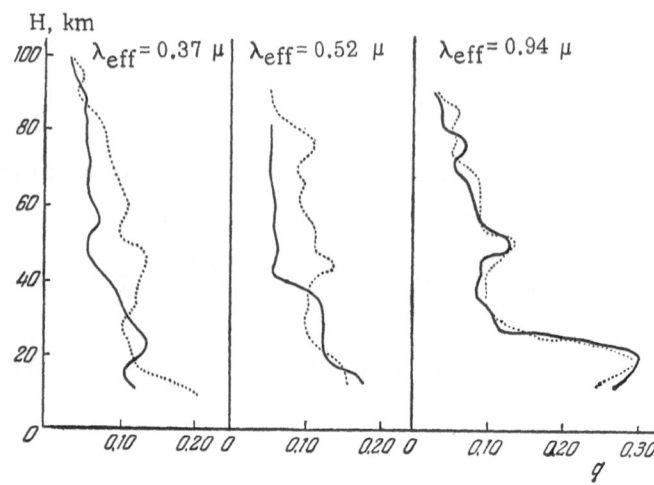

Fig. 123. Typical q(H) curves in three spectral regions, according to measurements by Megrelishvili at $z = 0°$ (solid curves) and $z = 70°$ (dotted).

On the other hand, we may start with the twilight-layer concept (§2), taking $I \sim D_{11}(h_{max})$ and once again $D_{11}(h) \sim e^{-kh}$; we will then have, as Shtaude [85, 238] has shown,

$$q = K(h_{max}) \frac{dh_{max}}{dH}. \qquad (IV.4, 5)$$

Fesenkov [77, 237] and Link [84, 227] had previously obtained the same result by somewhat different methods. Because of the small difference between $d\overline{H}/dH$ and dh_{max}/dH during the total-twilight period, as a comparison between Eqs (IV.2, 4) and (IV.3, 9) shows, the two formulas yield similar results. The main distinction is that in one case the formula contains the value of K at height $\overline{H}$, in the other case at height h_{max}. However, Shtaude, Fesenkov, and Link have all adopted $dh_{max}/dH = 1$, which as we have seen is realistic only in a very rough approximation, since even during the total-twilight period the quantities h_{max} and $\overline{H}$ depend on K(H) as well as H.

Nevertheless, if we neglect this dependence in the first approximation and take $d\overline{H}/dH = 1$, then $q \approx K(\overline{H})$, and by determining q from measurements of the I(ζ) relation we can evaluate K approximately at the level $\overline{H}$.

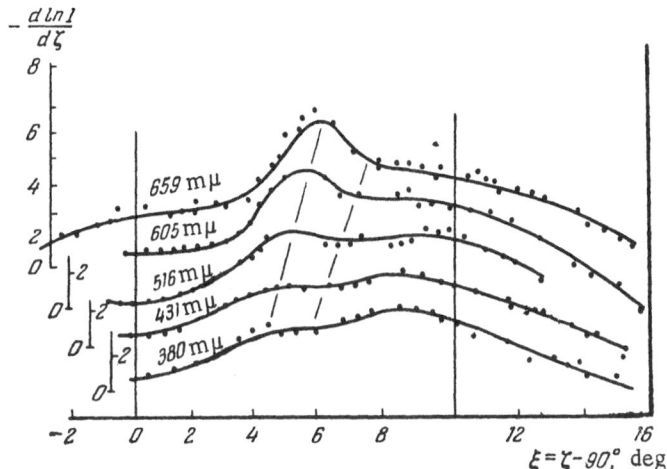

Fig. 124. The relation $-d(\ln I)/d\zeta$ for a single day and several wavelengths, according to measurements by Volz and Goody.

Following the work of Fesenkov, Shtaude, and Link, several authors therefore carried out extensive measurements of the function $q(\overline{H})$ [92, 95, 96, 101, and elsewhere]. Megrelishvili has conducted a particularly comprehensive program, measuring $q(H)$ over many years at the Abastumani Observatory. As an example, Fig. 123 shows q as a function of H according to simultaneous observations by Megrelishvili on the evening of October 14, 1955, in three spectral regions and along two lines of sight ($A = 0$, $z = 0°$ and $70°$), observations which she has kindly placed at the author's disposal. Figure 124 illustrates similar curves as obtained by Volz and Goody [124] on a single day, at several wavelengths, for $z = 70°$.

An especially noteworthy feature of these observations is the extreme variability of the $q(H)$ relation, both from day to day and from one value of z or λ to another. In other words, each $q(H)$ curve is highly individualistic, and hence very sensitive to changes in the atmospheric conditions. Again, individual $q(H)$ curves exhibit a capricious behavior, marked as they are by many small-scale kinks which definitely do not arise from errors of measurement. Bigg [111], in particular, has developed an instrument which measures directly the quantity

$$\frac{d \ln I}{dt} = - q \frac{dH}{d\zeta} \frac{d\zeta}{dt} \cdot$$

$$(IV.4, 6)$$

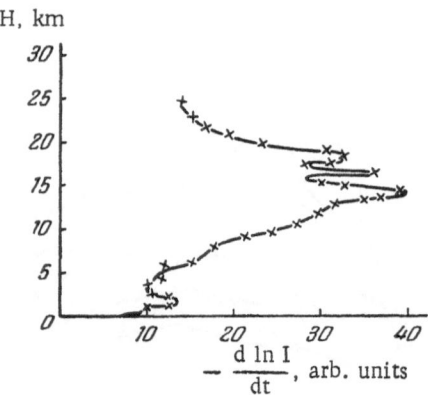

Fig. 125. A sample relation between $-d(\ln I)/dt$
and H for small H, as measured by Bigg.

Figure 125 shows a sample curve for $d(\ln I)/dt$ as a function of H, as obtained
by Bigg with allowance for refraction; it establishes that the small-scale variations in q are in fact real.

There are finally several peculiarities in the behavior of $q(H)$ that have
been observed more or less systematically, though with distinct variations. A
prominent feature in the overwhelming majority of cases is a sharp maximum
at $H = 5$-20 km; it is particularly strong and high at long wavelengths (see Figs.
123-125, as well as [92, 95, 96, 101, 111, 124, 282, and elsewhere]). Beyond
this maximum, which fluctuates strongly with atmospheric conditions (see,
for example, [111]), the value of q either stabilizes or declines weakly with
height. In the height range from $H \approx 30$ km to $H \approx 100$-150 km, the value of
q is only a few hundredths of $1\ km^{-1}$. Finally, for $H \gtrsim 150$ km, $q(H)$ again declines rapidly, reaching values of only thousandths of $1\ km^{-1}$ at $H \approx 250$ km.
In other words, if we adhere to the relations (IV.4, 5), we may infer a fairly
rapid systematic decrease of $K(H)$ with height, interrupted to be sure by several peculiarities which we shall discuss below (see also Chap. V).

We turn now to a discussion of the function $q(H)$ from the theoretical
standpoint. For this purpose we must revert to Eq. (IV.4, 3). First suppose that
$K(h)$ is constant and that the atmosphere is purely molecular, free from any
kind of inhomogeneities. In actual measurements of the sky brightness, the
quantity I will be evaluated as a function of the time t, or as a unique and,
in an adequate first approximation, a linear function of the zenith distance ζ

of the sun. It therefore suffices for us to examine the behavior of the quantity

$$-\frac{d\ln I}{d\zeta} = -\frac{d\ln I}{d\bar{H}}\frac{d\bar{H}}{d\zeta} .$$ (IV.4, 7)

[For small $\xi = \zeta - \pi/2$, and especially for $\xi < 0$, there is clearly no point in considering the quantity d(ln I)/dH.]

Again writing $\bar{H}$ in the form

$$\bar{H} = H + \bar{\eta},$$ (IV.4, 8)

where all the refraction effects and the relations between T and z, A, and ζ are absorbed by the quantity $\bar{\eta}$, we obtain

$$\frac{d\bar{H}}{d\zeta} = \frac{dH}{d\zeta} + \frac{d\bar{\eta}}{d\zeta} .$$ (IV.4, 9)

Over a quite broad region of the sky near the zenith, with $L = 0$ and refraction absent, Eq. (III.1, 23) implies

$$H = R(1 - \sin\zeta)\gamma,$$ (IV.4, 10)

whence

$$\frac{dH}{d\zeta} = -R\gamma\cos\zeta - R(1-\sin\zeta)\gamma\frac{\cos\varphi}{\cos\psi}$$ (IV.4, 11)

or, in the solar meridian,

$$\frac{dH}{d\zeta} = -R\gamma\left[\cos\zeta - (1-\sin\zeta)\operatorname{ctg}(\zeta - z)\right].$$ (IV.4, 12)

If z is not too large, this formula yields

$$\frac{dH}{d\zeta} \approx R\gamma\xi\left[1 - \frac{\xi}{2}\operatorname{ctg}(\zeta - z)\right]$$ (IV.4, 13)

and in the region near the zenith, where $\operatorname{ctg}(\zeta - z)$ is small,

$$\frac{dH}{d\zeta} = R\gamma\xi \qquad (\xi > 0).$$ (IV.4, 14)

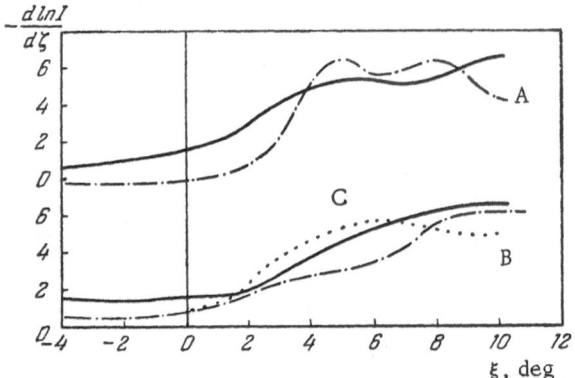

Fig. 126. The gradient $-d(\ln I)/d\zeta$ as a function of ξ, as computed by Volz and Goody. A) Molecular atmosphere; B, C) atmospheres with differing aerosol distributions.

On the other hand, in the preceding section we have found that $d\bar\eta/d\zeta = 0$ during the total-twilight period, so that

$$\frac{d\bar H}{d\zeta} = R\gamma\xi. \qquad \text{(IV.4, 15)}$$

During the bright stage of semitwilight, while the sun is still above the horizón and $\xi < 0$, $dH/d\zeta = 0$ and

$$\frac{d\bar H}{d\zeta} = \frac{d\bar\eta}{d\zeta}. \qquad \text{(IV.4, 16)}$$

Figures 121, 106, and 107 demonstrate that $d\bar\eta/d\zeta$ increases monotonically during this period.

Finally, during the dark stage of semitwilight when $\xi > 0$, both effects take place, that is,

$$\frac{d\bar H}{d\zeta} = R\gamma\xi + \frac{d\bar\eta}{d\zeta}. \qquad \text{(IV.4, 17)}$$

Thus when $\xi > 0$, the increase in $d\bar H/d\zeta$ continues until the rapid decrease in $d\bar\eta/d\zeta$ sets in at the border between semitwilight and total twilight (Figs. 121 and 106), whereupon $dH/d\zeta$ at once drops to the value (IV.4, 15),

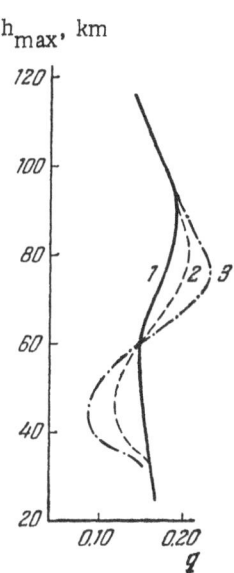

h_{max}, km

120

100

80

60

40

20

0.10   0.20
q

Fig. 127. The $q(h_{max})$ relation 1) for a standard molecular atmosphere, and for atmospheres with 2) one, and 3) two additional scattering layers.

subsequently rising again at an ever-accelerating rate. In other words, at some ξ just before total twilight sets in (for the zenith region at $\zeta \approx 93\text{-}94°$) there is a well-defined maximum in $dH/d\zeta$ (see Fig. 107), and hence in $-d(\ln I)/d\zeta$ also, insofar as K remains invariant with height. But as Fig. 106 indicates, at short wavelengths the maximum will be a weak one, and will fall at small values of ζ (see Figs. 123 and 124).

It is here that we find the explanation for the maxima in $-d(\ln I)/d\zeta$ or $-d(\ln I)/dH$ that almost always occur for small H or ξ (see Figs. 123-125). Thus Bigg's attempt [111] to interpret the maxima as evidence for the presence of an aerosol layer near the 20-km level is not on firm ground, as Megrelishvili [115] and Volz and Goody [124] have pointed out from other considerations. Nor is there any reason to ascribe the maxima to fluorescence, as Megrelishvili [112, 115] has sought to do. It is evident, however, that the state of the lower atmospheric layers will have a strong effect upon the quantity $\overline{\eta}$ and its ζ-dependence, and so upon the position and value of the maximum in $-d(\ln I)/d\zeta$, as Bigg [111] in fact points out.

These comments receive support from Fig. 126, which shows the ξ-dependence of $-d(\ln I)/d\zeta$ as computed by Volz and Goody [124] for three model atmospheres and several spectral regions.

It is all the more surprising, then, that in decisively refuting Bigg's interpretation, Volz and Goody [282] have used a very rough and approximate theory for the semitwilight period (neglecting to a considerable extent the very factors that are responsible for this effect) to analyze the results of their own measurements, and are in agreement with his conclusion that a maximum in the aerosol distribution exists at a height of about 20 km.

Bigg s interpretation [111] is untenable in other respects also, as Megrelishvili [115] has shown. One reason is evident from the preceding sections of this chapter, which indicate that $\overline{H}$ or h_{max} is considerably greater than H in the visible part of the spectrum, whereas Bigg regards these quantities as the same. The second argument is more subtle. It involves the resolving power of the twilight method, and deserves to be examined in some detail, especially since in certain other cases [90, 94, 102, 108, 282] its omission can lead to a fallacious interpretation of twilight observations.

The twilight-layer concept itself implies (see, for example, [96]) that even a thin layer of enhanced scattering power can produce an anomalous be- havior in the I(H) or q(H) twilight curves, with the anomaly extending over a height comparable to the thickness of the twilight layer. In other words, two layers separated from one another by heights less than the semithickness of the twilight layer would not be resolvable in the I(H) or q(H) relations.

To illustrate this general situation, Megrelishvili has performed com- putations of the function $q(h_{max})$ specifically for three model atmospheres (Fig. 127): a standard molecular atmosphere (solid curve), the same atmos- phere with a layer of twice the scattering power in the height range 60-70 km (dashed curve), and the same atmosphere again, but now with a second iden- tical layer located 5 km below the first one (dot-dash curve). The figure dem- onstrates that the layers actually are not resolved in the $q(h_{max})$ curves, and that in both cases the presence of the layers does not distort the curves locally, but instead perturbs them over a very wide range in height.

Volz and Goody [124] have made similar computations, with the same results.

Let us consider now the case of a thin scattering layer in a more general formulation. Suppose that the monotonic dependence $D_{11}(h) = D_{11}^0(h)$ breaks down at height $h = h_c$ because of a scattering layer there, with optical thick- ness τ_{11}^c and a δ-function distribution of scattering material with respect to height; thus

$$D_{11}(h) = D_{11}^0(h) + \tau_{11}^c \delta(h - h_c). \qquad \text{(IV.4, 18)}$$

We shall use the expression (IV.1, 1) for the sky brightness. If we now let $\tau'^{(0)}_{11}$ denote the value of τ'_{11} in the absence of the layer for the same value of h [see Eq. (IV.4, 1)], then with the layer present we easily see that

$$\tau'_{11}(\overline{H}) = \tau'^{(0)}_{11}(\overline{H}^0) + \tau_{11}^c T[y(h_c), \zeta, z, A, \lambda], \qquad \text{(IV.4, 19)}$$

where $\overline{H}^0$ is the effective height of the shadow for the same ζ in the absence of the layer, or

$$q(\overline{H}) = -\frac{d \ln \tau'_{11}}{dH} = -\frac{\dfrac{d\tau'^{(0)}_{11}}{dH} + \tau_{11}^c \left(\dfrac{dT}{dH}\right)_{h_c}}{\tau'^{(0)}_{11} + \tau_{11}^c T(h_c)}. \qquad \text{(IV.4, 20)}$$

If we restrict attention to the total-twilight period and to the region of sky near the zenith, then

$$T(y) = T\left(\frac{h_c - H}{\gamma}\right) \quad \text{and} \quad \left(\frac{dT}{dH}\right)_{h_c} = -\frac{1}{\gamma}\left(\frac{dT}{dy}\right)_{\frac{h_c - H}{\gamma}},$$

that is,

$$q = \frac{q_0 + \dfrac{\tau_{11}^c}{\tau_{11}^{\prime(0)}}\left(\dfrac{dT}{dy}\right)_{\frac{h_c - H}{\gamma}}}{1 + \dfrac{\tau_{11}^c}{\tau_{11}^{\prime(0)}} T\left(\dfrac{h_c - H}{\gamma}\right)} . \tag{IV.4, 21}$$

Since $h_c \gg H$, $\tau_{11}^c \ll \tau_{11}^{\prime(0)}$, and moreover,

$$T\left(\frac{h_c - H}{\gamma}\right) = 1 \quad \text{and} \quad \left(\frac{dT}{dy}\right)_{\frac{h_c - H}{\gamma}} = 0,$$

that is, $q = q_0$. As the boundary of the earth's shadow rises upward, the ratio $\tau_{11}^c / \tau_{11}^{\prime(0)}$ will increase and q will diminish. When $h_c - H$ falls to the point where perceptible attenuation of direct sunlight occurs in the vicinity of the layer [that is, to $h_c - H \approx \gamma(a + b)$], the variations in T(h) will still be insensitive, but dT/dy will increase, soon reaching the value 1/b. At the same time the drop in q will be retarded. As soon as dT/dy stabilizes, T will begin to fall rapidly (see Fig. 103), and the drop in q will give way to a rise. When

$$\left(\frac{d \ln T}{dy}\right)_{\frac{h_c - H}{\gamma}} = q_0$$

[that is, for $T \approx 1/K(y)b$], the equality $q = q_0$ will be re-established, but will immediately break down again because of the continuing decrease in T (see Fig. 126). When the illumination on the layer drops strongly ($T \ll 1$), as will happen when $h_c - H \approx \gamma a$, the quantity dT/dy will again begin to fall. This will at once cause q to diminish and soon after to revert to the value q_0 (see Fig. 127).

We can readily reproduce this pattern by starting with the idea of an effective shadow boundary. We have seen in the preceding section that, by Eq. (IV.3, 19), if a δ-function scattering layer is present the effective shadow boundary will first tend to approach the layer (corresponding to a decrease in

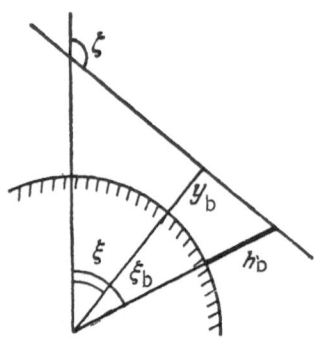

Fig. 128. The barrier effect.

q), then withdraw together with the layer (an increase in q), and finally break away from the layer to overtake its usual position (accompanied by another decrease in q).

If a layer with an enhanced (or reduced) scattering power has a sufficiently great horizontal extent, it will be observable along different lines of sight for the same $\bar{H}$, and hence at different ζ and at different times. This behavior may serve as a criterion for distinguishing effects caused by an intervening layer from effects of a different kind.

A theoretical analysis, then, provides a fairly clear understanding of the variations in q as a δ-function scattering layer passes through the penumbral region, as well as the contribution of the various factors responsible for these variations. At the same time it demonstrates that the numerous small-scale irregularities in the q(H) curves definitely are unrelated to the height structure of the function $D_{11}(h)$ or to local variations in K(H). These variations will also appear in the q(H) curves in a very diffuse form, since the quantity $\bar{\eta}$ will be governed by a value of K(H) averaged over the entire range of heights covered by the penumbra, that is, from $H + \gamma a$ to $H + \gamma(a + b)$.

Another reason for the kinks that appear in the q(H) curves could be the variations in $\bar{\eta}$ because of fluctuations in the T(y) relation. We consider first the case where light-obscuring formations intervene in the vicinity of the terminator, such as high mountains or systems of cloud, and for simplicity we confine attention to observing conditions in the solar meridian.

If we neglect refraction effects, the distance from the observer to the terminator, as measured along the earth's surface, will be $r_0 = R\xi$, where $\xi = \zeta - \pi/2$ as before (Fig. 128). Suppose now that at a distance $r = R\xi_b$ from the observer there is a barrier of height h_b, opaque to light. Then the perigee height y_b of a ray passing along the upper edge of the barrier will evidently be given by

$$y_b = (R + h_b) \cos(\xi_b - \xi) - R \approx h_b - \frac{R(\xi_b - \xi)^2}{2} . \qquad (IV.4, 22)$$

In §3 we found that if $y_b < a$, the presence of a barrier will have no influence on the effective shadow height, nor by the same token on the brightness of the twilight sky. Polyakova [253] has pointed out this fact. But if $y_b > a$, then we can use Eq. (IV.3, 18), taking $\Delta = y_b - a$. We readily see that during the total-twilight period $\bar{\eta}$ reaches a maximum at $\xi = \xi_b$, while during the semitwilight period the barrier effect is superposed on the refraction effect, and a more complex pattern results because of the dependence of T on z, A, and ζ. In both cases, however, the barrier effect depends very weakly on ζ. Indeed, Eq. (IV.4, 22) implies that a 1-km decrease in Δ with respect to its maximum value will entail a change of $\approx 1°$ in ζ, or a change of ≈ 110 km in r_b, while a 3-km change in Δ will correspond to a change of $1°.75$ in ζ and 200 km in r_b. This means that the presence of an opaque barrier near the ground will affect the $q(\bar{H})$ relations only in the sense of a reduction in q that is very diffuse with respect to height, and this only in the long-wave part of the spectrum. At short wavelengths, only stratospheric (for example, nacreous) clouds could be responsible for such an effect.

Another characteristic property of the barrier effect is that variations in Δ will induce simultaneous variations in $\bar{\eta}$ along the whole trajectory of the corresponding solar rays — that is, for given Δ and ζ, along the lines l sin z sin A = const, or in view of Eqs. (III.1, 8) and (III.1, 35), approximately along the lines $\varepsilon / \cos^2 \psi$ = const. But along lines with differing $\varepsilon / \cos^2 \psi$, in directions for which the perigees of the solar rays are displaced transversely, the barrier effects can differ. As a result, the transverse profile of a system of cloud or a mountain ridge will (in the event that $\Delta > 0$) be projected on the sky in the form of nearly parallel bands (converging slightly at the horizon) of greater or lesser brightness, since different $\tau'_{11}(\bar{H})$ will also correspond to lines of differing $\bar{\eta}$.

During total twilight, this effect will be a very feeble one, and can occur only at the longest wavelengths, since $\Delta < 0$ in the short-wave region of the spectrum. During semitwilight, however, a will be small, so that Δ will increase and the effect will become very distinct, especially in the orange and red spectral regions. We encounter here the so-called "rays of Buddha," mentioned in Chap. I, §2. Essentially they represent shadows of clouds or mountains that screen off direct sunlight. In the next section we shall consider their coloration.

We have shown, then, that during semitwilight, when changes in a produce rapid changes in Δ, barrier effects can induce variations in the sky brightness of substantial duration, and hence variations in the quantity q. But during the total-twilight period, to which most twilight measurements refer, the variations in q caused by the barrier effect will take place only very slowly and smoothly. In other words, barrier effects definitely cannot explain the small-scale variations in q that appear in Figs. 123 and 125, and that all observers have regularly commented upon (for example,[119]). They can be explained in only one way — in terms of atmospheric instability with respect to time.

We have already pointed out that the quantity actually measured is not q, but $\dfrac{d\ln I}{d\zeta} = \dfrac{d\ln I}{dt}\dfrac{dt}{d\zeta}$, and because of the very smooth and almost linear t-dependence of ζ, the derivative $dt/d\zeta$ cannot introduce any peculiarities into the dependence of $d(\ln I)/d\zeta$ on ζ or H. Since atmospheric instability with time can affect both τ'_{11} and $\bar H$ in Eq. (IV.4, 1), as well as the atmospheric transparency $P^m = e^{-m\tau^*}$, we can have

$$\frac{d\ln I}{dt} = -m\frac{d\tau^*}{dt} + \left(\frac{\partial\ln\tau'_{11}(\bar H)}{\partial t}\right)_{\bar H} + \left(\frac{\partial\ln\tau'_{11}}{\partial\bar H}\right)_{\tau'_{11}}\frac{d\bar H}{dt}\ .$$

(IV.4, 23)

We consider the three terms in this equation separately.

The time variations in τ^* (the first term) arise wholly from the variability of atmospheric aerosol. Measurements indicate [19] that the amplitude of the short-period variations in τ^* (only these are of interest to us here) usually comprise 20-30% of the magnitude of $\tau^{*,aeros}$, and occasionally reach 50% of this quantity. Thus if the air is highly turbid and $\tau^{*,aeros}$ is large, the time variations in τ^* will be of considerable importance. Since $\tau^{*,aeros}$ amounts to about 0.1-0.25 km^{-1} under ordinary conditions [19], the variations in τ^* will reach 0.05-0.5 km^{-1} after a few minutes or tens of minutes, and in the case of slow (such as high-latitude summer) twilights they can make a major contribution to the quantity $m\,d\tau^*/d\zeta$, particularly at points far from the zenith, where the factor m becomes sizable.

These remarks apply to daytime and night conditions. So far as we are aware, no measurements of the transparency during twilight hours have yet been made. One would expect, however, that because of the thermal phenomena accompanying the twilight process, the variations in τ^* would be intensified.

As the height of the observer changes, the variations in τ^* will diminish approximately in proportion to τ^*, and under high-mountain conditions the variations will fall to only ≈0.025 km^{-1} [19], about an order of magnitude smaller than at sea level. V. G. Kastrov [19] considers that purification of the air by a Föhn will have a similar effect. On high mountains, then, τ^* will be observed to vary much less widely, an important factor to consider when selecting a station from which to conduct twilight research.

The short-period transparency fluctuations under discussion here have one further important property which should not be overlooked. Since they are, to all appearances, produced by the presence in the atmosphere of relatively small-scale aerosol clouds, imperceptible to the eye, transported by the wind, and having a highly diverse character, these fluctuations will seem entirely

different, both in different parts of the sky and in different spectral regions [19].
Thus the short-period variations in the twilight-sky brightness, or in the quantity q, for different λ or z and A, need not be correlated with one another if they are caused by variations in τ^*.

Substantial revisions are necessary, however, in the arguments given above for the variations in the atmospheric transparency. Together with Eq. (IV.4, 1), they rest on the assumption of a point source of light.

But the twilight sky actually forms a source of very great extent, and it is only with extreme care that one can apply the Bouguer law to it. If the twilight layer were equally bright everywhere, the Bouguer law should be replaced by the relation (II.2, 58), so that for an observer on the ground we would have

$$I(z) = \pi I_c g(\cos z) \frac{l}{(1 - A_u) \tau^* + l} , \qquad (IV.4, 24)$$

where I_c is the brightness of the twilight layer itself, situated outside the lower and more turbid layers of the atmosphere, and A_u is the albedo of the underlying surface. (This formulation is fully adequate for the intrinsic night-sky glow also.) Equation (IV.4, 24) implies at once that I(z) is virtually insensitive to variations in τ^*, particularly for large A_u, since the attenuation of direct light from a given part of the sky is almost entirely compensated by the enhanced scattering of light from all the rest of the sky.

If now the sky brightness should vary from point to point, although not in a very strong manner, that is, if

$$I_c(z, A) = \bar{I}_c + \Delta I_c(z, A), \qquad (IV.4, 25)$$

where the quantity $\bar{I}_c$ is given by the condition

$$\bar{I}_c = \frac{1}{\pi} \int \int I_c(z, A) \cos z \, g(\cos z) \, d \cos z \, dA, \qquad (IV.4, 26)$$

then in the first approximation we may take, for an observer on the ground,

$$I(z, A) \approx \pi \bar{I}_c g(\cos z) \frac{l}{(1 - A_u) \tau^* + l} + \Delta I_c(z, A) e^{-m(z)\tau^*} \qquad (IV.4, 27)$$

(the quantity ΔI_c can be both positive and negative), and local variations in τ^* will affect the value of the last term almost exclusively. But under total-twilight conditions at the zenith, say, when an incomparably brighter twilight-glow segment persists at the solar horizon, the approximation (IV.4, 27) will already have become inapplicable, and can be regarded only as a basis for rough, qualitative considerations. Nevertheless, we may conclude immediately from Eq. (IV.4, 27) that a local rise in τ^* resulting in the attenuation of direct light from the twilight layer will simultaneously tend to increase the fraction of twilight-segment light, rescattered by the atmosphere, that is received by the measuring device. And if the segment is very bright, we would expect to find not merely a compensation of the direct light that has been lost, but the appearance of excess scattered light, and thereby a rise in the sky brightness with increasing τ^*.

Thus strong variations in the transparency are also hazardous in that they can produce substantial fluctuations in the multiple-scattering effects. But in §6 we shall find that variations in the scattering power of relatively high atmospheric layers are far more important here than variations in the ground layers.

In summary, we can assert that small-scale irregularities in the behavior of the q(H) curves will, to some extent, be produced by variations in τ^*, and that for a reliable interpretation, twilight measurements should be made, as far as possible, on clear days, along lines of sight at small zenith distances, and from high elevations. Moreover, it would be desirable for the observing program to be so designed that the interpretation process will eliminate the effects of τ^* fluctuations, especially since they can be responsible not only for small-scale, but also for relatively long-term variations in q(H).

As for the s e c o n d term in Eq. (IV.4, 23), all the discussion regarding the variations in τ^* (except the arguments dealing with the role of multiple scattering) will be largely valid. At present there is no reason to doubt that light scattering by aerosols makes a strong contribution at all heights, nor that the aerosol content of the air is distributed nonuniformly at great heights, in a more or less distinct cloud structure, changing with and transported by the wind. We would therefore expect the relative variations in τ'_{11} to have approximately the same scale at all heights [19]. The most important distinction from the case of the first term is that instead of a simple derivative of τ'_{11} we now have a logarithmic derivative; that is, the effect depends on relative rather than absolute variations in τ'_{11}, the relative variations naturally being considerably greater. Even if the aerosol component of τ'_{11} comprises only 10%, a very conservative estimate, and if its variations also remain within 10% we would still expect variations in $\Delta\tau'_{11}/\tau'_{11}$ of the order of 0.01 within a few minutes, a value comparable in scale with the short-period irregularities in the behavior of the q(H) curves. This effect should be further reinforced by the fact that as they pass through the instrumental field of view, the aerosol clouds will not only produce a change in the scattering coefficient, but in the scattering function itself, which also enters into the expression for τ'_{11}. For observations at the zenith, where the scattering angle φ is close to 90°, the effect should not be a strong one [19]. But it will become very marked for observations near the solar horizon, where φ is small and the aureole effect is large.

There is good reason to believe, then, that it is the second term in Eq. (IV.4, 23) which will be responsible for these irregularities in most cases.

But the possibility now arises that by measuring the duration of these irregularities in time, we can obtain some direct estimates of wind velocities, or of the dimensions of the inhomogeneities (aerosol clouds) at heights near $\bar{H}$ [96]. That we are concerned with just this height range (the penumbral region) follows from the rapid decline in τ'_{11} with height, that is, from the fact that most of the light comes from the twilight-layer region.

If t represents the "lifetime" of such a disturbance in the instrumental field of view (the least time necessary for displacing the twilight layer by its own thickness of about 25-30 km), d the extent of the cloud formation, $\bar{H}$ its height, and δ the angular size of the instrumental field of view, then the rate of displacement of the aerosol cloud, and hence the wind velocity at height $\bar{H}$, will evidently be given by

$$v \approx \frac{d + \bar{H}\delta}{t}. \qquad \text{(IV.4, 28)}$$

In the case of measurements carried out by the author [96], the quantity $\bar{H}\delta$ was about 5 km at a height $\bar{H} = 150$ km. The time duration of the twilight layer varied between 10 and 15 min. By adopting the maximum value of t for which the irregularities definitely persist (namely t = 10 min), and taking $d \leq 5$ km, we obtain as a lower limit for the wind velocity $v \gtrsim 30\text{-}60$ km/hr, which is not inconsistent with the results of direct measurements by various methods (see, for example, [185]).

We turn, finally, to the third term in Eq. (IV.4, 23). We have already seen that the factor $\partial(\ln \tau'_{11})/\partial\bar{H}$ cannot be responsible for the short-period variations in q. The second factor in this term can itself be resolved into the two terms

$$\frac{\partial\bar{H}}{\partial t} = \frac{dH}{dt} + \frac{d\bar{\eta}}{dt}. \qquad \text{(IV.4, 29)}$$

The first of these yields the regular variation in the height of the earth's geometric shadow with time, and is of no interest to us here. The second, however, comprises two effects. One effect is the variation in $\bar{\eta}$ as scattering structures appear in the instrumental field of view in the penumbral region. We have considered it above in connection with the problem of detecting a cloud with a δ-function structure in height, but now we should assume that the cloud, carried by the wind, will intersect the instrumental field of view in a horizontal direction.

Unlike the case considered previously, $\bar{H}$ will diminish when the cloud appears, and increase when it disappears. It is readily seen that this effect, invariably accompanied by the increase in $\tau'_{11}(\bar{H})$ at constant $\bar{H}$, as considered above, will always lead to an increased influence upon the $q(\zeta)$ relation for a cloud appearing in the instrumental field of view.

The other effect is the change in Δ in the perigee region, also discussed above. However, we have now to suppose that the barrier is not a fixed and opaque one, but a cloud barrier moving at a certain velocity. If the cloud moves along the solar meridian, then for plausible values of the velocity Eq. (IV.4, 22) implies that the motion cannot be expected to have any marked time effects. But if the cloud moves across the solar meridian, from one line $\varepsilon/\cos^2\psi = $ const to another, then we would expect to find time variations in Δ, and hence in $\bar{\eta}$. If we allow for the sun's own azimuthal motion, the lifetime of such a disturbance will be

$$t = \frac{d}{v} \sin A_0 \pm \frac{d}{R\xi\, dA/dt},$$
(IV.4, 30)

where, as previously, d is the extent of the cloud, v is its velocity, and A_0 is the azimuth of its velocity. We again obtain estimates of reasonable order of magnitude, so that the effect should be regarded as a probable source of short-period variations. There is, however, an important difference which enables us to distinguish one effect from the other.

The variations in $\tau'_{11}(\bar{H})$ are local in character, and will show no correlation in different parts of the sky. At the same time, they should be more or less uniformly distributed over all wavelengths, although judging from the variations in τ^*, it is difficult to expect a correlation between the variations in different parts of the spectrum. Still, we might expect to find a weak correlation between the $\tau'_{11}(\bar{H})$ variations in different parts of the sky for the same $\bar{H}$ (but different ζ and t) since when cloud structures appear at any level, they will normally cover a considerable area.

On the other hand, the variations in Δ as clouds pass through the perigee region should appear simultaneously (at the same ζ but different $\overline{H}$) along a given line with a constant value of $\varepsilon/\cos^2 \psi$ (for example, along the entire solar meridian), and should move regularly from one value of $\varepsilon/\cos^2\psi$ to the next (across the solar meridian). The existence of such a rule is a positive indication that the irregularities in the q(H) curves are to be explained in terms of the presence and motion of clouds in the terminator region. More-over, as we have already seen, the appearance of a cloud barrier in the term-inator region should manifest itself primarily in the long-wave spectral region, and for small ξ. As ξ increases and λ decreases, the barrier effect should regu-larly weaken. The criteria for the λ-dependence of the barrier effect will be especially sensitive here, since this relation will be free from all the uncer-tainties associated with the time variable.

We conclude this section by considering one further effect in connection with the problem of distinguishing between time variations of differing origin. In Chap. I, §4, we found that the degree of polarization p of the twilight sky is subject to short-term fluctuations to the same extent as the quantity q, if not more so (see Figs. 34 and 35). It is also noteworthy that neither variations in the transparency of the lower atmospheric layers nor changes in the circum-stances of illumination can have any effect on the degree of polarization, be-cause for single scattering the polarization of light will depend only on the conditions under which the scattering event itself occurs, and in this sense it will be a specifically local phenomenon.

In fact, Eqs. (II.2, 11) imply that

$$p = \frac{\sqrt{S_2^2 + S_3^2}}{S_1} , \qquad \text{tg } 2\psi_0 = \frac{S_3}{S_1} , \qquad q = \frac{S_4}{S_1} \quad \text{(IV.4, 31)}$$

or, reverting to Eq. (IV.3, 2),

$$p = \frac{\sqrt{\tau_{21}'^2 (\overline{H}) + \tau_{31}'^2 (\overline{H})}}{\tau_{11}' (\overline{H})} , \quad \text{tg } 2\psi_0 = \frac{\tau_{31}' (\overline{H})}{\tau_{21}' (\overline{H})} , \quad q = \frac{\tau_{41}' (\overline{H})}{\tau_{11}' (\overline{H})} , \quad \text{(IV.4, 32)}$$

so that all the polarization characteristics of the light from the twilight sky depend on $\tau_{11}'(\overline{H})$ only.

We may add that in most cases (see Chap. I, §4) it is not the true de-gree of polarization $p = (I_{max} - I_{min})/(I_{max} + I_{min})$ which is measured, but its apparent value

$$p_{app} = \frac{I_\perp - I_{||}}{I_\perp + I_{||}} = \frac{S_2}{S_1} = p \cos 2\psi_0, \quad \text{(IV.4, 33)}$$

where $I_\perp$ and $I_\parallel$ are the intensities of the light passed by the analyzer with the electric vector aligned respectively perpendicular and parallel to the scattering plane [19].

From Eq. (IV.3, 2), we have

$$P_{app} = \frac{\tau'_{21}(\bar{H})}{\tau'_{11}(\bar{H})} .$$

(IV.4, 34)

This means that only two effects can be responsible for the short-period variations in P_{app} (as well as in p, q, and ψ_0).

The first effect would occur if aerosol formations having a scattering matrix $f_{ik}(\varphi)$ different from that of the surrounding medium were to enter the portion of the cone of sight illuminated by the sun. In the second effect, the entry of clouds into the terminator region, in a direction crossed by solar rays, would cause fluctuations in the effective height of the earth's shadow. These fluctuations in $\bar{H}$ could affect the polarization of the light from the twilight sky only if the penumbral region (in the vicinity of the level $\bar{h} = \bar{H}$) happens to contain sharp changes of the scattering matrix with height — that is, if there are aerosol layers at this level with a scattering matrix different from that of the ambient medium.

Thus the presence or absence of a correlation between short-period variations in $q(\bar{H})$ [or $CE_{\lambda_2}^{\lambda_1}(\bar{H})$] and $p(\bar{H})$ (or other polarization properties of the scattered light of the sky) could serve as a sharp criterion for elucidating the nature of the variations.

§5. Dispersion in Effective Shadow Heights and the Color of the Twilight Sky

One of the most distinctive properties of the twilight sky is its sequence of color changes. Many authors have measured and described this process (Chap. I, §§2 and 3). Yet the theoretical side of this subject — the color of the sky, its relation to the structure of the atmosphere, and its ζ- and z-dependence — has hitherto received little study. Gruner [256] was the first to give a clear formulation, but the mathematical treatment he had developed explains only the general causes responsible for the color of twilight, without identifying explicitly the contributions of individual factors. Polyakova [253] has used particular model atmospheres to perform direct calculations of this type for the twilight-glow segment, and Shtaude's investigations [128, 240] can also be applied indirectly for such an analysis.

Finally, Divari [116], and subsequently Volz and Goody [282], have com-puted numerically the spectral variation of the twilight-sky brightness, as well as the color index

$$CE_{\lambda_2}^{\lambda_1}(\zeta) = -2.5\log\frac{I(\lambda_1, \zeta)}{I(\lambda_2, \zeta)} = -1.08 \ln \frac{I(\lambda_1, \zeta)}{I(\lambda_2, \zeta)} \qquad \text{(IV.5, 1)}$$

for a number of (λ_1, λ_2) pairs, as a function of ζ, for z = 70°, under various as-sumptions for the optical structure of the atmosphere.

In a general way, we can provide a graphic illustration of how the colora-tion of the sky relates to atmospheric structure and the parameters ζ, z, and A, by appealing to the concept of the effective height of the earth's shadow. In 1946, Rozenberg carried out the first such investigation (unpublished), for the total-twilight period and z = 0. After considerable refinement and extension to arbitrary z in the solar meridian, this study was later communicated to the Second Conference on Actinometry and Atmospheric Optics [130]. Divari [116] has advanced some general considerations based on the twilight-layer concept. In particular, he has examined the influence of atmospheric ozone on the color of the twilight sky, and has shown the effect to be very strong in the vicinity of the Chappuis bands (see also [124]).

As a basis for our discussion, we shall again start with Eq. (IV.4, 1), which after differentiation with respect to λ takes the form

$$\frac{d \ln I(z, A, \xi, \lambda)}{d\lambda} = \frac{d \ln I_0}{d\lambda} - m\frac{d\tau^*}{d\lambda} + \frac{d \ln \tau_{11}'(\overline{H})}{d\lambda} . \qquad \text{(IV.5, 2)}$$

The first two terms are of no special interest as they remain invariant throughout the twilight period (apart from the variations in τ^*, of course). But the last term can be written in the form

$$\frac{d \ln \tau_{11}'(\overline{H})}{d\lambda} = \left(\frac{\partial \ln \tau_{11}'(\overline{H})}{\partial\lambda}\right)_{\overline{H}} + \left(\frac{\partial \ln \tau_{11}'(\overline{H})}{\partial\overline{H}}\right)\lambda\frac{d\overline{H}}{d\lambda} . \qquad \text{(IV.5, 3)}$$

Once again, the first term here describes the spectral variation of the scattering power of the atmosphere above the level $\overline{H}$, and in all likelihood it will vary comparatively little with $\overline{H}$, that is, with ζ, z, and A.

But the situation is altogether different for the second term, which, since dH/dλ = 0, we shall write in the form

$$\left(\frac{\partial \ln \tau_{11}'(\overline{H})}{\partial\overline{H}}\right)_\lambda \frac{d\overline{H}}{d\lambda} = -Q(\overline{H}, \lambda)\frac{d\overline{\eta}}{d\lambda} , \qquad \text{(IV.5, 4)}$$

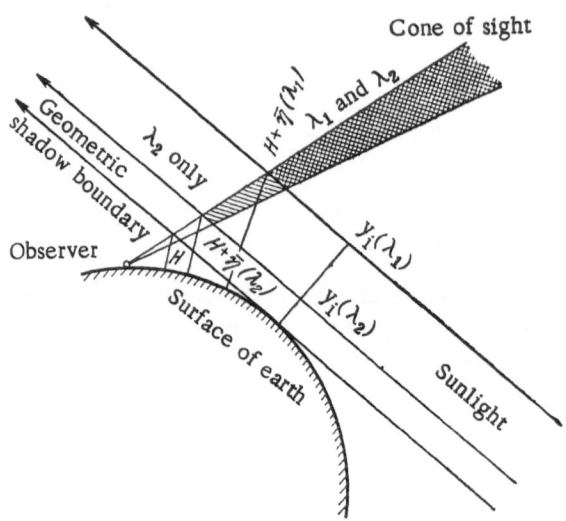

Fig. 129. The influence of the dispersion in the effective shadow heights on the color of the sky.

where, in view of Eq. (IV.1, 11), the logarithmic gradient of the optical thickness τ'_{11} is given by

$$Q(h, \lambda) = -\frac{d \ln \tau'_{11}(h, \lambda)}{dh} = -\frac{D_{11}(h, \lambda)}{\tau'_{11}(h, \lambda)} .$$

$$(IV.5, 5)$$

This term takes into account the dispersion in the effective heights of
the earth's shadow, and as we have seen in §3, the dispersion changes markedly during the semitwilight period, becoming very great during the stage of
total twilight. We shall demonstrate below that it is this term which is mainly
responsible for the ζ-, z-, and A-variations in the color of the sky (at least in
the region near the zenith) during the twilight period.

To begin with, we note that if Eq. (III.5, 5) holds at all heights, then
$Q = K = \text{const}$,

$$\frac{\partial \ln \tau'_{11}(\bar{H})}{\partial \lambda} = \frac{d \ln k_0}{d\lambda},$$

and, by Eqs. (III.1, 29) and (III.5, 13),

$$\frac{d\bar{\eta}}{d\lambda} = \frac{\gamma}{K} \frac{d \ln k_0}{d\lambda},$$

so that

$$\frac{d \ln \tau'_{11} (\bar{H})}{d\lambda} = (1 - \gamma) \frac{d \ln k_0}{d\lambda}.$$

With the condition (III.5, 5) and the value $\gamma = 1$, then, the color of the twilight sky during the total-twilight period will coincide with the color of direct
sunlight, for $z = \zeta$. Differences will appear only if $\gamma \neq 1$, or if the condition
(III.5, 5) is violated.

In order to illustrate the physical meaning of these processes more clearly,
let us consider the sky brightness at two wavelengths λ_1 and λ_2 so chosen that
$\bar{\eta}(\lambda_1) > \bar{\eta}(\lambda_2)$, and write Eq. (IV.4, 1) in the form

$$I(\lambda_1) = I_0(\lambda_1) \omega_0 P^m(\lambda_1) m \tau'_{11} [\bar{H}(\lambda_1), \lambda_1] \qquad \text{(IV.5, 6)}$$

and also

$$I(\lambda_2) = I_0(\lambda_2) \omega_0 P^m(\lambda_2) m \left\{ \tau'_{11} [\bar{H}(\lambda_1), \lambda_2] + \int_{\bar{H}(\lambda_2)}^{\bar{H}(\lambda_1)} D_{11}(h, \lambda_2) \, dh \right\}.$$

$$\text{(IV.5, 7)}$$

Figure 129 will serve to interpret the expression for $I(\lambda_2)$. The first term
in braces represents light scattered in the part of the cone of sight lying above
$\bar{H}(\lambda_1)$ and sending the observer light at both λ_1 and λ_2 [see Eq. (IV.5, 6)]. This
region is indicated by heavy shading. If the effective shadow boundaries co-
incided for the two wavelengths, the difference between Eqs. (IV.5, 7) and
(IV.5, 6) would depend only on the spectral variation of the scattering coefficient in the range $h > \bar{H}$, on the spectral dependence of the transparency of the
lower atmospheric layers, and finally, on the spectral variation of the sun's
brightness. However, because of the dispersion in the effective shadow heights
there is an additional portion of the cone of sight (light shading in Fig. 129)
where only light of wavelength λ_2 is scattered (at λ_1, this region falls in the
effective shadow). The second term in braces in Eq. (IV.5, 7) represents the
excess light of wavelength λ_2 scattered in the height range between $\bar{H}(\lambda_1)$ and
$\bar{H}(\lambda_2)$. Because of the rapid decline in the scattering coefficient with height,
the relative contribution of this excess light to the total sky brightness is a
major one, depending strongly on the logarithmic gradient K of the scattering
coefficient.

We turn now to a quantitative discussion. If we take logarithms in Eq. (IV.4, 1), expand $\ln \tau'_{11} [\bar{H}(\lambda_2), \lambda_2]$ in powers of

$$\Delta \bar{\eta} = \bar{\eta}(\lambda_1) - \bar{\eta}(\lambda_2)$$

$$(IV.5, 8)$$

(we have made the choice $\Delta \bar{\eta} > 0$), and, since $\Delta \bar{\eta}$ is not small, retain three terms in the expansion, we find, by Eq. (IV.5, 5),

$$\ln I(\lambda_1) = \ln \omega_0 + \ln m + \ln I_0(\lambda_1) - m\tau^*(\lambda_1) + \ln \tau'_{11}[\bar{H}(\lambda_1), \lambda_1],$$

$$(IV.5, 9)$$

$$\ln I(\lambda_2) = \ln \omega_0 + \ln m + \ln I_0(\lambda_2) - m\tau^*(\lambda_2) +$$

$$+ \ln \tau'_{11}[\bar{H}(\lambda_1), \lambda_2] + Q[H(\lambda_1)\lambda_2]\Delta\bar{\eta} - \frac{1}{2}\left(\frac{dQ(\lambda_2)}{dh}\right)_{\bar{H}(\lambda_1)} (\Delta\bar{\eta})^2$$

or in accordance with Eq. (IV.5, 1),

$$CE^{\lambda_1}_{\lambda_2} = C^{\lambda_1}_{\lambda_2} + 1.08 \left\{ Q[\bar{H}(\lambda_1), \lambda_2]\Delta\bar{\eta} - \frac{1}{2}\left(\frac{dQ(\lambda_2)}{dh}\right)_{\bar{H}(\lambda_1)} (\Delta\bar{\eta})^2 \right\},$$

$$(IV.5, 10)$$

where the term

$$C^{\lambda_1}_{\lambda_2} = 1.08 \left\{ \ln\frac{I_0(\lambda_2)}{I_0(\lambda_1)} - m[\tau^*(\lambda_2) - \tau^*(\lambda_1)] + \ln\frac{\tau'_{11}[\bar{H}(\lambda_1), \lambda_2]}{\tau'_{11}[\bar{H}(\lambda_1), \lambda_1]} \right\}$$

$$(IV.5, 11)$$

incorporates the effects arising from the spectral dependence of I_0, τ^*, and τ'_{11}. Leaving these effects aside for the present, let us examine the behavior of $\Delta \bar{\eta}$ during the twilight period. We shall confine attention to the case where selective absorption is absent and $\bar{\eta}(\lambda)$ increases monotonically with decreasing λ. If selective absorption is present, an analogous formulation will serve, except that the specific $\bar{\eta}(\lambda, \zeta, z, A)$ relation should properly be considered separately with reference to each special case.

At noon, when the sun is high above the horizon, $\bar{\eta}$ will essentially vanish over the entire visible spectrum (see §3, Fig. 121), so that $\Delta\bar{\eta} = 0$, and the color of the sky will be determined exclusively by the quantity $C^{\lambda_1}_{\lambda_2}$. As the sun drops toward the horizon, $\bar{\eta}(\lambda)$ will begin to rise, slowly at first, but then at an increasingly rapid rate. At the same time $\Delta\bar{\eta}$ will also increase, for all pairs of wavelengths. But this process will operate in a nonuniform manner. The rise in $\bar{\eta}$ will at first be perceptible only in the far ultraviolet, and only gradually will it extend toward longer wavelengths. Thus the dispersion in the effective shadow heights will initially develop in the violet and blue spectral regions, and not until ζ becomes large will it penetrate into the orange and red. In particular, at $\zeta \approx 80°$, $\overline{H}(\lambda = 0.4\ \mu)$ will already have reached 7-8 km, while $\overline{H}(\lambda = 0.6\ \mu)$ is still at the 1-1.5-km level (see Fig. 121). By the time the sun has fallen to the very horizon, $\overline{H}(\lambda = 0.4\ \mu)$ will have risen high into the stratosphere, but $\overline{H}(\lambda = 0.6\ \mu)$ will barely have reached 5-6 km. It is this behavior which explains the so-called "Alpine glow," the bright purple coloration of snow-covered mountain peaks still illuminated by the rays (but only those of longest wavelength) of the setting sun, and also the crimson-purple tint that is taken on by clouds at increasingly high levels.

The dispersion in $\overline{H}(\lambda)$ continues to develop and penetrate into the long-wave spectral region throughout the semitwilight period, and is responsible for the characteristic changes in the color of the sky that occur at this time. If we confine attention, as before, to the single wavelength pair (λ_1, λ_2), then we shall have a monotonic rise in $\Delta\overline{\eta}$ during semitwilight — a continual reddening of the sky, concluding only with the onset of total twilight, when $\Delta\overline{\eta}$ will have reached its limiting value

$$\Delta\overline{\eta}_{\text{lim}} \approx \gamma\Delta y_i. \qquad \text{(IV.5, 12)}$$

The color index of the twilight sky will reach its limiting value

$$CE_{\lambda_2}^{\lambda_1} = C_{\lambda_2}^{\lambda_1} + 1.08\left\{Q\gamma\Delta y_i - \frac{\gamma^2}{2}\frac{dQ}{dh}(\Delta y_i)^2\right\} \qquad \text{(IV.5, 13)}$$

at the same time.

One might have expected that the color index of the twilight sky would remain constant as ζ increases even further. But this is not so, for an entirely different factor now makes its appearance.

In the preceding section we found that if $d\overline{\eta}/dH = 0$, or $d\overline{H}/dH = 1$, as occurs during the total-twilight period, apart from the barrier effect and the time variations, then the logarithmic gradient in the sky brightness will equal the logarithmic gradient in $\tau'_{11}(h)$:

$$q(h) = Q(h). \qquad \text{(IV.5, 14)}$$

But we also found that during the total-twilight period the quantity $q(h)$ undergoes substantial variations, in general declining with increasing H (Figs. 123 and 124). If we turn now to the expression (IV.5, 13), we see at once that a decrease in Q will imply a decrease in $CE_{\lambda_2}^{\lambda_1}$, and the twilight will become bluer. This means that at the edge of total twilight, the reddening process in the twilight sky will give way to an increasing blueness, with a reflection of the height variations of Q. We encounter here an opportunity for direct experimental tests of the arguments advanced above.

In fact, if we measure simultaneously the ζ-dependence of $\ln I(\lambda_1)$ and $\ln I(\lambda_2)$, then $CE_{\lambda_2}^{\lambda_1}$ can be evaluated in either of two ways. First, the color index can be determined for all values of ζ, directly from Eq. (IV.5, 1). Secondly, from the ζ-dependence of $\ln I(\lambda_2)$ we can evaluate $q(\zeta, \lambda_2)$ and $dq(\zeta, \lambda_2)/dH$, and by Eq. (IV.5, 14) equate them to the quantities $Q(\zeta, \lambda_2)$ and $dQ(\zeta, \lambda_2)/dH$. Then if the values of $C_{\lambda_2}^{\lambda_1}$ and Δy_i are known, we can compute the function $CE_{\lambda_2}^{\lambda_1}(\zeta)$. Here in the first approximation we can take $C_{\lambda_2}^{\lambda_1} = \text{const}$, and can obtain the quantity Δy_i either from the values of $CE_{\lambda_2}^{\lambda_1}$ for two arbitrarily selected ζ, or else we can adopt theoretical estimates for Δy_i, either from Figs. 103 or 105, for example, or by using Eq. (III.5, 14).

Three attempts have been made to verify this analysis empirically. First, the author has taken a suitable mean over published data (for example, [92]) on the $q(H)$ relation and the quantity Δy_i, according to calculations for molecular scattering. Equation (IV.5, 13) was then used to compute (to within a constant

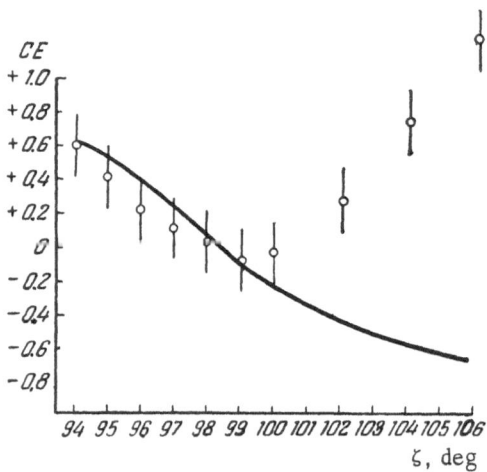

Fig. 130. Comparison between computed and measured functions $CE_{\lambda_2}^{\lambda_1}(\zeta)$ (mean data).

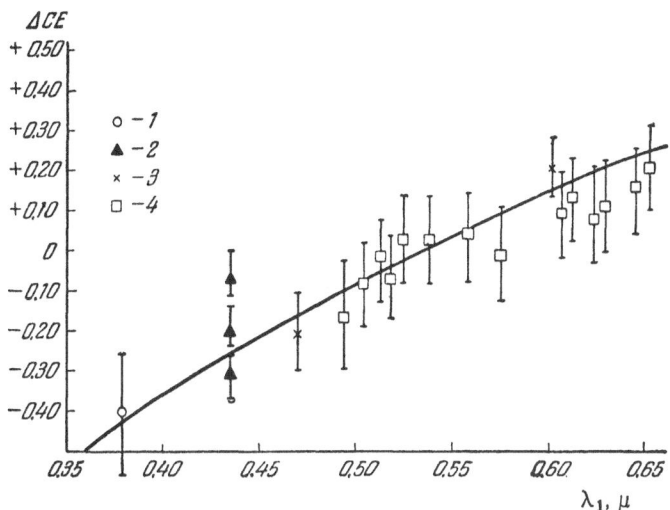

Fig. 131. Comparison between computed (solid curve) and measured (symbols) values of ΔCE as a function of λ_1. 1) [97]; 2) Fig. 28; 3) [113]; 4) [134].

term) the ζ-dependence of $CE_{\lambda_2}^{\lambda_1}$ for conditions corresponding to Megrelish-vili's measurements [97, 101] of $CE_{\lambda_2}^{\lambda_1}$. Figure 130 compares the computed function $CE_{\lambda_2}^{\lambda_1}(\zeta)$ (solid curve) with Megrelishvili's observations (circles). We have used here Megrelishvili's published mean $CE_{\lambda_2}^{\lambda_1}(\zeta)$ relation for spring twilights. The agreement is considerably poorer for other seasons, however, as would be expected since the turbidity of the air is minimal during the spring, and Δy_i has been selected on the assumption of pure molecular scattering. Note the good agreement between the curves in the range $94° \leq \zeta \leq 99°$, and the sharp divergence for $\zeta \geq 99°$. We shall return to this behavior in §§ 6 and 7.

In the second attempt, Rozenberg and Turikova [130] have employed measurements of the twilight-sky brightness at the zenith, as secured by them in the Crimea [122] in the two spectral regions $\lambda_{eff} = 0.44 \mu$ (blue filter) and 0.54μ (green filter) [19], during three different twilights. The corresponding $CE_{\lambda_2}^{\lambda_1}(\zeta)$ relations are shown by solid curves in Fig. 28 (Chap. I, §3). In this figure, the crosses indicate values of $CE_{\lambda_2}^{\lambda_1}(\zeta)$ as computed from Eq. (IV.5, 13), while Eq. (III.5, 14) has been used for Δy_i. In view of the uncertainty in the value of $C^{\lambda_1}_{\lambda_2}$ and the spectral sensitivity of the apparatus, the value

of $CE^{\lambda_1}_{\lambda_2}$ has been determined to within a constant only, as in the case of

Fig. 130. The experimental and computed relations were therefore displaced arbitrarily along the axis of ordinates until they coincided, so that it is not the

actual values of $CE^{\lambda_1}_{\lambda_2}$ which are being compared, but their ζ-dependence.

As Fig. 28 shows, despite substantial differences in the character of the function on different days, the agreement between observation and computation is perfectly satisfactory.

Finally, by starting once again with the values of $\Delta y_i\,(\lambda_1,\,\lambda_2)$ for a molecular atmosphere and with a suitable mean q(h) relation, Rozenberg has computed the λ_1-dependence of the quantity

$$\Delta CE = CE^{\lambda_1}_{0.53\ \mu}\ (\zeta = 98°20') - CE^{\lambda_1}_{0.53\ \mu}\ (\zeta = 95°20').$$

In Fig. 131, ΔCE is represented by a solid curve. The figure also shows the results of measurements by several authors, with an indication of the errors. Unlike the preceding case, it is not only the λ_1-dependence of ΔCE which is being compared here, but also the absolute values of ΔCE. We see that the computed curve accurately reproduces the $\Delta CE(\lambda_1)$ variation. Note that in the vicinity of the Chappuis bands (0.55-0.65 μ) there is a distinct dip in the experimental points, indicating as would be expected that the quantity $\Delta\bar\eta$ is sensitive to selective-absorption effects not included in the computation. However, since both measurements and computations refer to the zenith direction, the dip is a relatively weak one (compare Fig. 31, Chap. I, §3).

The comparisons we have made demonstrate that the dispersion in the effective heights of the earth's shadow indeed controls the basic pattern of color changes in the twilight sky during the semitwilight and total-twilight periods, and that it does yield to other effects in the transition to deep twilight (Fig. 130). The role of the other terms is considerably weaker, at least

if z is not too large; apart from generating the constant term $C^{\lambda_1}_{\lambda_2}$, these

terms should be responsible for diverse fluctuations in $C^{\lambda_1}_{\lambda_2}$, just as for the

daytime sky.

Turning now to the z- and A-dependence of the twilight-sky coloration, we first recall that at any instant different twilight phases will be in progress over the sky, and that not only $\bar H$ [and hence $Q(\bar H)$] and γ will vary, but

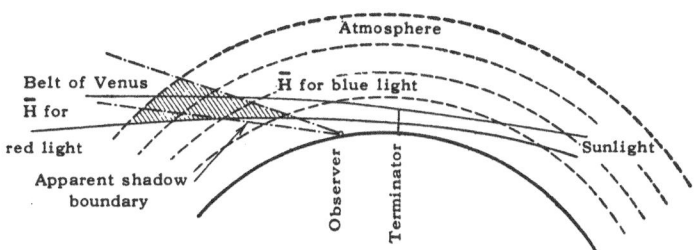

Fig. 132. Formation of the Belt of Venus.

also Δy_i. However, it is at once evident that the distribution of $CE_{\lambda_2}^{\lambda_1}$ along the solar meridian for given ζ will mimic the relation $CE_{\lambda_2}^{\lambda_1}(\xi)$ at constant z and A = 0, although with a certain amount of distortion.

We therefore shall not pause to trace this distribution, but shall consider only three directions where the color effects are particularly striking.

After the sun has sunk below the horizon and the round shadow of the earth has risen above the opposite horizon, the shadow is bordered by a rose-colored rim. This feature is the so-called "Belt of Venus," mentioned in Chap. I, §2. We need only turn to Figs. 84 and 85 (Chap. III, §1) to see that in the region where the Belt of Venus appears, total twilight has already set in by this time, and that the value of γ is very large there. This latter fact implies that the T(y) relation is projected on the line of sight ($\eta = \gamma y$) in a highly magnified form; that is, the dense layers of the atmosphere are illuminated exclusively by rays with very small values of y_i, while rays with large y_i illuminate only such rarefied layers of air that we receive practically no light from them (Fig. 132). In other words, the dispersion in the effective shadow heights along the line of sight is so great that the short-wave spectral region is completely excluded from participating as a source of the sky brightness (apart from multiple scattering, of course). However, as z decreases the quantity γ will decline rapidly (Fig. 85b), as will the height of the geometric shadow (Fig. 84b), so that the dispersion in the effective shadow heights will also fall off sharply, resulting in a strong enhancement of the blueness of the sky. The Belt of Venus, then, essentially represents a projection of the T(y) relation on the perpendicular to the line of sight; hence its small angular size. As ζ increases, together with the zenith distance at which the Belt of Venus appears, the Belt spreads out into a broad and ill-defined band.

Moreover, shortly after sunset (or before sunrise) a light, diffuse, rose-colored glow appears at $z \approx 65\text{-}75°$ near the solar vertical; this is the "purple light" (Chap. I, §2), and it reaches its maximum development when $\zeta \approx 94\text{-}95°$ (see Fig. 25). A number of special investigations have been made of this phenomenon. It has usually [297] been attributed to an aerosol layer (Chap. I, §3) that allegedly occurs at a level of about 20-25 km, as the boundary of the earth's geometric shadow arrives there (Fig. 84). We have already seen that a purple light should occur in a molecular atmosphere also; and while the presence of aerosols will naturally affect the intensity of the glow, in particular because of the altered scattering function, it is not at all obligatory for the aerosol constituent to be distributed in stratified form in the atmosphere.

In fact, at the very time when $\zeta \approx 94\text{-}95°$, the transition from semitwilight to total twilight takes place at $z \approx 65\text{-}75°$ in the solar vertical, the geometric shadow height having risen above ≈ 15 km (Fig. 84). But we already know that $CE_{\lambda_2}^{\lambda_1}$ reaches a maximum at just this time — the sky has its reddest color. At smaller ζ or larger z, the reddening process still occurs in the sky, for $\Delta\bar{\eta}$ has not yet reached its limiting value. But at larger ζ or smaller z, a progressively stronger blueness occurs because of the decline in $Q(\bar{H})$. In a sense, the purple light therefore represents a projection of the $CE_{\lambda_2}^{\lambda_1}(\bar{H})$ relation on the celestial sphere.

Evidently it is this effect that Volz and Goody [282] erroneously regarded as evidence for an aerosol layer at the 20-km level. Figure 154 (Chap. V, §4) distinctly shows that we are concerned here with the purple-light phenomenon.

We finally consider the region of the twilight-glow segment. Just as everywhere else in the sky, we encounter here the same progressive sequence of colors and the same smooth change in the sky brightness, a result of the varying effective shadow height and its dispersion. The inevitable changes induced by the varying conditions of projection of the perigee region on the line of sight will ultimately affect only the stage of development and the time scale of any given process. However, if $\alpha_z = \pi/2 - z$ is small, large values of the air mass m_z will arise along the line of sight, so that the factor $P^m(\lambda)$ in Eq. (IV.4, 1) will make an important contribution, or the corresponding terms $m\tau^*(\lambda)$ and $m[\tau^*(\lambda_2) - \tau^*(\lambda_1)]$ in Eqs. (III.5, 9) and (III.5, 11). At the same time the optical state of the lower atmospheric layers will naturally take on great importance, as they will begin to exert complete control over the aspect of the twilight sky at the very horizon. In particular, because of the rapid rise in m as the line of sight nears the horizon, the brightness of the twilight segment should decline; that is, it should pass through a maximum at some value α_z (see Figs. 21a and 22a), as Hulburt [91] has pointed out.

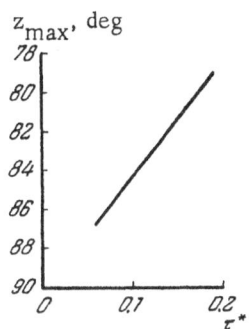

Fig. 133. Position of maximum brightness of the twilight segment at sunset or sunrise, as an approximate function of τ^*.

Let us first take the case where the sun lies at the horizon itself, so that ξ is very small. Then the twilight-segment region will still fall in the daytime hemisphere, and in evaluating $\tau_{11}(\bar{H})$ we can, in the first approximation, appeal directly to Eq. (III.5, 29), as we have seen that $\bar{H}$ will not differ greatly from h_i, in general.

We should further note that the required maximum in the brightness of the twilight segment usually falls well above the horizon (Figs. 21 and 22, and [91]), so that $\xi \leq \alpha_z \ll 1$. Under these circumstances we may take

$$m \approx \frac{1}{\alpha_z} \quad \text{[cf. Eq. (III.5, 16)]},$$

$$\frac{d\alpha}{dy} \approx \frac{1}{R\,(\xi+\alpha_z)} \approx \frac{1}{R\alpha_z} \quad \text{[cf. Eq. (III.5, 26)]},$$

$$\gamma \approx \frac{\alpha_z}{\xi+\alpha_z} \approx 1 \quad \text{[cf. Eqs. (III.1, 26), (III.1, 7), and Fig. 85]}, \quad \text{(IV.5, 15)}$$

to obtain, upon substitution into Eq. (III.5, 29),

$$\tau'_{11}(\bar{H}) \approx \frac{1}{m_\xi-m} \left[\, 1 + \left(1 + \frac{KR}{mm_\xi}\right)^{-2}\right], \quad \text{(IV.5, 16)}$$

where we have neglected m/m_ξ in comparison with 1.

Thus under our stipulated assumptions, Eq. (IV.5, 9) reduces to the approximate form

$$\ln I = \ln I_0 + \ln \omega_0 + \ln m - m\tau^* - \ln(m_\xi - m) +$$

$$+ \ln \left[1 + \left(1 + \frac{KR}{mm_\xi}\right)^{-2}\right], \quad \text{(IV.5, 17)}$$

and the condition for maximum brightness, $d(\ln I)/d\alpha_z = -m^2[d(\ln I)/dm] = 0$, becomes, after some simple transformations and with m/m_ξ neglected in comparison with 1,

$$\tau^* = \frac{1}{m}\left[1 - \frac{\dfrac{2KR}{mm_{\xi}}}{\left(1+\dfrac{KR}{mm_{\xi}}\right)+\left(1+\dfrac{KR}{mm_{\xi}}\right)^3}\right] . \qquad (IV.5, 18)$$

Figure 133 shows the resulting τ^*-dependence of $z_{max} \approx 90° - 1/m_{max}$ for $K = 0.15$ and $m_{\xi} = 40$. We readily see that the relation agrees qualitatively with that of Fig. 22b (Chap. I, §3). We would not have expected to find a closer agreement, both because our approximations are such rough ones, and because of the large uncertainties in the measurements on which Fig. 22b is based.

As ξ increases, Eq. (III.5, 29) should be replaced by Eq. (III.5, 35). But we shall follow a different procedure, and consider at once the case where ξ is so large that the twilight-segment region already falls within total twilight.

If α_z is not too small, we may then replace $\bar{H}$ by $H + \gamma y_i$, where y_i is already independent of α_z and is approximately equal to the perigee height of the rays forming the effective shadow boundary. On this assumption, the condition implied by Eq. (IV.5, 9) for the maximum brightness of the twilight segment takes the form

$$\frac{1}{m} - \tau^* = Q\left(\bar{H}\right)\frac{d\bar{H}}{dm} , \qquad (IV.5, 19)$$

and since

$$\gamma \approx \frac{1}{1+m_{\xi}} ,$$

$$H = R\left(1 - \sin \zeta\right)\gamma \approx \frac{R\xi^2}{2\left(1+m\xi\right)} , \qquad (IV.5, 20)$$

we have the relation

$$\tau^* \approx \frac{1}{m} + Q\left(\bar{H}\right)\frac{\bar{H}\xi}{\left(1+m\xi\right)^2} . \qquad (IV.5, 21)$$

Figure 134 presents curves, as computed from this formula, for the zenith distance z_{max} of maximum brightness of the twilight segment as a function of the depression ξ of the sun below the horizon, for $\tau^* = 0.1, 0.2,$ and 0.3. For small τ^*, z_{max} is almost independent of y_i, but a perceptible dependence sets in at large τ^*. For $\tau^* = 0.3$ in Fig. 134, we therefore give separate $z_{max}(\xi)$ curves for $y_i = 5$ km and 25 km. A comparison shows good qualitative agreement between Figs. 134 and 21b.

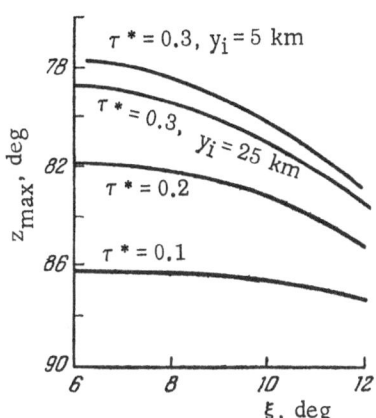

Fig. 134. Position of maximum brightness of the twilight segment as an approximate function of ξ, for large ξ and selected τ^* and y_i.

We recall that the value of y_i is determined essentially by the transparency of relatively high layers in the atmosphere, while for large m the quantity $m\tau^*$ depends mainly on the state of the lower layers. Since the vertical and horizontal transparency of the atmosphere are usually uncorrelated [19], the quantities $\bar{H}$ (or y_i) and τ^* in Eq. (IV.5, 21) should be regarded as independent functions of wavelength. Nevertheless, in the absence of selective absorption both quantities will increase monotonically with advancement into the short-wave spectral region. Figure 134 therefore indicates that as the wavelength decreases, the zenith distance z_{max} will increase, generally speaking. It is this circumstance which ultimately governs the color of the twilight-glow segment, and its variability with weather conditions. Thus our analysis of single scattering of light under twilight conditions has accurately reproduced, at least qualitatively (and in many cases quantitatively), the pattern of the chief twilight phenomena. We have been able to identify the dispersion in the effective shadow heights $\bar{H}$ as the basic factor responsible for the colors of the twilight sky, and to conclude that the concept of the effective shadow height enables us to analyze the twilight phenomena in the graphic manner necessary to formulate an approximate theory.

§6. Multiple Scattering

Our theoretical discussion of the course of the twilight phenomenon has so far treated the scattering of direct sunlight only, and has neglected effects arising from the intrinsic glow of the atmosphere. Lambert [257] seems to have been the first to note that these effects might make an important contribution to the development of twilight, at least in the part of the sky spanned by the projection of the earth's shadow (in the antisolar vertical). In his very first paper on this topic, Fesenkov [5] emphasized the need to include these effects when applying twilight sounding to a study of the upper atmosphere. But if the formulation of a theory capable of describing accurately just single-scattering phenomena under twilight conditions is already a rather complex problem, and one by no means fully solved, as we have seen, then the introduction of multiple scattering, not to mention reflection from the earth's surface, raises as yet almost insuperable difficulties. With the development of computer techniques for solving equations of transfer, we should in fact be able to obtain illustrative data before long which would furnish us with a reasonably good idea of the state of affairs, at any rate for certain model atmospheres. At present, however, we are restricted to working out manifold considerations to cover an exceptionally wide range of possibilities — from a complete neglect of secondary and higher-order scattering, to the idea that scattering of direct sunlight is wholly lost in the background of multiple scattering, particularly during the deep-twilight period. We have therefore to weigh the persuasiveness of arguments advanced by various authors without attempting to claim a decisive solution to the problem. Nevertheless, we shall find that an analysis of the observational data can yield fairly definite semiquantitative estimates.

We begin with some qualitative considerations regarding the general appearance of the twilight sky (see Chap. I, §2). Certainly one must agree with Lambert [257] that the pronounced brightening in the region where the earth's shadow is projected on the antisolar horizon, a brightness comparable with that where the sky is illuminated by direct sunlight, already indicates that multiple scattering makes a significant contribution to the total brightness of the sky. In this regard, the contrast sensitivity of the eye and the fact that the brightness of multiply scattered light should not depend strongly on the direction of the line of sight (see below) imply that the fraction of multiple scattering in the over-all balance of the sky brightness amounts to approximately 10-20% in the antisolar vertical, for small depressions of the sun (see Figs. 19 and 24). In particular, the very existence of a more or less sharp apparent boundary to the earth's shadow and an easily observable Belt of Venus suggest that during the total-twilight period, already in progress in the vicinity of the Belt, the brightness of multiply scattered light cannot exceed the brightness of singly scattered light by any substantial margin. Yet the presence of a distinct brightness maximum in the zenith region [7, 58, 67, 81] (see Figs. 18, 19, and 24) must unquestionably be interpreted as a consequence of considerable amounts of multiply scattered light, since the single-scattering theory would imply a monotonic $I(z)$ relation. Furthermore, the presence during all twilights of such a relatively bright source of light as the twilight-glow segment near the solar horizon cannot help but generate an impressive contribution from secondary scattering ([103] and Fig. 12).

Finally, detailed computations of the brightness of multiply scattered light for various model terrestrial atmospheres under daytime conditions [12-16] indicate that the contribution of multiple scattering to the brightness balance of the daytime sky amounts to a few tens of percent, and that it increases both with the turbidity of the atmosphere (with τ^*) and with the zenith distance of the sun. We have encountered examples in Chap. III, §§3 and 4. The computations fail, however, to provide any explanation of how the multiple-scattering contribution varies after sunset occurs.

Thus we would expect at the outset that the role of multiple scattering should not be a negligible one during the semitwilight and total-twilight periods, nor should it be too great. That is to say, it will be necessary to allow for multiple scattering when interpreting twilight measurements, but the presence of multiple scattering should not impede applications of twilight measurements to the study of the upper atmosphere. The problem of correctly estimating its contribution and its ζ- and z-dependence thereby becomes particularly urgent, and it is no wonder that despite the difficulties involved, attempts have been made again and again to obtain such estimates.

We shall not pause here to consider the ideas advanced in the earlier work on this subject [5, 62, 84, 258-260], although we shall return to some of them later, but instead we shall review at once the fundamental paper of Hulburt [91], who has long entertained doubts as to the legitimacy of applying the twilight method in stratospheric research, and has carried out an extensive series of theoretical investigations. His paper is also of much interest as an example of the risks that one encounters when making theoretical calculations of the brightness of doubly scattered light under twilight conditions.

From a number of absolute measurements of the sky brightness at solar depressions $\xi = 0$ to $13°$ below the horizon, both at the zenith and in the twilight-segment region, Hulburt established that the $I(\xi)$ relations at the zenith and the horizon are similar, depending very little on weather conditions, and he determined quantitatively the degree to which the radiation of the twilight sky is concentrated toward the twilight-segment region. (Khvostikov [261] subsequently showed that Hulburt had made a mistake in reducing his observational data, with a substantial effect on the reported results.) Assuming the structure of the atmosphere to be isothermal, molecular, and of uniform composition [262], Hulburt further computed the sky brightness at the zenith resulting from scattering of direct sunlight. Then, regarding the twilight segment as a source of light and adopting the measured values for its brightness and size, he computed the sky brightness at the zenith due to additional scattering of the singly scattered light coming from the twilight segment. A combination of the two brightnesses yielded a ξ-dependence of the total sky brightness at the zenith that differs little from the observed relation. Hulburt regarded this agreement as confirmation of the techniques he had used. Finally, Hulburt reached the following conclusions by comparing the brightnesses of the singly and doubly scattered light. So long as the height of the earth's

TABLE 15. Ratio of Brightness I_2 of Doubly Scattered
Light to Brightness I_1 of Singly Scattered Light
at the Zenith (Hulburt)

ζ	94°34′	96°26′	97°52′	99°36′	100°91′	102°2′	102°52′
$\dfrac{I_2}{I_1}$	0.01	0.11	1.3	18	85	$2.3 \cdot 10^4$	$2.9 \cdot 10^5$

geometric shadow remains below about 60 km, primary scattering was found to
be dominant, with secondary scattering extremely weak. But this result clearly
conflicts with the observed sky brightness in the vicinity of the earth's shadow
along the antisolar vertical, as well as with the role of secondary scattering in
the daytime hemisphere, an effect which can be followed to some extent even
at small ζ. Shtaude [263] has shown that the disagreement results from Hulburt's neglect of the light received from the sky as a whole in comparison with
the twilight-segment radiation. Moreover, Hulburt found a sharp change in the
state of affairs at $H \approx 60$ km, corresponding to the characteristic bend that is
always observed in the log I(ζ) curves. At $H \gtrsim 60$ km, the contribution of primary scattering was found to diminish rapidly until the entire twilight-sky
brightness was produced exclusively by secondary scattering, and perhaps by
higher-order scattering also. Table 15 shows how completely the multiple scattering comes to dominate the single scattering.

These results also contradict a number of other facts, particularly the presence of short-period variations in the brightness and of polarization of the twilight sky, as discussed in §4, and observed up to very considerable shadow
heights (Chap. I, §§3 and 4). As it is essentially a three-dimensional effect,
secondary scattering would not exhibit such fluctuations [96].

Another very important result of Hulburt's investigation was that secondary scattering does not arise mainly near the earth's surface, but at some height
above. Its sources are concentrated in the form of a well-defined layer extending over a relatively narrow range in height, from about 10-15 km to 30-40 km,
with a strong brightness maximum at the 20-30-km level. There is every reason to believe that this finding is the only positive result of Hulburt's work, for
all his other conclusions are erroneous; indeed, it is on Hulburt's authority that
stratospheric research by the twilight method has been thought a hopeless pur-
suit, and for a long time this claim has ruinously affected further progress.

Where, then, lies the source of Hulburt's error? Quite apart from the numerical miscalculation detected by Khvostikov, serious objections can be raised,

in our view, against Hulburt's general approach to the problem [96, 261]. The
fact is that the brightness of singly and doubly scattered light are rapidly declining functions of ζ, varying over a factor of at least 10^7 during the twilight
process. According to Hulburt's ideas, the brightness of the doubly scattered
light is essentially determined by the product of the sky brightness actually observed at the horizon and the density of the atmosphere at the 15-40-km level,
a quantity that is comparatively well known (at least to within a factor of
about 2), with the doubly scattered component varying during twilight by a
factor of $\approx 10^7$. For the brightness of the primary scattering, which as we have
seen is determined by the vertical optical thickness (in the first approximation,
by the air pressure) at level $\bar{H}$, Hulburt a priori selects a far more rapid
variation with ζ, so that during the twilight process the value of the pressure
at level $\bar{H}$ declines by a factor of fully 10^{17}, according to the isothermal model
atmosphere that he adopts. Since $I_2/I_1 \approx 10^{-1}$ at the start of twilight, one needs
neither measurements nor intricate calculations to reach Hulburt's conclusion.
In other words, the exceptionally high values that Hulburt finds for the ratio
I_2/I_1 are a direct consequence of comparing the brightness of doubly scattered
light as extracted from the observations, and thus close to the true values, with
a hypothetical brightness for singly scattered light based exclusively on
a law for the height variation of the scattering coefficient in an arbitrarily
prescribed atmosphere, as proposed at one time by Humphreys [262] but
wholly refuted by subsequent measurements (Chap. II, §1). We may add that
Hulburt's calculation, like most of the theoretical work that followed, was
based on the assumption that the scattering in the upper atmosphere is molecular in character. And this too conflicts with current views.

The need had naturally arisen to perform a series of theoretical calculations under various assumptions for the structure of the upper atmosphere, the
more so following the acquisition in the early 1940's of considerable data af-
fording independent refutations of Humphreys' model. We refer here to the results of acoustic and radiosonde observations of the atmosphere, the valuable
studies of the processes involved in meteoric ablation (for example, [264]),
the photometric investigation of polar aurorae [265], and soon afterward the
initial data from rocket research. On the other hand, it also became necessary
to examine the behavior of the twilight process itself, in order to obtain any
possible evaluation of multiple scattering. We begin with the first group of
purely theoretical papers.

In 1942, Link [266] applied a method developed by Tikhanovskii [53] to
compute the fraction of doubly scattered light in the brightness of the zenith
region of the sky, for $\zeta = 100°$ and $\lambda = 5300$ A, and he found that it comprises
only 28% of the total sky brightness. Shtaude [263] and Yudalevich [267, 261]
published the results of much more extensive approximate calculations in 1948.

TABLE 16. Theoretical Estimates of the Ratio I_2/I_1

ζ, deg	92	94	96	98	100	102	103	104	105	107
Shtaude, twilight atmosphere	0.20	0.23	0.33	0.62	0.63	—	—	—	—	—
Shtaude, isothermal atmosphere	0.20	0.23	0.53	3.65	97.5	—	—	—	—	—
Yudalevich	—	0.30	0.49	0.76	0.78	0.87	0.97	1.16	1.68	2.40

For her investigation, Shtaude selected an atmospheric structure in two variants, one agreeing with that adopted by Hulburt (an isothermal atmosphere), and the other corresponding to the atmospheric structure obtained by analyzing observed $I(\zeta)$ relations for the brightness of the twilight sky at the zenith, under the assumption that secondary scattering is absent [98] (a twilight atmosphere). It turned out that while the brightness I_1 of singly scattered light depends entirely on the choice of model atmosphere, the brightness I_2 of doubly scattered light is almost insensitive to this choice. But it was also established that the ratio I_2/I_1 depends critically on the structure of the atmosphere, with the immediate implication, in particular, that theoretical calculations of this type can only be very provisional ones, and that to apply the twilight method in upper-atmosphere research, one must measure experimentally, for each individual case, the contribution of multiple scattering to the sky brightness. Table 16 gives Shtaude's estimates for the quantity I_2/I_1 (these are lower bounds, as the author points out).

Also included in the table are the results of more precise calculations performed by Yudalevich[267] for $\lambda = 0.55~\mu$, on the basis of an atmospheric structure with respect to height (pure molecular scattering) that seemed most likely at the time, and with an ingenious geometrical device [268] to simplify the operations. We see that under reasonable assumptions for the structure of the atmosphere, the contribution of secondary scattering is incomparably smaller than found by Hulburt, and agrees far better with the qualitative description outlined above.

Nevertheless, the assumption of a scattering coefficient varying exponentially with height — an isothermal atmosphere — continued to be adopted by many authors, mainly because of its simplicity. It was, for example, used as a basis for calculation by Robley [269], Dave [270], and Sato [271], and inescapably led these authors to conclusions differing little from those of Hulburt. Their work remains of special interest in that the authors derived the

polarization of doubly scattered light as well as its brightness. We shall return to this aspect of their results later. Here we wish only to make two remarks of a procedural character. Robley's investigation [269] rests entirely
on the method developed by Chandrasekhar for solving the matrix equation
of transfer (including polarization of the radiation) for a plane-parallel atmosphere, so that his solution can apply to twilight conditions only in an artificial
and very rough manner. In other words, while the method of solution is perfectly correct in itself, the way in which it is extended to twilight conditions
of illumination raises decisive objections and makes any definitive conclusions
completely untenable, as Dave has pointed out [270].

On the other hand, Dave [270] and Sato [271] have applied a method in
which successive calculations are made of the brightness and polarization of
singly and doubly scattered light, and have carefully taken into account the
effects due to curvature of the atmosphere (although Dave, as well as Shtaude
and Yudalevich, have neglected refraction). However, instead of using the
Stokes parameters and scattering matrix (Chap. II, §2), which would have en-
sured an accurate allowance for the polarization effects, they have resorted
to resolving the light beams into two noncoherent, alternately polarized components, and tracing the course of each component separately. This procedure,
at one time applied by Soret [51], Ahlgrimm [52], and Tikhanovskii [53] for
computing the polarization of the daytime sky, actually involves a significant
systematic error [96], namely the failure to recognize that the noncoherence
of the alternately polarized components is already violated in the first scattering event. The results obtained in this way must therefore be revised, even
within the scope of the molecular model atmosphere that was used. Yudalevich [272] has recently suggested another plan for calculating the brightness
and polarization of doubly scattered light under twilight conditions for an atmosphere of arbitrary optical structure, but the numerical work has not yet
been carried out.

Returning now to Dave's investigation [270], we note that according to
his calculations the height range mainly responsible for generating doubly
scattered light (for observations at the zenith) does not remain invariant, but
moves upward as the sun sinks below the horizon. Figure 135 illustrates this
result, which differs markedly from Hulburt's findings, but approximates the
results of Shtaude's calculations.

Fesenkov [273] has considered the very interesting problem of how the
zodiacal light is singly scattered in the atmosphere. In many respects it is
analogous to the problem of secondary scattering of sunlight during twilight
hours.

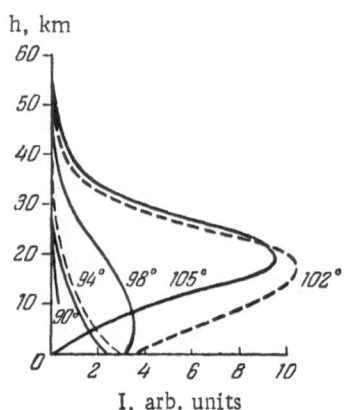

h, km

I, arb. units

Fig. 135. The vertical height distribution of the brightness of doubly scattered light for zenith observations, in relative units, for selected ζ (Dave [270]).

There are two further effects concerning scattering under twilight conditions that have not yet been investigated. First, light that is multiply scattered at small angles (in the aureole region) in the lower layers of the atmosphere, along the earth's surface, can produce a very noticeable effect if the directivity of the atmospheric scattering diagram is taken into account. However, one would expect true absorption of light in the atmosphere to nullify this effect. Secondly, there can be scattering of atmospheric light that has been reflected from the earth's surface (especially from snow-covered ground). The considerable illumination of the earth's surface during twilight, mainly because of the twilight-glow segment, might render this effect comparable with the effect of secondary scattering in the atmosphere. In particular, the enhanced brightness of the twilight sky in winter, as noted by Volz and Goody [282], might possibly arise to some extent from the presence of snow cover.

To conclude our brief survey of theoretical work, let us inquire how the role of multiple scattering depends on wavelength, that is, on the color of multiply scattered light. According to the Rayleigh theory (Chap. II, §2), the scattering coefficient for gases is proportional to λ^{-4} outside the absorption bands. Since a similar factor enters in each new scattering event, it might seem that the brightness of light that has suffered k-fold scattering should be proportional to λ^{-4k}. This argument has in fact been generally accepted, and has often been used in an attempt to evaluate the contribution of doubly scattered light under both daytime and twilight conditions, on the basis of experimental data for the color of the sky. In particular, the increasing blueness of the sky that occurs as ζ increases during the total-twilight period has been treated as the result of an increased contribution of secondary scattering to the twilight-sky brightness, in apparent accord with the theoretical estimates cited above for this contribution. (We have already shown, however, that the enhanced blueness of the sky actually arises from other factors.) The only difficulty that has been thought to occur is that if aerosols are present, the λ^{-4} law should be replaced by a weaker λ^{-n} relation, with n < 4. But this correction would not alter the essential feature that for k-fold scattering only a relation of the form λ^{-kn} would be expected; that is, provided n > 1, the higher the multiplicity of the scattering, the "bluer" the scattered light should be.

Proceeding on this basis, Link [266] estimated the spectral dependence of the ratio I_2/I_1 of the brightness of doubly and singly scattered light, and reached the obvious conclusion that the ratio should diminish with increasing wavelength. Yudalevich [274] used the same arguments to compute the spectral dependence of I_2/I_1, and found that in the near infrared region the ratio I_2/I_1 should amount to only a few percent, for all ζ.

However, the claim that a λ^{-kn} factor describes the color of k-fold scattered light is unrealistic, since it omits the effects of extinction. Ashburn [103] and Rozenberg [130] have pointed out this fact in connection with the colors of the twilight sky. A rigorous extinction correction for doubly scattered light would quite properly be the main obstacle in the brightness calculations. Until such calculations can be made with the necessary confidence, any final judgment as to the color of doubly scattered light will therefore be impossible. Still, one can advance arguments to show that doubly scattered light will have a color differing in a relatively minor way from that of singly scattered light, and in the sense that the longer wavelengths will predominate — a "reddening" rather than the enhanced blueness that a λ^{-kn} law would imply. By the same token, it would follow that the observed increase in the blueness of the twilight sky with respect to ζ during the total-twilight period indicates that multiple scattering plays a comparatively unimportant part at that stage. We remark in passing that the whitening of the daytime sky that occurs when the turbidity of the atmosphere increases likewise arises not merely from the weakened wavelength dependence of the scattering coefficient, but also from the excess of doubly and multiply scattered light less blue in color.

To demonstrate these claims, let us first consider the combined effect of scattering of all multiplicities in a layer having a sufficiently great optical thickness τ^*, illuminated by diffuse light on one side, and at rest above an underlying surface of albedo A. We may then apply Eq. (II.2, 56) for the transmissivity, and if we regard absorption as absent and the quasidiffuse illumination as orthotropic for $\beta = 0$, we will have, for the intensity of the light scattered by the layer,

$$I_{trans} = \frac{\Phi_0}{\pi} g(\cos z) \frac{l}{(1-A)\tau^* + l} \qquad (IV.6, 1)$$

for transmitted light and

$$I_{refl} = \frac{\Phi_0}{\pi} \frac{(1-A)\tau^* + Al}{(1-A)\tau^* + l} \qquad (IV.6, 2)$$

for reflected light, where Φ_0 is the luminous flux irradiating the layer, and l is a constant, depending on the form of the scattering matrix. We readily see that if $A = 1$, that is, if the layer lies above a nonabsorbing, orthotropic surface, then for both reflected and transmitted light the intensity will not depend on τ^*, nor thereby on wavelength. The same effect will occur (although for different reasons) for reflected light if τ^* is large enough, that is, if $(1 - A)\tau^* \gg l$. Indeed, it has long been well known that a sufficiently thick layer of a nonabsorbing scattering medium will appear white in reflected light, independently of the spectral dependence of the scattering coefficient (such as clouds or snow). In transmitted light, if $(1 - A)\tau^* \ll l$ a layer of the scattering medium will also appear white, independently of the scattering law. In the opposite case, $(1 - A)\tau^* \gg l$, we find from Eq. (IV.6, 1) that $I_{trans} \approx 1/\tau^*$, so that if the scattering coefficient is proportional to λ^{-n} we will have $I \sim \lambda^n$. In other words, the intensity of multiply scattered light penetrating the thickness of the layer will rise, not decline, with increasing wavelength; thus multiply scattered light will be considerably "redder" than singly scattered light, and in fact redder than direct light. But the average multiplicity of the scattering will increase with increasing τ^*, or decreasing λ.

As another example to illustrate qualitatively the physical meaning of the phenomenon under specific twilight conditions, consider the secondary scattering of light in the layer near the ground, for $\xi = 8°$. At this stage the twilight segment has clearly become predominant as the main source of illumination of the lower atmosphere, with its radiation concentrated into a relatively narrow and well-defined brightness maximum (Fig. 21a, Chap. I, §3). In a first and fairly rough approximation, we may therefore assume that all the light of the segment comes only from this maximum, whose position depends on τ^* and is shown graphically in Fig. 134.

We further assume that the scattering is purely molecular in character, so that the values $\tau^*_1 = 0.2$ and $\tau^*_2 = 0.1$ correspond to wavelengths of $\lambda_1 = 0.46$ μ and $\lambda_2 = 0.54$ μ (Table 4, Chap. 2, §3), and the values of m_{max}, in accordance with Fig. 134, are approximately $m_1 = 6.6$ and $m_2 = 12.3$. Moreover, from Fig. 103, and the fact that for molecular scattering Eq. (III.5, 13) gives $y_i(\lambda_1) = y_i(\lambda_2) + (4/K) \ln(\lambda_2/\lambda_1)$, we find $y_i(\lambda_1) = 22$ km and $y_i(\lambda_2) = 17$ km. Finally, since Eq. (IV.5, 20) yields $\gamma(m_1) \approx 0.48$ and $\gamma(m_2) \approx 0.37$, we have $\bar{H}(\lambda_1) = 41$ km and $\bar{H}(\lambda_2) = 30$ km, while at the zenith ($\gamma \approx 1$) we have $\bar{H}_{zen}(\lambda_1) = 86$ km and $\bar{H}_{zen}(\lambda_2) = 81$ km.

To compute $CE^{\lambda_1}_{\lambda_2}$ for singly scattered light coming from the zenith, we may apply Eqs. (IV.5, 10) and (IV.5, 11) directly. But if we wish to evalu-

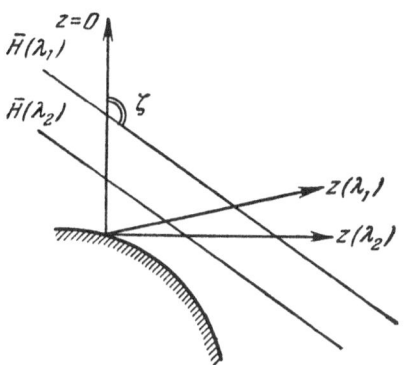

Fig. 136. Geometry for estimating $CE_{\lambda_2}^{\lambda_1}$

for doubly scattered light.

ate $CE_{\lambda_2}^{\lambda_1}$ for light that has been doubly scattered in the ground layer of the

atmosphere, with the illumination coming from the twilight segment, then we must add two terms to Eq. (IV.5, 11). One of them, equal to 1.08 ln [m(λ_2)/ /m(λ_1)], allows for the difference in the values of m appearing in Eq. (III.5, 9) because of the spectral dispersion in the positions of the brightness maximum of the twilight segment. We recall that the corresponding term in Eq. (IV.5, 9), namely ln m, takes into account the inclination of the line of sight to the boundary of the earth's shadow, that is, the length of the line of sight within the height range from $\overline{H}_2$ to $\overline{H}_1$, say (Fig. 136). A second term, equal to 1.08 · · 4[ln (λ_2) − ln (λ_1)], must be included to allow for the change in the spec- tral composition of the light occurring in the multiple-scattering event itself.

If we return to Eq. (IV.5, 10) with these corrections and neglect the com- paratively small term proportional to $(\Delta \overline{\eta})^2$, we obtain as an approximate esti-

mate for $CE_{\lambda_2}^{\lambda_1}$ for singly scattered light at the zenith,

$$CE_{\lambda_2,\,sing}^{\lambda_1} \approx 5.5Q - 0.65 + const, \qquad (IV.6, 3)$$

and for doubly scattered light in the ground layer,

$$CE_{\lambda_2,\,doub}^{\lambda_1} \approx 12Q - 0.8 + const, \qquad (IV.6, 4)$$

where the terms

$$\text{const} = 1.08\,[\ln I_0\,(\lambda_2) - \ln I_0\,(\lambda_1)].$$

Since at heights $\bar{H} = 30\text{-}40$ km the values of $Q \approx q$ are usually close to 0.1, while at heights $\bar{H} = 80\text{-}90$ km, $Q \approx q = 0.02\text{-}0.03$ (Fig. 123, §4), we obtain

$$CE^{\lambda_1}_{\lambda_2,\,\text{sing}} \approx -0.55 + \text{const} \quad \text{and} \quad CE^{\lambda_1}_{\lambda_2,\,\text{doub}} \approx 0.38 + \text{const},$$

so that singly scattered light is "redder" than would follow from a λ^{-4} law (for which $CE^{\lambda_1}_{\lambda_2} = -0.7 + \text{const}$), while doubly scattered light is much redder than singly scattered. In fact, by Eq. (IV.5, 1) we have

$$\frac{I\,(\lambda_2)}{I\,(\lambda_1)} = 10^{\,0.4CE^{\lambda_1}_{\lambda_2}},$$

so that

$$\left(\frac{I_{\text{green}}}{I_{\text{blue}}}\right)_{\text{doub}} \Big/ \left(\frac{I_{\text{green}}}{I_{\text{blue}}}\right)_{\text{sing}} \approx 10^{0.37} \approx 2.3.$$

In other words, for identical brightness in the blue spectral region, doubly scattered light will be 2.3 times as bright in the green as singly scattered light. If we take the less favorable but less probable values $Q(85 \text{ km}) = Q(35 \text{ km})$, then the ratio will approach unity, but it still will not fall to the value 0.5 that a λ^{-4} law would imply.

This estimate should be regarded as a qualitative illustration only. The diffuseness of the brightness maximum of the twilight segment will somewhat weaken the amount of reddening of doubly scattered light. On the other hand, as Hulburt and Dave have shown, doubly scattered light is not produced in the ground layer, but in a layer located at some finite height, and the inevitable dispersion in these heights will tend to strengthen the reddening effect for doubly scattered light.

To avoid confusion, we should emphasize that even though doubly scattered light is "redder" than singly scattered, it is still "blue" radiation, for the short-wave spectral region remains predominant. This means that the brightness of doubly scattered light should rise monotonically with decreasing wavelength.

We offer, finally, a qualitative explanation for the brightness minimum in the zenith region of the sky. For this purpose we shall apply the concepts of the twilight layer and the effective layer where secondary scattering occurs

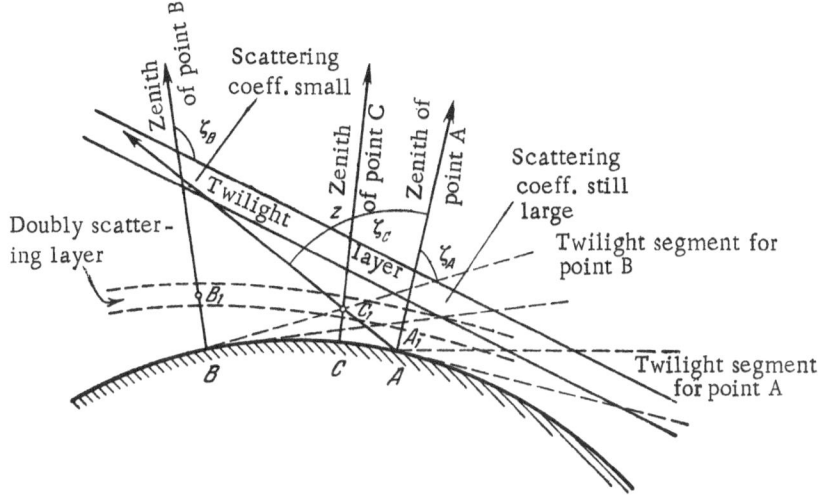

Fig. 137. Geometry for estimating the contribution of doubly scattered
light in the solar and antisolar verticals.

(Fig. 137). The behavior of the intensity of singly scattered light as a function
of z for given ζ has been considered in detail in §§2 and 3. Near the antisolar
vertical, it declines monotonically with increasing z, because of the increasing height of the twilight layer (the decreasing value of the scattering coefficient) and also the increasing value of mτ* (the decreasing atmospheric transparency), although the lengthening of the portion of the line of sight within
the twilight layer somewhat delays the fall in intensity. As for the secondary
scattering produced at low heights, Dave's calculations [270] indicate that in
the antisolar vertical we should expect only a very slight rise in the effective
scattering layer with increasing z. The illumination on the layer will decline,
for as z increases the twilight segment will be produced by progressively higher
portions of the twilight layer (Fig. 137). In particular, the illuminance E at
the observed point C_1 in the doubly scattering layer for $ζ = ζ_A$ will be the same
as at the point A_1 for $ζ = ζ_C$. If h_{doub} is the height of the effective scattering
layer, then we readily find that

$$ζ_C - ζ_A \approx \frac{h_{doub}}{R} \text{tg} z, \qquad \text{(IV.6, 5)}$$

and since $h_{doub} = 10\text{-}30$ km, $ζ_C - ζ_A \lesssim 5 \cdot 10^{-3}$ tg z radian. The corresponding relative change in the illuminance on the doubly scattering layer will be
approximately

$$-\Delta \ln E \lessgtr Q \frac{d\overline{H}}{d\zeta} (\zeta_C - \zeta_A) \lessgtr Q \frac{d\overline{H}}{d\zeta} 5 \cdot 10^{-3} \, \mathrm{tg} \, z,$$

$$\text{(IV.6, 6)}$$

and since $Q \approx 0.1$ and $d\overline{H}/d\zeta \approx \overline{H} (d\gamma/d\zeta)$ in the vicinity of the illuminated layer of the twilight segment, where $\overline{H} \approx 20$-40 km and $d\gamma/d\zeta \approx 2$-3, we obtain $-\Delta \ln E \lessgtr 5 \cdot 10^{-2}$ tg z, in order of magnitude; thus if z is not too large, the change in the illuminance will not exceed a few percent. The change in the transparency of the lower atmospheric layers will also be of little importance if z is not too large. However, the length of the line of sight within the effective scattering layer will increase approximately in proportion to sec z. The brightness of doubly scattered light in the zenith region should therefore increase in the direction toward the antisolar horizon approximately in proportion to sec z $(1 - \Delta \ln E)$, or somewhat more slowly because of the increasing h_{doub}. And since the brightness of singly scattered light declines rapidly with increasing z, the sky brightness should pass through a minimum at a certain value of z, marking the beginning of a gradual transition of the main contribution from single to double scattering. If secondary scattering played an important part in the zenith region of the sky, the minimum would fall at the zenith [263]. Actually, however, the minimum is displaced noticeably from the zenith along the antisolar vertical (Figs. 18 and 19, Chap. I, §3) [298], at least when total-twilight is in progress at the zenith. This fact indicates that single scattering makes the main contribution to the twilight-sky brightness during the total-twilight period (see Chap. V, §4). As further evidence, we note that the theory of single scattering correctly describes the color changes of the sky during the semitwilight and total-twilight periods.

In order to estimate quantitatively the displacement of the brightness minimum relative to the zenith, let us suppose that for small z we may neglect the scattering-diagram effect and take the brightness of doubly scattered light in the solar meridian to be

$$I_2(z) = I_2(z = 0) \, P^{\sec z} \sec z \, (1 + c \, \mathrm{tg} \, z),$$

where $c < 5 \cdot 10^{-2}$, and z is considered positive in the solar vertical and negative in the antisolar vertical. Thus the brightness minimum in the doubly scattered light will be displaced from the zenith by no more than 2° in the direction away from the sun.

If we now appeal to Eq. (IV.4, 1) for the brightness of singly scattered light, the condition $d(I_1 + I_2)/dz = 0$ for a minimum in the sky brightness will yield the approximate relation

$$z_{\min} \approx -\frac{Q(\bar{H}_{\min})\,\bar{H}_{\min}\,\xi + c\left[I_2\,(z_{\min})/I_1\,(z_{\min})\right]}{(1 + \ln P)\left[1 + I_2\,(z_{\min})/I_1\,(z_{\min})\right]},$$

where we have considered that during the total-twilight period we have, in the zenith region of the sky,

$$\frac{d\,\ln r'_s\,(\bar{H})}{dz} \approx \frac{d\,\ln r'_s\,(\bar{H})}{d\bar{H}}\,\bar{H}\,\frac{dy}{dz} \approx Q\,(\bar{H})\,\bar{H}\,\xi.$$

This expression for $z_{\min}$ enables us to relate the displacement from the zenith of the brightness minimum in the twilight sky directly to the ratio I_2/I_1 at $z = z_{\min}$; thus

$$\frac{I_2\,(z=0)}{I_1\,(z=0)} \approx \frac{I_2\,(z_{\min})}{I_1\,(z_{\min})} \cdot \frac{1 - Q\,(\bar{H})\,\bar{H}\,\xi\,|\,z_{\min}\,|}{1 - c\,|\,z_{\min}\,|}.$$

We also gain here the opportunity to obtain quantitative estimates of the ratio I_2/I_1 from measurements of $z_{\min}$ for various ζ, and we can test the theory quantitatively if we can evaluate I_2/I_1 from independent measurements (at different z, for example).

We come now to the attempts that have been made to derive the contribution of light doubly scattered in the atmosphere to the brightness of the twilight sky through analysis of observational data. Let us begin with Robley's work [172], an isolated effort against which there are serious objections. It was Robley's idea that if scattering were purely molecular in character, with the polarization of the scattered light given by the Rayleigh law with the Cabannes correction (Chap. II, §2), and if the doubly scattered light were completely depolarized, then the degree of polarization observed in the light of the twilight sky could inform us about the brightness of the secondary-scattering component. By no means new, this idea has been applied in several attempts to derive the fraction of doubly scattered light under daytime conditions. Actually, however, it meets with some inescapable difficulties.

To begin with — and this applies equally to daytime and twilight conditions — the presence of aerosols will change the degree of polarization of light scattered in the atmosphere (see Chap. II, §2); usually, but certainly not always, it will tend to reduce the polarization, even to the point where so-called negative polarization sets in [17-19]. This effect will evidently render uncertain any estimate of the contribution due to secondary scattering, primarily in the sense of materially overestimating that component. Secondly, there are no grounds for claiming that doubly scattered light is completely

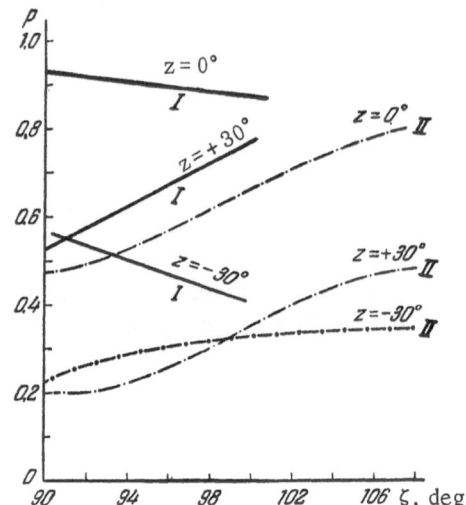

Fig. 138. Degree of polarization of I) singly
and II) doubly scattered light from the twilight sky as a function of ζ at selected z.

depolarized. While under daytime conditions this assertion may perhaps be
regarded as a fairly rough approximation to reality, it simply is wrong during
twilight hours, when the main source of secondary scattering is the directed
illumination of the atmosphere by the twilight-glow segment. Thus, according to provisional calculations by Rozenberg [39, 96], the degree of polarization of doubly scattered light for observations at the zenith during twilight is
very high, amounting to about p = 0.7 if molecular scattering is assumed (and
if only the light coming from the twilight segment is considered, p will reach
values of 0.82-0.86). This means, incidentally, that the decrease in the degree of polarization of the light from the twilight sky to a value p = 0.3-0.5
or less can in no way be explained as a secondary scattering effect; on the
contrary, it actually argues for the smallness of the contribution from secondary scattering, despite the opinion of Ginzburg and Sobolev [298]. Yet at
the same time, if we were nevertheless to accept Robley's treatment, we would
have to regard his estimates of the secondary-scattering contribution as much
too low. We can hardly place any confidence, then, in his estimates for the
quantity I_2/I_1, values that are often quoted. Indeed, Robley himself afterward
came to the same conclusion [269] when his computations of the degree of polarization of doubly scattered light yielded a value of 50%. The considerably
more accurate computations of Dave [270] have materially improved Robley's

TABLE 17. The Ratio $I(A = \pi)/I(A = 0)$ as Computed by Fesenkov [83]

ζ, deg	92	94	96	98
Single scattering	0.64	0.26	0.054	0.01
Double scattering.	0.67	0.50	0.42	0.16

value, bringing it close to the estimate derived by Rozenberg [39, 96], and revealing a systematic rise in the degree of polarization with increasing ζ. Figure 138 presents the ζ-dependence of the degree of polarization of singly and doubly scattered light, as obtained by Dave on the assumption of molecular scattering, for $z = 0°$ and $\pm 30°$ in the solar meridian.

Of some interest is Dave's attempt to compare the computed behavior of the degree of polarization of the combined light from the twilight sky, as based on obviously exaggerated estimates of the role of secondary scattering (isothermal atmosphere), with the results of observations by Dave and Ramanathan [110]. In all cases ($z = 0°$, $\pm 30°$), the observed degree of polarization was found to be considerably lower than the computed value. The qualitative character of the $p(\xi)$ relation was similar, however. Two circumstances clearly emerged here. First, the relative displacement of the experimental and theoretical curves along the ζ-axis distinctly points to an overestimate of the contribution of doubly scattered light. Secondly, the sharp drop in the degree of polarization for $\zeta \gtrless 102°$ can be explained by the gradually increasing translucence of the night sky as seen through the haze of scattered sunlight. We shall return to this question in the next section.

In 1934, Fesenkov [83] became the first to secure adequate semi-empirical estimates of the role of secondary scattering, and the method he developed is incomparably more correct than the method to which Hulburt later resorted. It was based on measurements of the sky brightness at two points having the same zenith distance $z = 67°$ in the solar ($A = 0$) and antisolar ($A = \pi$) verticals. By applying the twilight-layer concept, measurements of the sky brightness, and the assumption of molecular scattering in the stratosphere (the belief generally held at the time), Fesenkov derived the density of the atmosphere up to the 60-km level. He also computed theoretically the brightness of singly scattered light at a number of points scattered over the whole sky. Finally, with these data and with estimates of the scattering power in the lower layers of the atmosphere (in particular, the scattering diagram), as obtained from the results of measurements of the daytime-sky brightness, he performed calculations for the brightness of doubly scattered light at various ζ, for both the points at $z = 67°$. Table 17 shows the resulting values of the ratio $I(A = \pi)/I(A = 0)$ for singly and doubly scattered light. We see that at $z = 67°$ the brightness of

TABLE 18. Relative Variations in I_2/I_1 as a Function of ζ, as Computed by Fesenkov [83]*

ζ, deg	92	94	96	98
A = 0	1	0.82	0.90	0.99
A = π	1.02	1.54	6.8	> 14

*The ratio is arbitrarily taken as unity at $\zeta = 92°$ for the point with A = 0.

doubly scattered light is about half as great in the antisolar vertical as in the solar vertical (including a correction for the scattering-diagram effect), and that the ratio $I(A = \pi)/I(A = 0)$ varies comparatively little with ζ (by a factor of about 4 in the range of ζ considered). On the other hand, for singly scattered light the same ratio varies by some two orders of magnitude. Table 18 indicates the manner in which the ratio I_2/I_1 varies with ζ for the two points.

The data of Table 18 show unambiguously that while the contribution of secondary scattering remains almost constant at A = 0 and $z = 67°$ (a region where, as we have seen, semitwilight and the bright stage of total twilight are still in progress at the values of ζ listed), it rises rapidly with ζ at A = π and the same z, until at $\zeta = 96°$ it predominates decisively over single scattering (in this region of the sky, the dark stage of total twilight and primarily deep twilight are in progress for the ζ listed).

The most important result of this study by Fesenkov was that at sufficient solar depressions below the horizon (for $\zeta \geqslant 96°$, say), the sky brightness at the point $z = 67°$, A = π arises almost wholly from the intensity of doubly scattered light, with the latter intensity differing comparatively little from its value at the corresponding point on the solar vertical (A = 0), although the brightness of singly scattered light is very much higher at A = 0. This finding immediately afforded a method for estimating the brightness of doubly scattered light at the point $z = 67°$, A = 0; one had merely to multiply $I(A = \pi)$ for the same value of ζ by the slowly varying ratio $I(A = 0)/I(A = \pi)$ for doubly scattered light. Several authors, including Megrelishvili [109], have subsequently applied this method in order to free measurements of the twilight-sky brightness from the contribution of secondary scattering.

Fesenkov's method is subject to a certain amount of refinement, as follows. Suppose that observations are taken at two points in the solar meridian, corresponding to lines of sight at zenith distances of +z and −z, where z is great enough for one of the points to fall in the twilight-segment region, and the other in the region of the earth's rising shadow ($z = -70°$, say). The measured sky brightness consists of the brightness I_1 of singly scattered light and the brightness I_2 of doubly scattered light:

$$I(\zeta, z) = I_1(\zeta, z) + I_2(\zeta, z). \tag{IV.6, 7}$$

We have seen that for given ζ, I_2 depends only weakly on z, so that without serious error we may take

$$I_2(\zeta, +z) = cI_2(\zeta, -z), \tag{IV.6, 8}$$

where c depends weakly on ζ. On the other hand, $I_1(\zeta, z)$ varies regularly with ζ, and there will exist a ζ' for which

$$I_1(\zeta, -z) = I_1(\zeta', +z). \tag{IV.6, 9}$$

By Eq. (IV.4, 1), this correspondence will occur if

$$\overline{H}(\zeta, -z) = \overline{H}(\zeta', +z), \tag{IV.6, 10}$$

that is, if

$$\frac{[R(1-\sin \zeta)+\bar{y}]}{\sin (\zeta+z)} = \frac{[R(1-\sin \zeta')+\bar{y}]}{\sin (\zeta'-z)} . \tag{IV.6, 11}$$

Here the total sky brightness in the direction +z for the value ζ' will be

$$I(\zeta', +z) = I_1(\zeta', +z) + I_2(\zeta', +z) \tag{IV.6, 12}$$

or

$$I_1(\zeta', +z) = I(\zeta', +z) - cI_2(\zeta', -z). \tag{IV.6, 13}$$

Using Eq. (IV.6, 9), and substituting Eq. (IV.6, 13) into Eq. (IV.6, 7) for $I(\zeta, -z)$, we find

$$I(\zeta, -z) = I(\zeta', +z) - cI_2(\zeta', -z) + I_2(\zeta, -z), \tag{IV.6, 14}$$

whence

$$I_2(\zeta, -z) = I(\zeta, -z) - I(\zeta', +z) + cI_2(\zeta', -z). \tag{IV.6, 15}$$

But if z is great enough, $I_2(\zeta', -z) \ll I_2(\zeta, -z)$, so that if c is not too large we shall have

$$I_2(\zeta, \ -z) \approx I(\zeta, \ -z) - I(\zeta', \ +z), \qquad \text{(IV.6, 16)}$$

where the right-hand member contains only measurable quantities. The main difficulty with this procedure consists in determining ζ', since the value of $\bar{y}$ is poorly known. But this problem will not be too important for values of ζ and z such that $I(\zeta, -z) \gg I(\zeta', +z)$, as will be the case with the onset of deep twilight in the vicinity of the antisolar vertical.

Fesenkov [275] has recently introduced substantial further improvements into the method (see also Chap. V, §4).

We also wish to mention the semi-empirical work of Rozenberg [130] and Divari [116], who have shown in various ways that calculations of the color index $CE_{\lambda_2}^{\lambda_1}$, if only single scattering is included, will agree well with the observations only up to the close of total twilight; there will be complete disagreement during the deep-twilight period, when the increasing blueness of the sky is replaced by a reddening. Divari suggests that the incipient transition to night is responsible for the reddening — that the intrinsic airglow, which is more intense at red wavelengths, begins to play a part (§7). However, we have seen that the sky can also be reddened because of the increased contribution of doubly scattered light, and we are inclined, though not adamantly so, to ascribe the reddening effect to this factor (§7). We would then have a positive indication that doubly scattered light makes only a minor contribution during the total-twilight period, but a sharply increased one during the deep-twilight stage, in full accord with Fesenkov's views and the results of the estimates he has made [83].

We turn, finally, to two experimental investigations of another character [100, 276], which provide completely independent and very strong evidence for the relatively small role of multiply scattered light during the total-twilight period. This research is based on very abundant material acquired through many years of observation by T. G. Megrelishvili, under the direction of I. A. Khvostikov, in a program conducted since 1942 at the Abastumani Astrophysical Observatory of the Georgian Academy of Sciences, and it will be discussed fully in the next chapter. At this point we merely wish to review those results that relate to the multiple-scattering problem.

The principal result obtained by these authors is that the data on the density and pressure of the atmosphere, as derived by analyzing the ζ-dependence of the twilight-sky brightness, agree quite well with the findings of

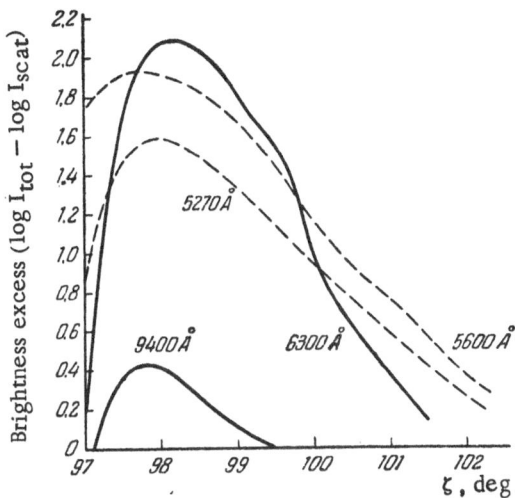

Fig. 139. Excess of the true sky brightness over
that expected for single scattering.

other types of research (including rocket techniques) so long as total twilight
prevails in the region under observation, but that pronounced discrepancies
between the twilight data and the other methods appear as soon as deep twilight sets in. Megrelishvili and Khvostikov properly regard this result as
demonstrating that the contribution of secondary scattering, which was not included in the reduction of the twilight measurements, cannot be of much importance during the total-twilight period. Link [266] had earlier reached the
same conclusion, but from relatively meager observational material.

In her demonstration, Megrelishvili [109] used the rocket data on the
structure of the atmosphere as given in Kallmann's survey [277], assumed that
the scattering is molecular in character, and then computed the $I(\zeta)$ relation
for the zenith (with restriction to single scattering only) at $\lambda = 0.52$, 0.56,
0.63, and 0.94 μ. The values of $\log I_{scat}(\zeta)$ derived in this way were then
subtracted from mean $\log I_{tot}(\zeta)$ curves covering many days (at the corresponding wavelengths) of direct measurement of the twilight-sky brightness
at the zenith. Measured values of the night-sky brightness were used for transformation to an absolute brightness scale.

Figure 139 shows the resulting differences — the excess in the actual sky
brightness as compared with that expected for single scattering. If the entire
excess in brightness is to be ascribed to multiple scattering, the values of I_2/I_1

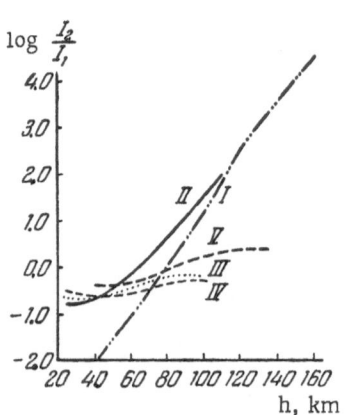

Fig. 140. Comparison of several estimates for the ratio I_2/I_1 as a function of the maximum height of the twilight layer [109]. I) Hulburt; II) Shtaude, isothermal; III) Shtaude, twilight; IV) Megrelishvili; V) Yudalevich.

will be considerably smaller than found by Hulburt, yet considerably larger than indicated by the calculations of Shtaude, Yudalevich, and Link. In the next section and in Chap. V, we shall discuss the extent to which the presence of aerosols and other factors might be responsible for this excess.

Megrelishvili also reports that preliminary results from a reduction of observations obtained by the Fesenkov method, but now with subsequent correction for the effect of doubly scattered light, indicate, on the one hand, a satisfactory agreement with the estimates of Yudalevich and Shtaude (for a twilight atmosphere), as shown in Fig. 140, while on the other hand there is good agreement with rocket data on the density of the atmosphere, at least up to a height of 120-130 km. On this basis, Megrelishvili suggests that the brightness excess she has found (Fig. 139) should be ascribed mainly to multiple scattering. But this would conflict with Fig. 140. In our own view, the explanation for the high brightness excess lies in defects in the model atmosphere that was adopted, and possibly in shortcomings in the computation procedure, which was based on the twilight-layer concept. In this regard, we recall that the only directly measured quantity involved in the comparison is just the estimated contribution of secondary scattering according to the Fesenkov method. We also point out that for reasons discussed above, the excess in brightness at $\overline{H} \approx 100$-150 km definitely cannot be due to the presence of an aerosol layer at that level, as Shtaude and Link have believed.

It seems to us, then, that Fig. 140 convincingly establishes the moderate (though not negligible) contribution of doubly scattered light to the general sky brightness during the total-twilight period. We may add in conclusion that a thorough familiarity with Megrelishvili's observational data has enabled the author to offer certain modifications to the Fesenkov method, which permit one to evaluate the ratio I_2/I_1 experimentally with fully adequate accuracy (Chap. V, § 4). Preliminary and none too reliable measurements by this modified procedure on one day of observation have shown good agreement with the estimates obtained at Abustumani, and illustrated in Fig. 140.

§7.   Deep Twilight and Nightfall

Let us return now to the color index $CE_{\lambda_2}^{\lambda_1}$ for a given part of the sky, (z, A), and again follow its ζ-dependence. During the daytime, when $\zeta \ll 90°$, $CE_{\lambda_2}^{\lambda_1} = C_{\lambda_2}^{\lambda_1}$, a quantity given by Eq. (IV.5, 11). As the sun nears the horizon, a dispersion begins to develop in the effective heights of the earth's shadow, so that $CE_{\lambda_2}^{\lambda_1}$ increases, the sky reddens, and the semitwilight (dusk or sun-set) phase commences (Fig. 141). At a certain value of ζ, $\Delta\bar{\eta}$ ceases its rise and the height dependence of $Q = - d(\ln \tau'_{11})/dh$ begins to dominate; the sky therefore becomes progressively bluer during this total-twilight phase. Then, when the height of the earth's geometric shadow reaches ≈ 100 km, the con-tribution of doubly scattered light begins to rise rapidly, so that the twilight sky reddens again as deep twilight sets in. Soon afterward the night sky be-gins to emerge through the haze of scattered sunlight, and the deep-twilight phase culminates with the transition to night.

Figure 141 illustrates schematically the sequence of twilight phases at the observer's zenith (see Fig. 27, Chap. I, §3). The times of transition be-tween phases and their duration depend on the wavelength band selected and on the line of sight. Note that throughout the twilight period the sky is redder than during the day.

Since doubly scattered light is rather highly polarized, though less so than singly scattered light, the beginning of deep twilight will be evinced in the behavior of the $p(\zeta)$ curves, although not so strongly as in the $CE_{\lambda_2}^{\lambda_1}(\zeta)$ curves. This is the explanation given by Ginzburg [259], Ginzburg and Sobolev [260], Robley [269], Dave [270], and Divari [113] for the polarization mini-mum or "step" at angles $\zeta = 96$-$100°$ (Figs. 34, 35, 37, 38, 40, 41). However, a serious objection may be raised at once against this explanation. As we have seen, the weight of information available compels us to recognize that the contribution of doubly scattered light does not begin to rise until ζ has reached larger values than the above, as indicated by the transition from a progressively bluer to a redder sky (since doubly scattered light is redder than singly scat-tered). It is true that on some days the sky begins to redden slightly (some-times considerably) earlier than usual, at times when $\zeta = 96$-$97°$. But in all likelihood this exception arises from an enhanced turbidity in the lower layers of the stratosphere. We must therefore seek another explanation for the mini-mum in the polarization.

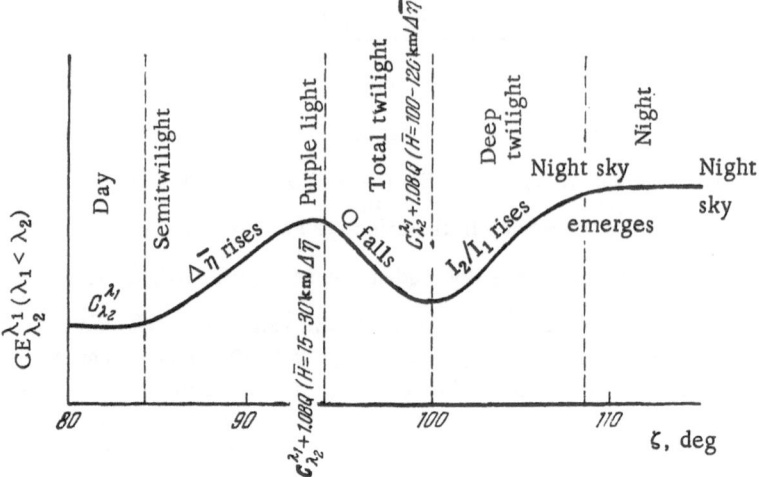

Fig. 141. Schematic sequence of twilight phases for the zenith region of the sky.

Some authors are inclined to explain the reduced degree of polarization at $\zeta \approx 96°$ by a rise in the relative contribution of the nearly unpolarized intrinsic glow of the upper atmospheric layers, as if this component were already beginning to emerge against the bright background of scattered sunlight. However, one will readily find (Figs. 14, 16, 17) that the intrinsic airglow does not become perceptible (except in the near infrared spectral region) until ζ is considerably larger, and the longer the wavelength, the earlier it appears. But the drop in polarization proceeds in the opposite sense, beginning in the short-wave region (Figs. 40 and 41, Chap. I, §4), so that the explanation cannot be accepted.

On the other hand, if the depolarization is caused by the presence of an aerosol layer located at heights of about 80-100 km, in the region where noctilucent clouds and the temperature inversion with height (mesopause) occur, then its initial appearance should come at short wavelengths, because the dispersion in the effective shadow heights will ensure that the twilight ray reaches the aerosol-layer region for a value of ζ increasing with λ.

It should be added that at all heights the degree of polarization remains smaller than for secondary scattering alone; that is, aerosol scattering essentially prevails throughout the height range. Thus preference should be given to the hypothesis that explains the dip in the polarization curves for angles ζ in the range 96-100° through the presence of an aerosol layer at the corresponding heights. This interpretation also helps us to understand the correla-

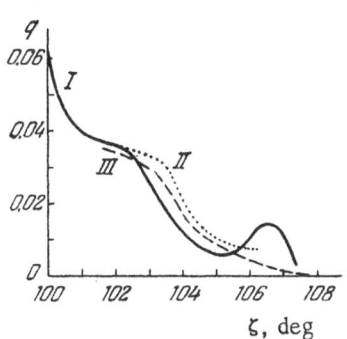

Fig. 142. The gradient q(ζ) for
ζ > 100°, at z = 0°, for three days
of observation: I) August 9; II)
August 13; III) August 30.

tion between the degree of depolarization and the critical frequency for
radio-wave reflection (Fig. 36). Further arguments in support of this hypothesis are provided by Dietze's polarization measurements [123], and by
the fact that as ζ continues to increase, the degree of polarization will
often begin to rise (Figs. 34, 35, 37),
a behavior which could not be understood at all if one were to hypothesize
an increased contribution from the intrinsic airglow or from secondary scattering. In this last case, some rise in
the degree of polarization could in
fact occur, because of the reduced
angular size of the twilight segment;
but it would be a very weak effect, and would essentially be overwhelmed by
an effect operating in the opposite sense — the increased relative intensity of
doubly scattered light.

We may close by adding that Ginzburg [298] has shown the original idea
of Khvostikov to be untenable, namely that the polarization minimum for ζ ≈
96-100° might arise from the properties of light scattering by ions. Rozenberg
[96] has also demonstrated that the properties of scattering by molecules arranged in a force field cannot be responsible.

When ζ ≈ 102-104°, that is, for geometric shadow heights at the zenith
of H ≈ 200-250 km, we again observe a sharp change in the behavior of the
p(H) or p(ζ) curves (Figs. 34, 35, 37, 38, 40, 41), beginning sooner at long
wavelengths, and manifestly having an entirely different cause. Calculations
by Dave [270] show convincingly that the reason for the rapid decline in p for
ζ ≳ 102-104° is that the nearly unpolarized glow of the night sky is already
beginning to emerge through the background of scattered sunlight. That the
night-sky glow has become perceptible (for the zenith, at the above values of
ζ) also appears distinctly in the ζ-dependence of the sky brightness and especially the gradient q = − d(ln I)/dH [96]. In fact, Fig. 15 (Chap. I, § 2) shows
the variation in the slope of the I(ζ) curve for this range of angles ζ (at the
boundary of astronomical twilight). In Fig. 142, typical q(ζ) curves for ζ >
100° are presented, as derived from measurements by the author at the zenith
on three days [95, 96].

One can readily show that the decline in q for $\zeta \gtrsim 102\text{-}104°$ does arise from the superposition of the relatively constant night-sky brightness on the diminishing brightness of scattered sunlight. To this end, one need merely turn to Figs. 15, 16, or 17, which show at once that the brightness of the night-sky glow comprises about 10% of the twilight-sky brightness at just these values of ζ, but at smaller ζ in the red spectral region than in the blue.

Figure 39 also provides a convincing demonstration of this interpretation, for it clearly shows that at angles $\zeta \approx 102\text{-}103°$ the emergence of the constant background of night-sky glow begins to affect the background of the fainter component $I_{\parallel}$, whereas the stronger component $I_{\perp}$ experiences this change only when $\zeta \gtrsim 104°$. We can also understand in this way the correlation between the degree of depolarization in the sky for this range in the angle ζ and the critical frequency for radio-wave reflection (Fig. 36), because the brightness of the night-sky glow is closely associated with the state of the ionosphere. Furthermore, in addition to the night-sky glow there is often a significant contribution from the background consisting of light from terrestrial sources that has been scattered in the atmosphere (such as from large cities). In this connection, we note that the attempt by Ginzburg and Sobolev [260] to explain through secondary scattering the strong rotations in the plane of polarization (Fig. 42) for $\zeta \gtrsim 104°$, as detected by Rozenberg, cannot be regarded as convincing (see also [39, 96, 171]).

We arrive, then, at the conclusion that during the semitwilight and total-twilight periods the main factor responsible for the brightness of the sky is single scattering of direct sunlight above the level of the effective boundary of the earth's shadow. During the deep-twilight period, however, a change occurs twice in the mechanism responsible for the luminosity of the sky. At first, multiply scattered light assumes the dominant role, and then the night-sky glow. Thus from the standpoint of extracting information from twilight observations on the structure of the atmosphere, only semitwilight and total twilight remain of interest to us, and it is with these periods that we shall be concerned in Chap. V.

CHAPTER V

THE INVERSE THEORY:
TWILIGHT SOUNDING OF THE
MESOSPHERE

§ 1. Formulation of the Problem

Up to this point we have endeavored to explain and calculate the principal twilight phenomena, and to clarify how the various properties of the atmosphere's optical structure influence the course of these processes. The discussion has established that while secondary scattering of light by the atmosphere does contribute significantly during the bright portion of twilight – the
semitwilight and total-twilight periods –it is not the predominant factor, and
the distinctive features of twilight phenomena result from single scattering of
direct sunlight that has been attenuated along the way to the scattering site.
Secondary scattering is allotted a substantial role, but does not.have a decisive impact.

We have also found that during the total-twilight period the main contribution comes from scattering at relatively great heights, over a range corresponding approximately to the mesosphere and in part to the lower ionosphere adjacent to the mesopause: from 20-30 km to 100-120 km. This distribution affords us with the opportunity of utilizing observations of the twilight sky to extract information relating to these and lower heights, a region
which, in spite of developments in rocket technology, has remained comparatively inaccessible to investigation. We are confronted, then, with the problem of how twilight observations are to be interpreted and efficiently organized – twilight theory in its inverse form. The scope of this problem, at least
in its present stage, is defined by the need to answer the following three ques-
tions: 1) What information can be gleaned by analyzing the behavior of the
twilight sky, and with how much reliability? 2) What type of information
and how much of it will ensure a unique interpretation, reliably free of any
disturbing effects from properties of the atmosphere's optical structure or op-

275

tical instabilities? 3) What specific methods of acquiring and reducing this information can be recommended for extracting the data we need on the mesosphere?

These are the topics we shall consider in this chapter, insofar as is possible at the present time.

To open the discussion, we should return to Eq. (IV.1, 1), adding now a term $S'_i(z, A, \zeta, \lambda)$ to represent the contribution of second- and higher-order scattering, as well as the intrinsic airglow and scattered light from terrestrial sources; thus in taking

$$S_i(z, A, \zeta, \lambda) =$$

$$= I_0 \omega_0 P^m(\lambda) \int_0^\infty D_{i1}[h, \varphi(z, A, \zeta), \lambda] T[y(h), \zeta, z, A, \lambda] dh +$$

$$+ S'_i(z, A, \zeta, \lambda), \qquad (V.1, 1)$$

we shall regard the function $S_i(z, A, \zeta, \lambda)$ as known, and the function $D_{i1}[h, \varphi(z, A, \zeta), \lambda]$ to be determined.

To clarify the basic characteristics of this formulation of the problem, we restrict attention first to observations in the solar meridian ($A = 0, \pi$), and neglect the contribution of higher-order scattering, taking $S'_i(z, A, \zeta, \lambda) = 0$. In a fairly rough approximation this simplification is permissible during the bright portion of twilight, since during that period the quantity S_i varies over a factor of about 10^5, and inclusion of the term S'_i would introduce a change of no more than a factor of 2. Equation (V.1, 1) will then become

$$S_i(z, \zeta, \lambda) = I_0 \omega_0 P^{m(z)}(\lambda) m(z) \times$$

$$\times \int_0^\infty D_{i1}(h, \zeta - z, \lambda) T[y(h), \zeta, z, \lambda] dh.$$

$$(V.1, 2)$$

If we regard Eq. (V.1, 2) as a function of ζ for $z = $ const, or a function of z for $\zeta = $ const, then mathematically it will represent a Volterra integral equation of the first kind with respect to D_{i1}, with kernel T, and if the function T were known the equation could be solved by the usual techniques of integral-equation theory. The inverse twilight theory is a difficult problem precisely because the kernel T is unknown, and depends on the optical state of the atmosphere to the same extent as the required function $D_{i1}(h, \zeta - z, \lambda)$.

In principle, one might so design an experiment as to evaluate the func-tion T from independent measurements, such as through observations of the brightness of direct sunlight at various heights by means of balloons, rockets, or satellites; through measurements of the illumination on a satellite as it emerges from the earth's shadow, a technique exploited by Venkateswaran et al. [184]; or, finally, through indirect measurements, such as a determina-tion of the height dependence of $\sigma(h, \lambda)$ by probing the atmosphere with searchlights [19].

Such measurements are absolutely essential, to serve as a foundation for the twilight method of sounding the atmosphere, that is, as a direct test of the method in a variety of actual cases. However, one must remember that the function T is subject to substantial fluctuations, including short-period ones. The measurements would therefore only yield random, or at best, average or typical data for the function T. But information of this type would not suffice to solve the inverse problem; in each particular case the function T will have to be measured specially, with the data having to be extracted from the pat-tern of the twilight sky itself. The possibility of doing so is afforded by the fact that the relations (V.1, 2) or (V.1, 1) essentially represent a system of equivalent integral equations for different z or ζ, and if the form of the func-tion T is known merely in a general way, the values of the parameters char-acterizing its behavior in a particular case can then be extracted from an analysis of the system. It is the search for efficient procedures in this direc-tion that constitutes the main problem of the approximate twilight theory.

The physical meaning of the problem becomes particularly evident if we approach Eq. (V.1, 2) from another point of view. Tracing the dependence of S_i on ζ or z may in fact be regarded as a process of scanning the function $D_{i1}(h)$ by means of an instrumental function T, where unlike ordinary scanning situations, the form of the instrumental function here varies regularly during the scanning process, and moreover is subject, in a manner unknown to us, to fluctuations with atmospheric conditions. However, the scanning proceeds si-multaneously in two interrelated directions ζ and z, and if we know how the behavior of the instrumental function differs for the ζ-scan as compared with the z-scan, we can infer certain information from Eq. (V.1, 2) on the manner in which the instrumental function T itself varies.

This procedure will be possible in principle for any type of relation be-tween the deformations of the instrumental profile as projected on the ζ- and z-axes. The operation can be performed realistically, however, only in the simplest cases, and with the proviso that the interrelation be described by a minimal number of unknown parameters (in our case, parameters specifying the state of the atmosphere).

From this standpoint, let us consider the behavior of the function T examined in detail in Chap. III, §5. We note at the outset that the way in which refraction effects and the air mass depend on the path of direct sunlight from the site of the observed point will have an appreciable effect on the form of the function T only during the semitwilight period; the dependence will be almost completely obliterated during the total-twilight period, particularly in the zenith region of the sky, where Eqs. (III.1, 28) and (III.1, 29) hold, so that we will have

$$h = H + \gamma y$$

(V.1, 3)

and $T = T(y, \lambda)$ will not depend on ζ or z. Furthermore, during the semitwilight period the scattering process of direct sunlight develops in approximately the same height range as the attenuation process; thus the functions D_{i1} and T are related to each other since they are determined by the optical structure of the same atmospheric layers. An entirely different situation occurs during the total-twilight period, outside the selective-absorption regions. For although the form of the function T is still controlled by the structure of the lower atmospheric layers, the form of the function D_{i1} depends only on the structure of the higher layers; as a result, the functions D_{i1} and T are independent.

It follows at once that the most favorable period for attacking the inverse problem of twilight theory — carrying out a twilight sounding of the atmosphere — is during total twilight, since the problem is complicated by too many extraneous factors during the semitwilight period. Since we have far less interest in applying the twilight method for an investigation of the lower, better studied, and comparatively accessible layers of the atmosphere, especially if we recognize the low resolving power of the twilight method, we shall henceforth concentrate our attention primarily on the investigation of the mesosphere, during the total-twilight phase. But we shall exclude from consideration the region of sky near the horizon, since at large z, as we have seen, the magnitude and variations of the transparency of the layers of air near the ground assume decisive significance.

§2. The Twilight Layer and the Principal Results of Twilight Sounding

The first systematic attempt to solve the inverse problem of twilight theory was that undertaken by Fesenkov [5], who used his own measurements of the sky brightness at the zenith as a function of ζ. The solution was based on the assumption, a natural one at the time, that the scattering of light in the upper atmosphere is exclusively molecular in character, with a scattering coefficient (equal to the extinction coefficient) proportional to the density of the air. In place of Eq. (V.1, 2), one then obtains for the sky brightness at z = 0 the expression

$$I(\zeta) = \text{const} \int_0^\infty \rho(h) T(h, \zeta) \, dh,$$

$$(V.2, 1)$$

where ρ is the air density at height h. The form of the instrumental function T(h, ζ) is computed from a certain model for the structure of the lower atmosphere, with a correction for refraction. Fesenkov then expresses the ρ(h) relation in the form

$$\rho(h) = \rho_0(h) [\alpha + \beta h + \gamma h^2],$$

$$(V.2, 2)$$

where ρ_0(h) is adopted arbitrarily on the basis of general knowledge about the atmosphere, while the coefficients α, β, and γ are quantities to be determined. Since ρ_0(h) and T(h, ζ) are known, substitution of Eq. (V.2, 2) into Eq.(V.2, 1) yields the relation

$$I(\zeta) = A(\zeta) \alpha + B(\zeta) \beta + C(\zeta) \gamma,$$

$$(V.2, 3)$$

where A, B, and C are known functions of ζ. If observations have provided us with a series of values of I(ζ) for different ζ, we can readily apply the method of least squares to find the parameters α, β, and γ, and thereby ρ(h). The resulting function ρ(h) can then be used in place of ρ_0(h) in Eq. (V.2, 2), enabling us, after recomputing the functions A, B, and C, to obtain the next approximations to ρ(h). In this way the solution of the inverse problem provides a determination of the height distribution of the density of the atmosphere.

However, Shtaude [85] has shown that the formulation (V.2, 2) leads to contradictory results upon comparison with experiment. Moreover, the atmosphere is essentially described by a function T(h, ζ) of particular form, in no way incorporated into the process of solving the inverse problem. Hence the specific results obtained by Fesenkov cannot be regarded as realistic. But his detailed analysis of the twilight process has nevertheless remained of primary importance as groundwork for subsequent developments in the twilight theory. He not only demonstrated the possibility of solving the inverse problem, but he explained how one might effectively obtain approximate solutions to the inverse problem on the basis of the twilight-layer concept (Chap. IV, §2). Referring the reader to the original papers [77, 83, 85, 98, 237-240] for details, we shall review here only some of the principal implications of the twilight layer and its use in solving the inverse problem.

Since the twilight ray is refracted by only a few minutes of arc during the total-twilight period, according to Fesenkov's calculations [5] (see also Chap. III, §1, and Chap. IV, §2], we may neglect refraction. Moreover, Fesenkov assumes [77] that within the narrow confines of the twilight layer,

$$\rho(h) = \rho(H)\,e^{-K(h-H)},$$ (V.2, 4)

where $H = H(\zeta)$ is the height of the earth's geometric shadow, and K is constant within the layer. Evidently if the layer is well defined and there is no cloud structure in the upper atmosphere, there will be no serious objections to this assumption. We remark that Shtaude's suggestion [85] of replacing Eq. (V.2, 4) by the expression

$$\rho(h) = \rho(0)\exp\left[-\int_0^h K(h)\,dh\right]$$

introduces nothing essentially new, since this expression represents the limit of Eq. (V.2, 4) as the thickness of the twilight layer approaches zero, while $K(h)$ is here specified as a mean for the twilight layer, and cannot be defined more accurately within the scope of the twilight-layer concept.* For the instrumental function T(h, ζ), Fesenkov suggests, on the basis of calculation [5], adopting the approximation

$$T = \exp\left[-\alpha + \beta\sqrt{y} + \gamma y\right],$$ (V.2, 5)

where y is the perigee of a ray passing the zenith at height h, $\beta = 3.83$, and $\gamma = 0.352$. By substituting Eqs. (V.2, 4) and (V.2, 5) into Eq. (V.2, 1), Fesenkov [77] obtains the approximate relation

$$\frac{d\ln I}{d\zeta} = \frac{\cos\zeta}{\sin^3\zeta}\,RK$$ (V.2, 6)

between K and the measured quantity $d(\ln I)/d\zeta$, not depending on β and γ, that is, on the form of the function T. (The more exact solution obtained by Fesenkov has not yet been applied for twilight analysis because of its complexity.) As Shtaude has shown [85], the same relation follows directly from the twilight-layer concept if we adopt Eq. (V.2, 4), and assume that the twi-

---

*We note in passing that Shtaude's arguments, as set forth in her Chaps. III and IV [85], suffer to a large extent from a confusion of interpretation (see Chap. II, §2).

light-sky brightness depends on neither h nor ζ. In this way we obtain a physical explanation of the first approximation (V.2, 6), and of why its form is independent from that of the instrumental function. Its meaning is further clarified if we use Eq. (V.1, 3), and recall that γ is practically independent of ζ in the zenith region of the sky. We can then write Eq. (V.2, 4) in the form $\ln \rho(h) = \ln \rho(H) - K\gamma y$, whence for constant $y = y_{max}$, where y_{max} is the perigee height of the ray corresponding to the center of gravity h_{max} of the twilight-layer brightness, we will have

$$\left(\frac{d \ln \rho}{d\zeta}\right)_{\nu_{max}} = \left(\frac{d \ln \rho}{dh}\right)_H \cdot \frac{dH}{d\zeta} = K \frac{\cos \zeta}{\sin^2 \zeta} R = \frac{d \ln I}{d\zeta} ,$$

$$(V.2, 7)$$

so that

$$I(\zeta) = \text{const} \cdot \rho(h_{max}).$$

$$(V.2, 8)$$

This means that the brightness of the twilight sky at any moment will be determined, to within a constant factor, by the density (more accurately, by the scattering coefficient) of the air at the height where the center of gravity of the twilight layer falls at that time. But the form of the function T governs only the quantity y_{max}, that is, the actual value of the height h_{max}. In other words, the relation (V.2, 6) is a mathematical formulation of the twilight-layer concept, and is valid, as we have seen in Chap. IV, §2, only if the thickness of the layer is small in comparison with the scale of the mesosphere. The requirement that T be independent of h and ζ limits applicability of the concept to the total-twilight period.

If scattering were actually purely molecular in character in the mesosphere region, with aerosols absent at those heights, as most authors of papers on twilight sounding of the atmosphere have assumed, then a knowledge of the height dependence of K would yield a determination of the air temperature over a range of heights. In fact, if we regard the molecular composition of the air as constant with height, and revert to the equation of state

$$p = \frac{R_0}{M} \rho T,$$

$$(V.2, 9)$$

where p is the pressure, R_0 is the universal gas constant, M is the molecular weight, and T is the absolute temperature of the air, then we would obtain

$$\frac{M}{R_0} \frac{dp}{dh} = T \frac{d\rho}{dh} + \rho \frac{dT}{dh}$$

$$(V.2, 10)$$

or, with the hydrostatic equation

$$\frac{dp}{dh} = -g\rho,$$

(V.2, 11)

where g is the acceleration of gravity,

$$-\frac{d\ln\rho}{dh} = \frac{Mg}{R_0 T} + \frac{d\ln T}{dh}.$$

(V.2, 12)

On the other hand, by Eq. (V.2, 4) we have $d(\ln \rho)/dh = -K$, so that

$$K(h) = \frac{Mg}{R_0 T(h)} + \frac{d\ln T(h)}{dh}.$$

(V.2, 13)

If we discard the simplified twilight-layer concept and turn to the more correct relation (IV.4, 1), corresponding to a refined concept for the twilight layer [128], then for the case of pure molecular scattering Eq. (II.3, 12) would imply

$$\tau'_{11}(h) = \sigma_0^{\text{mol}} \frac{R_0 T_0}{gM} \frac{p(h)}{p_0}$$

(V.2, 14)

(the subscript 0 means that the quantities are referred to sea level), corresponding, for $z = 0$, to

$$I = I_0 \omega_0 P \sigma_0^{\text{mol}} \frac{R_0 T_0}{gM} \frac{p(h)}{p_0}.$$

(V.2, 15)

Then in the event that M is constant with height and statistical equilibrium obtains,

$$\frac{d\ln I}{d\zeta} = \left(\frac{d\ln p}{dh}\right)_{h_{\text{max}}} \frac{dh_{\text{max}}}{d\zeta} = \frac{gM}{R_0 T} \frac{\cos\zeta}{\sin^2\zeta} R_0,$$

(V.2, 16)

whence we can determine the air temperature at height h_{max} (or more accurately, as we have seen in Chap. IV, §4, at height $\bar{H}$); and we may also write

$$\frac{dI}{d\zeta} = I_0 \omega_0 P \sigma_0^{\text{mol}} \frac{R_0 T_0}{gM p_0} \left(\frac{dp}{dh}\right)_{h_{\text{max}}} \frac{dh_{\text{max}}}{d\zeta} = I_0 \omega_0 P \sigma_0^{\text{mol}} \frac{\cos\zeta}{\sin^2\zeta} R_0 \frac{p(h_{\text{max}})}{p_0},$$

(V.2, 17)

which yields the air density at height h_{max} (or $\bar{H}$).

The question unavoidably arises as to how much of our discussion remains valid if we recognize that scattering of light at all heights actually is essentially aerosol in character — that it is governed not so much by the air density as by the content of scattering constituents. At first glance, it might seem that this circumstance would simply annul any opportunities for twilight sounding of the atmosphere, or at any rate limit them merely to a study of atmospheric aerosol. Actually, however, the situation is altogether different.

In the first place, the role of aerosol scattering depends materially on wavelength, diminishing rapidly as the wavelength shortens. Thus, in particular, we have an opportunity in principle to separate the effects of molecular and aerosol scattering to some extent, and by taking measurements in the short-wave spectral region we can substantially reduce the aerosol effects. Secondly, apart from the case where cloud structure is present (noctilucent clouds, for example), both observations and theoretical estimates show that the relative concentration of optically active aerosol varies comparatively weakly with height. So far as can be judged from available data ([19]; Part II of [13]), the aerosol scattering coefficient in a typical section of the visible spectrum is of the same order of magnitude at all heights as the molecular scattering coefficient, and fluctuates from a fraction of the latter to about twice the molecular value. Particularly suitable from this standpoint are observations at the zenith, where the contribution of aerosol scattering is at a minimum because of scattering-diagram effects. Thus, the presence of aerosols will alter the scattering coefficient of air by no more than a factor of 2-3, as a rule, and even less in the short-wave spectral region. Meanwhile the air density will vary, over the height range of interest to us, by at least four orders of magnitude; even over the thickness of the twilight layer, across which one actually averages the data of a twilight sounding, the density will vary by almost one order. As a result, the aerosol constituent will produce an apparent increase in the air density, but generally speaking the increase will fall within the limits of error in the interpretation of data on the twilight-sky brightness. In particular, the uncertainty in the height h_{max} to which the data refer will amount to several kilometers, corresponding to an uncertainty in density close to the error arising from the fluctuations in atmospheric aerosol. In other words, the presence of aerosols (provided they do not collect in well-defined layers of cloud) will cause an additional roughening of the data from twilight soundings of the atmosphere, but will not deprive them of their significance, even in comparison with the data of direct rocket experiments.

In this connection, three papers seem especially important, both for the observational material on which they are based and for their findings. In 1948, Megrelishvili and Khvostikov [276] reported the initial results of a protracted and comprehensive program for measuring the brightness of the twilight sky, to acquire data on the density and pressure of air in the upper atmosphere.

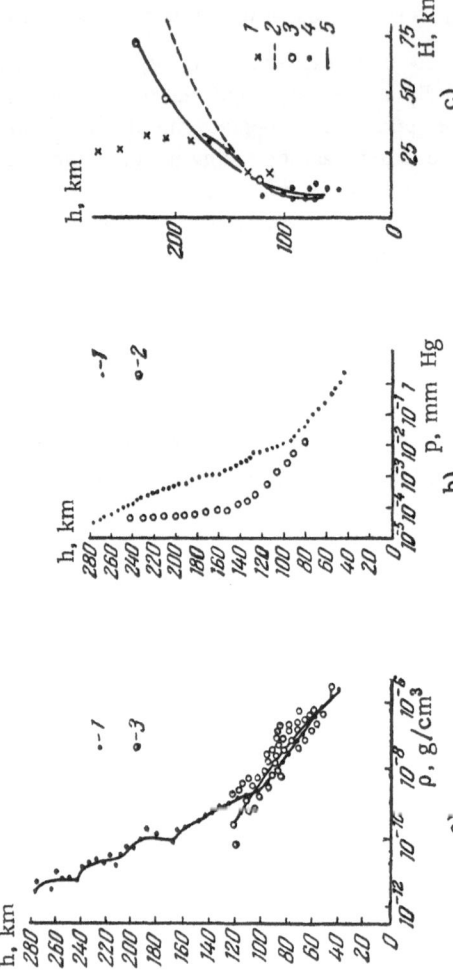

Fig. 143. Density ρ, pressure p, and height H of a homogeneous atmosphere as mean functions of the height h of the atmospheric layers. 1) Data from twilight sounding [276]; 2) observations of polar aurorae [265]; 3) meteor observations [264]; 4) results of ionospheric measurements; 5) mean curve in panel c.

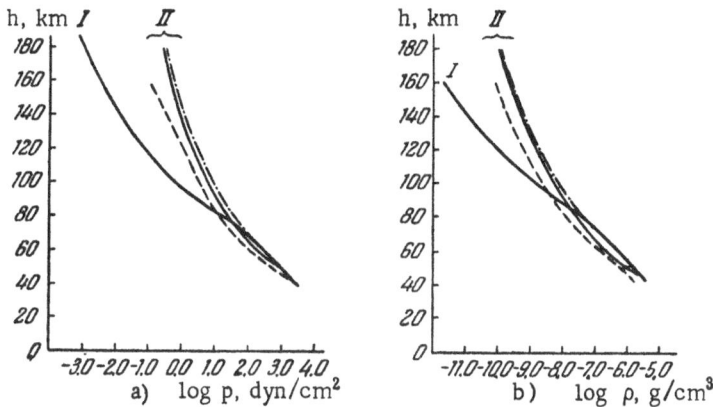

Fig. 144. Air pressure and density as functions of height, for various techniques of measurement. I) Rocket data; II) twilight data (λ = 3790, 5270, 9400 A).

The authors utilized a mean ζ-dependence for the zenith brightness of the twilight sky, as derived from the results of observations of 250 individual twilights, with the effects of the night-sky background removed. With this mean brightness curve, the expressions (V.2, 15) and (V.2, 17), and the relation

$$\ln \frac{\rho_0}{\rho} = \frac{h}{H(h)},$$

(V.2, 18)

where $H(h) = R_0 T/gM$ is the height of a homogeneous atmosphere at the level h, the authors proceeded to obtain mean climatological curves for $p(h)$, $\rho(h)$, and $H(h)$. Figures 143a, b, and c present their results. We readily see that up to heights of about 100-120 km, the data of the twilight measurements agree well with the findings of other authors — data obtained by different methods and representing the whole complex of experimental information available at that time on the upper layers of the atmosphere. In fact, the mean of all the previous data on the height distribution $H(h)$ practically coincides with the curve derived from the analysis of the twilight-sky brightness (Fig. 143c). The twilight data were reduced without any correction for secondary scattering, and the authors were justified in regarding the comparison as demonstrating that the correction must be a relatively small one.

Megrelishvili [109] published a second paper in 1956, in which she analyzed the possibilities of reconciling the twilight method of atmospheric re-

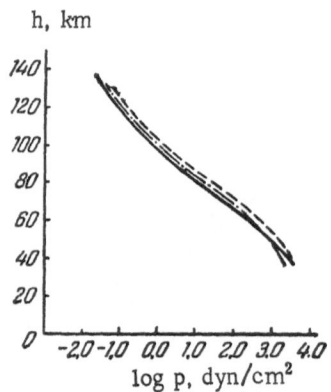

h, km

log p, dyn/cm²

Fig. 145. Air pressure as a function of height, according to rocket meas-urements (solid curve) and twilight data obtained on October 13 and 27, 1955 (broken curves), corrected for secondary scattering.

search with other types of data. A new comparison of the height distri-bution of pressure and density, as de-rived from twilight observations in different parts of the spectrum, with rocket soundings of the atmosphere, as reviewed by Kallmann [277], in-dicates good agreement up to heights of about 90 km, but a sharp disparity at greater heights (Figs. 144a and b).

But the picture changes com-pletely if we introduce corrections for the background of secondary scat-tering. To this end, Megrelishvili used a procedure developed by Fesen-kov [83, 231, 232], involving simul-taneous observations at two points of the solar meridian, namely $z = \pm 70°$ (Chap. IV, §6; also Fig. 24). Figure 145 compares the rocket data with the results of twilight observations on two days; the reduction included a sec-ondary-scattering correction. The agreement is strikingly improved as com-pared with Fig. 144. And the good agreement persists up to a height of at least 120-130 km.

The third paper to which we wish to refer was published in 1952 by Ash-burn [103]. A distinctive feature of Ashburn's work, like that of Megrelishvili, is that in addition to the careful attention paid to several factors sometimes ignored by other investigators, the author has measured the sky brightness not only in the zenith, but also in the twilight-glow segment, and has simultane-ously measured the degree of polarization of the light from the zenith. On the basis of these data, together with computations of the degree of polarization of doubly scattered light under molecular scattering, Ashburn was able to elim-inate the brightness of doubly scattered light and to derive the concentration of air molecules as a function of height. Unfortunately, the presentation given [103] does not permit one to follow the author's considerations in full and to compare them with other theoretical developments. In Fig. 146, Ashburn's re-sults are compared with values he has deduced from several other programs of twilight measurements, as well as with the data from rocket measurements and the standard NASA atmosphere. As in Megrelishvili's work [109], good agree-ment is found up to the 120-km level, if secondary-scattering effects are in-corporated. But if these effects are neglected, the agreement definitely breaks down as low as the 80-km level.

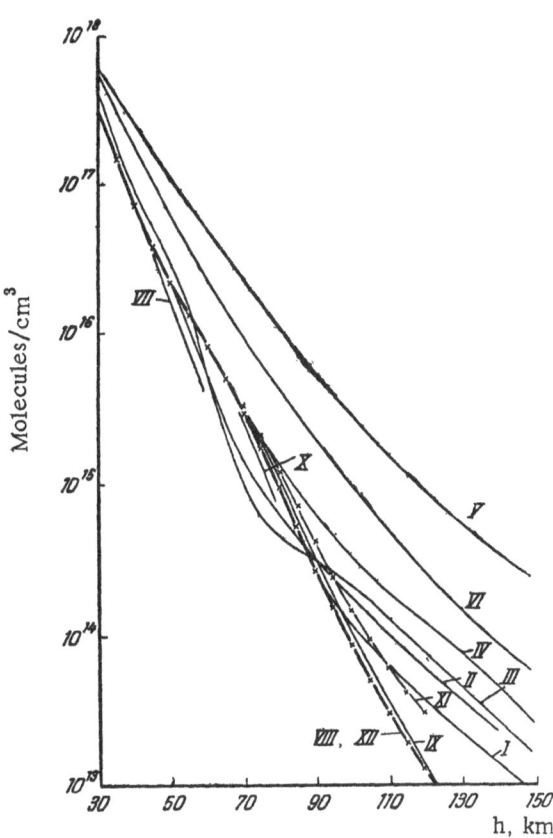

Fig. 146. Density of the atmosphere as a function of
height, compared for twilight and rocket measurements.
I) Megrelishvili and Khvostikov [276]; II) Link, 1949
[266]; III) Fesenkov [5]; IV) Ljunghall [67]; V) Link
[227]; VI) Chiplonkar [106]; VII) Hulburt [91]; VIII)
standard NASA atmosphere; IX) Grimminger; X)
Heavens et al. (V-2 rockets); XI) Ashburn, first approximation; XII) Ashburn, second approximation.

Thus on the one hand the available data indicate without question that the twilight method of probing the atmosphere indeed deserves to be applied. It has become perfectly clear, however, that the disparity already observed at heights $\gtrless 90$ km arises mainly from the influence of multiple scattering. We may conclude, in particular, that at lower heights the effect is comparatively minor, and does not become dominant until at least 120-130 km, in accord with the findings of the preceding chapter. Finally, a certain systematic excess in the values of the twilight data should, in all probability, be ascribed to the influence of atmospheric aerosol, which as we have seen will cause an apparent rise in p and ρ, but by no more than a factor of 2-3.

There is every reason to believe, then, that if they are appropriately organized and reduced, twilight observations are capable of yielding reliable data on the upper atmosphere. In fact, as we shall see presently, the possibilities of the twilight method are far from being exhausted, and substantial refinements can be anticipated in the quality of the material.

§3. A Differential Solution

Any approximate solution to the inverse problem of twilight theory that utilizes the twilight-layer concept must meet the requirement of a rapid decline in the scattering coefficient with height. However, in the upper atmosphere this decline becomes far milder, with the twilight layer extending over tens of kilometers; thus its height is more uncertain, and the resolving power of the twilight-sounding method is reduced. In its original form, moreover, the method cannot be applied to twilight photoluminescence (of sodium, say), for the distribution of luminescent material with height will differ completely from the distribution of air density. A more effective approximate solution to the inverse problem can be obtained by another procedure suggested by Rozenberg [96]; among other advantages, the calculation automatically eliminates the constant term in the twilight-sky brightness — the contribution of the night-sky glow and the illumination from terrestrial sources.

Let us first confine attention to the case of total twilight in the zenith region of the sky, where T is independent of ζ and z, and let us neglect the weak dependence of the scattering matrix on the scattering angle $\varphi = \zeta - z$ for values of φ close to 90°; thus we set $(\partial D_{i1}/\partial\zeta)_h = 0$. Then by differentiating Eq. (V.1, 2) with respect to ζ, we obtain the formula

$$\frac{dS_i}{d\zeta} = I_0\omega_0 P^m m \int\limits_0^\infty D_{i1}(h)\, T'\left(\frac{h-H}{\gamma}\right)\left(\frac{dy}{d\zeta}\right)_h dh,$$

$$(V.3, 1)$$

where $T' = dT/dy$. Since Eqs. (V.1, 3) and (IV.4, 12) imply that

$$\left(\frac{dy}{d\zeta}\right)_h \approx -\frac{1}{\gamma}\frac{dH}{d\zeta} \approx R\cos\zeta,$$

(V.3, 2)

we may write Eq. (V.3, 1) in the form

$$\frac{dS_i}{d\zeta} = I_0\omega_0 P^m mR\cos\zeta\int\limits_0^\infty D_{i1}(h)\,T'\left(\frac{h-H}{\gamma}\right)dh.$$

(V.3, 3)

Link [84, 294] was the first to derive this formula for the sky brightness $I = S_1$ (see also [160, 161]). Returning to Eq. (V.1, 2), Link now integrated by parts and, setting $T(0) = 0$, obtained

$$S_i = I_0\omega_0 P^m m\int\limits_0^\infty \tau'_{11}\left(\frac{dT}{dh}\right)_\zeta dh,$$

(V.3, 4)

and since

$$\left(\frac{dT}{dh}\right)_\zeta = T'\left(\frac{dy}{dh}\right)_\zeta = \frac{1}{\gamma}T',$$

(V.3, 5)

the function

$$S_i(\zeta) = I_0\omega_0 P^m\frac{m}{\gamma}\int\limits_0^\infty \tau'_{i1}T'\,dh.$$

(V.3, 6)

Finally, dividing Eq. (V.3, 3) by Eq. (V.3, 6), Link found

$$\frac{d\ln S_i}{d\zeta} = R\gamma\cos\zeta\frac{\int\limits_0^\infty D_{i1}T'\,dh}{\int\limits_0^\infty \tau'_{i1}T'\,dh},$$

(V.3, 7)

which yields Eq. (V.2, 6) if D_{i1} varies exponentially with h, so that we revert to the twilight-layer concept.

However, we shall follow another route [96] to Eq. (V.3, 3), differing from that of Link in that our treatment will rest on the behavior of the function $T(y)$, as investigated in Chap. III, §5.

In Fig. 147, the solid curve represents the $T'(y)$ relation for the case illustrated in Fig. 104. The figure indicates that this function reaches a sharp maximum at the inflection point y_i in the $T(y)$ curve. As shown in Chap. III,

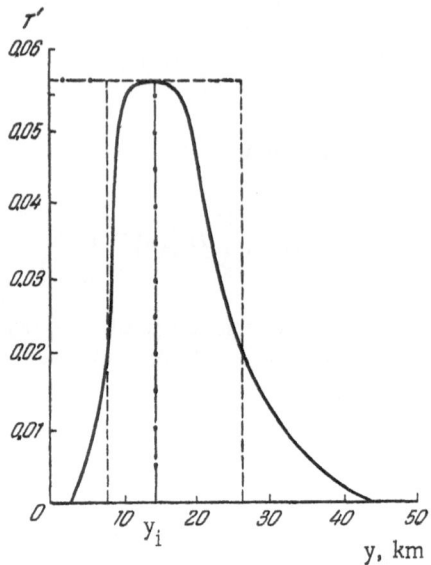

Fig. 147. A typical T'(y) curve outside the
selective-absorption region, corresponding
to Fig. 104.

§5, the optical structure of the atmosphere (in the absence of selective absorption) will affect the value of y_i almost exclusively, but not the shape of
the T(y) curve, nor thereby T'(y). The halfwidth of the maximum in T'(y)
is approximately 15 km, slightly smaller than the halfwidth of the twilight
layer, and it does not depend on K(H), so that there is an improvement in the
resolving power.

Returning now to the approximation (III.5, 12) to T(y) (see Fig. 104),
we obtain

$$T'(y) \approx \begin{cases} 0 & \text{for} \quad y \leqslant y_i - \dfrac{1}{K}, \\[2mm] e^{-1}K & \text{for} \quad y_i - \dfrac{1}{K} \leqslant y \leqslant y_i + \dfrac{e-1}{K}. \\[2mm] 0 & \text{for} \quad y > y_i + \dfrac{e-1}{K}. \end{cases} \qquad (V.3, 8)$$

In Fig. 147 this approximation is shown by dashed lines and, as the figure indicates, is quite satisfactory.

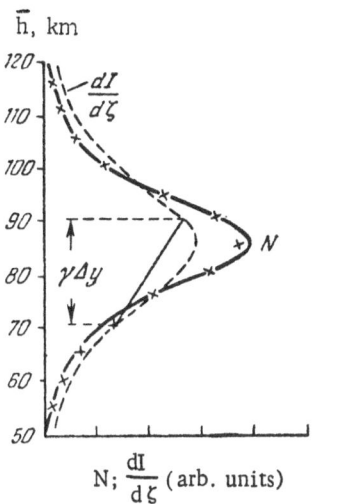

Fig. 148. A typical application of the chord method for refining the height distribution $D(\bar{h})$.

If we use our approximation in Eq. (V.3, 3), we obtain the formula

$$\frac{dS_i}{d\zeta} = I_0\omega_0 P^m mR \cos \zeta e^{-1}K \int_{H+\gamma\left(y_i-\frac{1}{K}\right)}^{H+\gamma\left(y_i+\frac{e-1}{K}\right)} D_{i1}(h)\, dh \qquad (\text{V.3, 9})$$

or, applying the theorem of the mean,

$$\frac{dS_i}{d\zeta} = I_0\omega_0 P^m R \cos \zeta \cdot \gamma D_{i1}(\bar{h}), \qquad (\text{V.3, 10})$$

where K refers to the ground layers of the atmosphere ($K \approx 0.15$),

$$\bar{h} = H + \gamma\bar{y}, \qquad (\text{V.3, 11})$$

and where, under reasonable assumptions for the height variation of D_{i1} in the absence of selective scattering, $\bar{y} \approx y_i$, with a difference of no more than ± 5 km between the two values. Thus, to the uncertainty indicated here in the determination of $\bar{h}$,

$$D_{11}(\bar{h}) = \frac{\dfrac{dS_i}{d\zeta}}{I_0\omega_0 P^m m R \cos \zeta \gamma}. \qquad (V.3, 12)$$

This expression admits of further improvement. We readily see that the transition from Eq. (V.1, 2) to Eq. (V.3, 3) corresponds to a transition from scanning with an instrumental function of the type of Fig. 104 to scanning with an approximately symmetric instrumental function of the type of Fig. 147, characterized by a halfwidth Δy. The general theory of instrumental profiles(see, for example, [278]) states that space frequencies with a period less than Δy are suppressed in the scanning process. However, if high space frequencies vanish without a trace, the first harmonic may be restored to a certain extent. As applied to Eq. (V.3, 3), the theory leads to the formula (see [278])

$$D_{11}(\bar{h}) = \frac{\dfrac{dS_i}{d\zeta}(H) + \dfrac{4\sigma^2}{\gamma^2(\Delta y)^2}\left[\dfrac{dS_i}{d\zeta}(H) - \dfrac{\dfrac{dS_i}{d\zeta}\left(H+\gamma\dfrac{\Delta y}{2}\right)+\dfrac{dS_i}{d\zeta}\left(H-\gamma\dfrac{\Delta y}{2}\right)}{2}\right]}{I_0\omega_0 P^m m R \cos \zeta \gamma \displaystyle\int_0^\infty T'(y)\,dy},$$

$$(V.3, 13)$$

where

$$\sigma^2 = \gamma^2 \frac{\displaystyle\int_{-\infty}^{+\infty} T'(z)\,z^2\,dz}{\displaystyle\int_0^\infty T(y)\,dy}, \qquad (V.3, 14)$$

with

$$z = y - y_i. \qquad (V.3, 15)$$

$$\bar{h} = H + \gamma y_i + \gamma \frac{\displaystyle\int_{-\infty}^{+\infty} T'(z)\,z\,dz}{\displaystyle\int_0^\infty T'(y)\,dy}. \qquad (V.3, 16)$$

Returning now to the approximation (V.3, 8) to T'(y), we have

$$\Delta y = \frac{e}{K}, \quad \int_0^\infty T'(y)\,dy = 1, \quad \sigma^2 = \frac{\gamma^2 e^2}{12K^2},$$

and $\bar{h} = H + \gamma y_i$, whence

$$D_{i1}(H + \gamma y_i) = \frac{\dfrac{dS_i}{d\zeta}(H) + \dfrac{1}{3}\left[\dfrac{dS_i}{d\zeta}(H) - \dfrac{\dfrac{dS_i}{d\zeta}\left(H + \gamma\dfrac{\Delta y}{2}\right) + \dfrac{dS_i}{d\zeta}\left(H - \gamma\dfrac{\Delta y}{2}\right)}{2}\right]}{I_0\omega_0 P^m m \cos \zeta \gamma}.$$

(V.3, 17)

The first term here coincides with Eq. (V.3, 12), but the second term represents a correction that can easily be shown graphically. In fact, if we draw a graph of $dS_i/d\zeta$ as a function of H or $\bar{h}$, and then draw the chord joining the points with $H + \gamma\Delta y/2$ and $H - \gamma\Delta y/2$, the distance between the $dS_i/d\zeta$ curve at the point H and the midpoint of the chord will be given by the bracketed expression in Eq. (V.3, 17). Figure 148 exemplifies how the correction can be made by the chord method for the case of the twilight flash of sodium [161]. By adopting a height distribution for the concentration N of sodium atoms (solid curve) and a function T'(y), the author computed $dI/d\zeta$ as a function of $\bar{h}$ (broken curve), and then used the chord method, Eq. (V.3, 17), to recover the height distribution of the sodium concentration (crosses). We see that the correction is a very effective one, approximately doubling the resolving power of the twilight-sounding technique.

§4. The Method of Effective Shadow Heights

The author has developed a variation on the differential solution described above to the inverse problem of twilight theory; this is the method of effective heights of the earth's shadow, and in our view it has definite advantages. The method is based on Eq. (IV.3, 2), with an added term to represent higher-order scattering:

$$S_i(z, A, \zeta, \lambda) = I_0\omega_0 P^m m \tau'_{i1}(\bar{H}) + S'_i(\zeta, z, A, \lambda). \quad \text{(V.4, 1)}$$

The left-hand member of this equation is a measured quantity. On the right-hand side, the unknown quantities are $P(\lambda)$, $\tau'_{i1}(h, \varphi, \lambda)$, $\bar{H}(z, A, \zeta, \lambda)$, and $S'_i(\zeta, z, A, \lambda)$. Evidently the amount and character of the information needed to solve the inverse problem will be governed by the relationships these quantities obey.

In Chap. II, §3, we found that P is subject to substantial fluctuations, and is quite difficult to measure accurately. We are therefore confronted at once with the need to reduce as much as possible the influence of the factor P^m, such as by observing from high mountains and in the zenith region of the sky. The air mass m will also be evaluated more accurately under these conditions (Chap. III, §3). Moreover, since P can vary over the sky, it is best to concentrate the observations in a relatively small area.

The contribution S'_i of doubly scattered light in Eq. (V.4, 1) depends strongly on ζ and λ, but comparatively little on z and A. We have shown in Chap. IV, §6, that the variation in the illuminance on the doubly scattering region of the atmosphere, and thereby in the brightness of doubly scattered light, as a function of z along the portion of the solar meridian near the zenith in comparison with the illuminance at the zenith itself does not exceed $|\Delta \ln E| \approx 5 \cdot 10^{-2}$ tg z [Eq. (IV.6, 6)], or about 1% at z = 10°. At large z, however, there is a substantial difference in the illuminance, further enhanced by the scattering-diagram effect, which is very considerable because of the high directivity of the primary-scattering process. But at small z, where the angle for secondary scattering is close to 90°, the diagram effect will be relatively weak.

Next, since secondary scattering occurs mostly in a layer located within and above the troposphere, the brightness of doubly scattered light will evidently be proportional to the quantity mP^m, like the brightness of singly scattered light; thus

$$S'_i (\zeta, z, A, \lambda) = I_0 (\lambda) \omega_0 P^m (\lambda) m b_i (\zeta, z, A, \lambda),$$

(V.4, 2)

where b_i depends only weakly on the line of sight, because of the variation in the illuminance E on the doubly scattering layer of the atmosphere, as well as the scattering-diagram effect. Specifically, $b_i(z) \approx b_i(0) (1 + 5 \cdot 10^{-2}$ c·tg z), where c < 1.

Finally, as was shown in Chap. III, §5, and Chap. IV, §3, the effective shadow height $\bar{H}(z, A, \zeta, \lambda)$ varies in a complicated manner during the semi-twilight period, but during total twilight in the region near the zenith, where h and y are related in a simpler way, we have the relation $\bar{H} = H + \gamma \bar{y}$, a convenient formula for the analysis; here $\bar{y} \approx y_i$. This circumstance clearly specifies the conditions under which the solution of the inverse problem will materially simplify: observations in the zenith region of the sky during the total-twilight period. We shall therefore confine our treatment to this case alone.

We consider first the conditions obtaining in the verticals A = ± π/2, that is, in the meridian normal to the solar meridian. According to Eq. (III.1, 25), we will have in this case

$$\bar{H} = \frac{R (1 - \sin \zeta) + \bar{y}}{\sin \zeta},$$

(V.4, 3)

so that $\overline{H}$ will be independent of z, and the z-dependence of $\tau'_{11}(\overline{H})$ will result exclusively from that of the scattering diagram, with the scattering angle φ given, in accordance with Eq. (III.1, 1), by the formula

$$\cos\varphi = \cos z \cos \zeta.$$

$$(V.4, 4)$$

Since $\cos \zeta \approx - \xi$ during the total-twilight period, where $\xi \ll 1$, we have $\varphi \approx \pi/2 + \xi \cos z$, and for $z \lesssim 60°$, φ will vary by no more than 5°, which is not a serious amount from the standpoint of generating a diagram effect. The conditions for appearance of secondary scattering are the same at all points of the meridian $A = \pm \pi/2$; that is, the quantity b_i will also be independent of z. Thus if we incorporate Eq. (V.4, 2), Eq. (V.4, 1) will take the form

$$S_i \left(z, A = \pm \frac{\pi}{2}, \zeta, \lambda \right) = I_0(\lambda) \omega_0 P^m(\lambda) m \left[\tau'_{i1}(\overline{H}) + b_i \right],$$

$$(V 4, 5)$$

where the bracketed quantities in the right-hand member do not depend on z. We may conclude that measurements along the meridian $A = \pm \pi/2$ will not yield any additional information about $\tau'_{11}(\overline{H})$, $\overline{H}$, or b_i as compared with measurements at the zenith. Nevertheless, for this very reason observations in the meridian can furnish an independent determination of $P(\lambda)$. In fact, Eq. (V.4, 5) implies that

$$\ln \frac{S_i \left(z, A = \pm \frac{\pi}{2}, \zeta, \lambda \right)}{S_i (z=0, \zeta, \lambda)} = \ln m - (m-1)\tau^*$$

$$(V.4, 6)$$

or

$$\tau^* = \frac{\ln m + \ln S_i (z=0, \zeta, \lambda) - \ln S_i \left(z, A = \pm \frac{\pi}{2}, \zeta, \lambda \right)}{m-1},$$

$$(V.4, 7)$$

where it is desirable for $m - 1$ to be small ($m = 1.5$-2), and for the measurements to be carried out in both of the verticals $A = \pm \pi/2$.

We turn next to the solar meridian, where if z is not too large we have, by Eq. (III.1, 24),

$$\overline{H} = [R(1 - \sin \zeta) + \overline{y}] \frac{\cos z}{\sin(\zeta - z)}$$

$$(V.4, 8)$$

and

$$\varphi = \zeta - z.$$

$$(V.4, 9)$$

Let us introduce the notation

$$B_i = \frac{S_i}{I_0\omega_0 p^m m} .$$

(V.4, 10)

Then

$$B_i(z, \zeta, \lambda) = \tau'_{i1}(\bar{H}, \lambda) + b_i(\zeta, \lambda),$$

(V.4, 11)

so that differentiation yields, with Eq. (IV.1, 11),

$$\left(\frac{\partial B_i}{\partial \zeta}\right)_z = - D_{i1}(\bar{H})\left(\frac{\partial \bar{H}}{\partial \zeta}\right)_z + \frac{\partial b_i}{\partial \zeta} ,$$

(V.4, 12)

where Eq. (V.4, 8) gives

$$\left(\frac{\partial \bar{H}}{\partial \zeta}\right)_z = - \frac{R \cos \zeta \cos z}{\sin(\zeta - z)} - \frac{\bar{H} \cos(\zeta - z)}{\sin(\zeta - z)} ;$$

(V.4, 13)

again, we have

$$\left(\frac{\partial B_i}{\partial z}\right)_\zeta = - D_{i1}(\bar{H})\left(\frac{\partial \bar{H}}{\partial z}\right)_\zeta + \frac{\partial b_i}{\partial z} ,$$

(V.4, 14)

where

$$\left(\frac{\partial \bar{H}}{\partial z}\right)_\zeta = \bar{H}\,\frac{\cos \zeta}{\sin(\zeta - z)\cos z} .$$

(V.4, 15)

For the zenith region of the sky, the second term in Eq. (V.4, 13) is much smaller than the first, and may be neglected. If $\partial b_i/\partial \zeta$ should also be small, Eq. (V.4, 12) will become

$$\frac{\partial B_i}{\partial \zeta} = D_{i1}(\bar{H})\,\frac{R \cos \zeta \cos z}{\sin(\zeta - z)} ,$$

(V.4, 16)

which is equivalent to Eq. (V.3, 12), and the whole challenge of the inverse problem of twilight theory will reduce to determining the height $\bar{H}$, that is, the quantity $\bar{y}$. It is upon Eq. (V.4, 16) that the accepted methods of solution essentially rest, although in the best practice one would retain the additional term $\partial b_i/\partial \zeta$, which can be evaluated independently by Fesenkov's method, say, or by theoretical estimates. Since the quantity $\partial b_i/\partial \zeta$ actually is not small, the quality of its estimation will control the accuracy of the method.

At the author's request, T. G. Megrelishvili has reduced the data of her own measurements, covering a number of days, by means of Eqs. (V.4, 14) and (V.4, 15). With her kind permission, we present herewith some of the re-

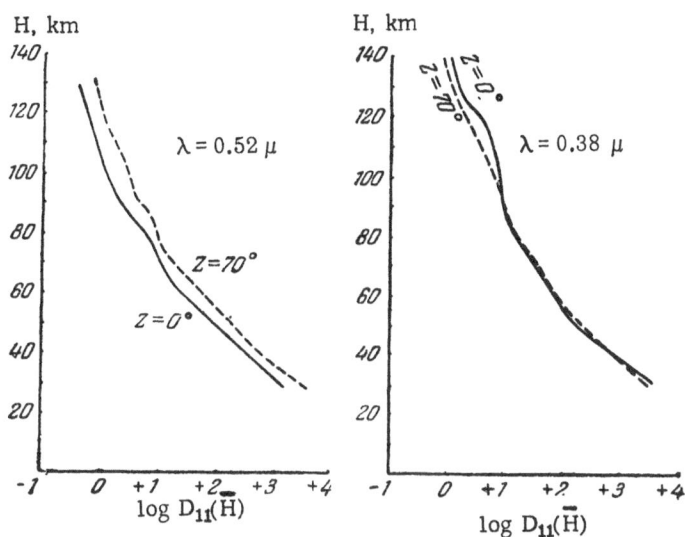

Fig. 149. Typical curves for log $D_{11}(\overline{H})$ as simultaneous functions of the height H of the geometric shadow at $z = 0°$ and $70°$, for two spectral regions. In this and succeeding figures, $D_{11}(\overline{H})$ is expressed in arbitrary units.

sults that were obtained. As initial data, the results were adopted of simultaneous measurements at three points of the solar meridian ($z = 0, \pm 70°$); one example appears in Fig. 24 (Chap. I, §3). Moreover, a slight modification of Fesenkov's method, Eq. (IV.6, 16), was used to determine the brightness of doubly scattered light for $z = 0°$ and $z = + 70°$ (with scattering-diagram effects neglected), in order that Eq. (V.4, 16) could be applied directly to evaluate $D_{11}(\overline{H})$ as a function of the height H of the earth's geometric shadow (without allowance for refraction).

Figure 149 shows the values of $D_{11}(\overline{H})$ obtained in this way on one morning, for two spectral regions (the measurements were taken with interference filters). Note the rather good agreement of the data obtained from the measurements at $z = 0°$ and $z = 70°$. Sometimes there is an almost perfect merging of the curves (to within the accuracy attained). In other cases, however, the agreement is considerably poorer, with a tendency for the curve at $z = 70°$ to exhibit a milder variation than at $z = 0°$. One would expect this effect to be caused partly by an insufficiently accurate correction for secondary scattering because of the neglect of diagram effects, and partly by a difference in the diagrams for singly scattered light at $z = 0°$ and $70°$.

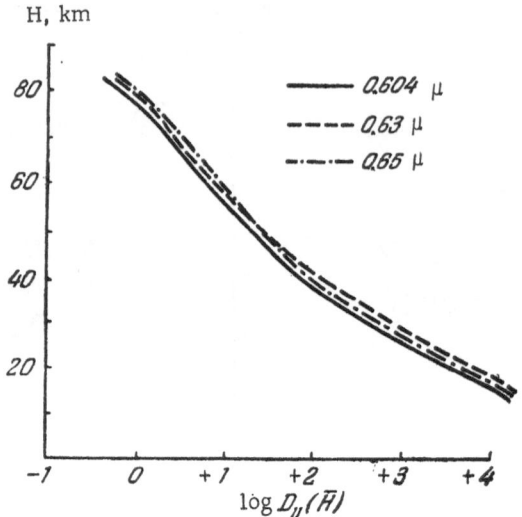

Fig. 150. Simultaneous curves for log $D_{11}(\overline{H})$ as functions of H, for three adjacent spectral regions.

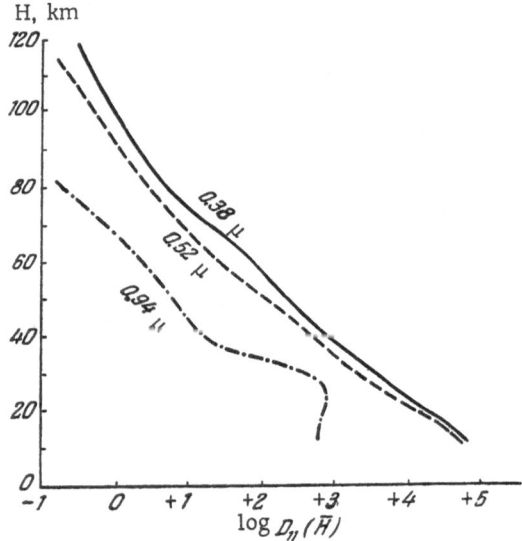

Fig. 151. Simultaneous curves for log $D_{11}(\overline{H})$ as
functions of H, for three widely separated spectral
regions.

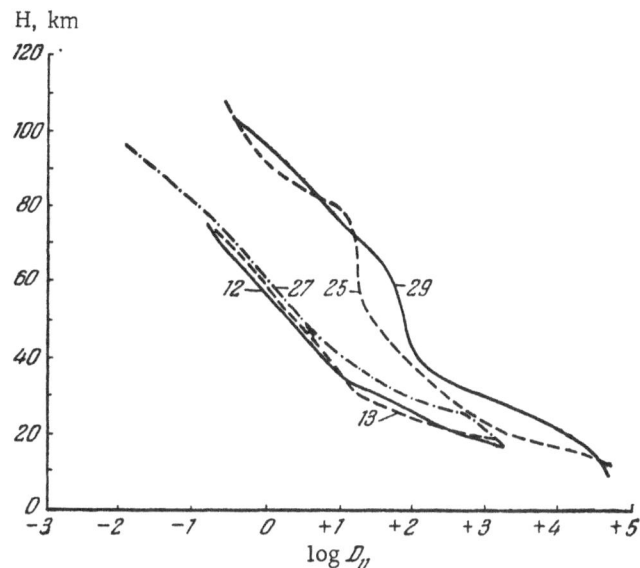

Fig. 152. Curves for log $D_{11}(\bar{H})$ as functions of H at $\lambda = 0.94$ μ, for several dates in a single month.

For one morning of observation, Fig. 150 presents the H-dependence of log $D_{11}(\bar{H})$, as obtained at the zenith in three adjacent spectral regions. The good mutual agreement of the curves indicates satisfactory accuracy in both the measurements and the computing procedures. The fact that all these measurements refer to the vicinity of the Chappuis absorption bands of ozone is of no great importance because, as we have seen, the bands are quite weak at the zenith during the semitwilight period. Moreover, the agreement of the curves in Fig. 150 furnishes evidence that the marked disparity between the curves representing well-separated spectral regions (Fig. 151) is a real effect. The maximum at $H \approx 20$ km in the curve for $\lambda = 0.94$ μ (Fig. 151) very likely represents a typical case of the effect of Bigg (the purple light), since the data were reduced under the assumption that total twilight had already been established. Thus we should not attach any real importance to it.

In comparing curves obtained on different days, the first feature we notice is their extreme individuality, both in the behavior of the function log $D_{11}(\bar{H})$ and in its absolute values. As an illustration, Fig. 152 presents curves for $\lambda = 0.94$ μ, as observed on several different dates during the same month.

Figure 153 shows similar curves for three successive days, as observed simultaneously in two nearby spectral regions. Clear evidence emerges of an

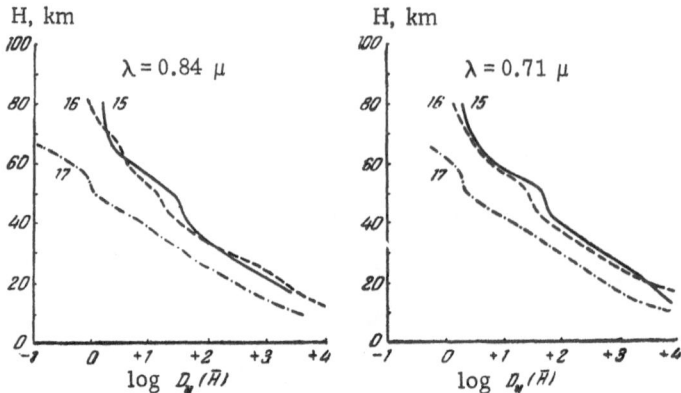

Fig. 153. Curves for log $D_{11}(\bar{H})$ as functions of H on three successive dates, revealing the presence of an aerosol layer.

aerosol layer at the level H $\approx$ 50-55 km, remaining almost stationary over the
three days. If we take into account the wavelength and the possibility of strong
turbidity in the lower atmospheric layers, we would expect that the aerosol
layer was actually located at a height of about 60-65 km. Note also the extension of the layer in height, which corresponds closely to the estimates we
have given of the resolving power of the twilight-sounding method.

Another example of a well-defined aerosol structure in the upper atmosphere appears in Fig. 152, in the curves for the 25th and 29th of the month.
But here we find that during the interval between these dates, on the 27th, the
structure is missing. This would appear to indicate that the aerosol component was arranged in the form of large-scale cloud systems, now entering, now
leaving the field of view. Since the true height of the aerosol layer was about
$\bar{H} \approx$ 80-90 km, although probably covering a considerable range in height(Fig.
153), we cannot exclude the possibility that noctilucent clouds had been observed in this case (the observations were carried out in the Caucasus
in October 1956).

The amplitude of the day-to-day variations in $D_{11}(\bar{H})$ may reach about
two orders of magnitude, especially in the near infrared spectral region. This
fact would not be surprising in itself, particularly in view of the large amplitude of the density and temperature fluctuations in the air at these heights, as
revealed by rocket-sounding data (see, for example, Figs. 48 and 49, Chap. II,
§1). A special, more detailed investigation would be desirable, however. Of
particular interest, in our view, is the case illustrated in Fig. 153, where between the 16th and the 17th a drop occurred in the scattering coefficient by

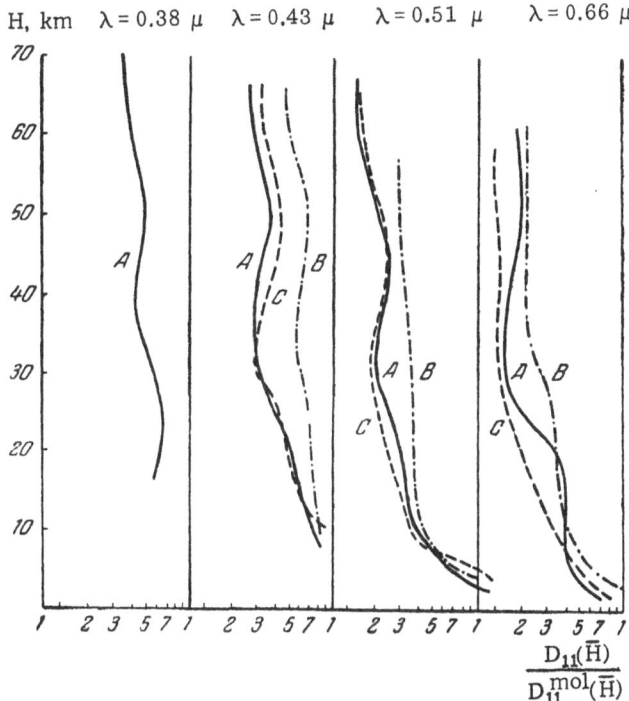

Fig. 154. Mean curves for $D_{11}(\overline{H})/D_{11}^{mol}(\overline{H})$ as functions of
H, according to Volz and Goody [282].

a factor of 5-7 over a wide range in height, and at the same time the aerosol
layer rose by 5-7 km with only a relatively slight change in its characteristics.

In this connection, we should return to the work of Volz and Goody
[282]. Adopting a certain mean distribution of atmospheric density with height,
and the results of absolute measurements of the twilight-sky brightness, these
authors determined, for a number of days, the ratio of $D_{11}(\overline{H})$ to the scattering
coefficient $D_{11}^{mol}(\overline{H})$ of a gaseous atmosphere at the same height. Figure 154
shows the results of averaging these ratios over three groups of days (A, B, C)
differing in the behavior of their twilight curves. Since the authors made no
correction for secondary scattering, they regard the decrease in $D_{11}(\overline{H})/D_{11}^{mol}(\overline{H})$
with increasing λ, shown in Fig. 154, as caused by secondary scattering alone,
and they subject only the data for $\lambda \approx 0.66\ \mu$ to further analysis. They reach
two significant conclusions in this way: they claim that the contribution of
aerosol scattering is usually small in the upper atmosphere, and that no well-
defined aerosol layers were detected.

It would appear, however, that both these conclusions are unfounded. To begin with, an inspection of Fig. 154 shows that even in the mean curves there is a well-defined minimum at heights in the range 30-40 km, with its position shifting systematically with wavelength. This last effect strikes us as particularly noteworthy. It evidently indicates that the theoretical derivation of the wavelength dependence of $\overline{H}$ (or h_{max} in Volz and Goody's treatment) is not exact enough, and that the values of $\overline{H}$ actually become too low at longer λ. But then the values of $D_{11}^{mol}(\overline{H})$ would diminish further at longer wavelengths, resulting in a considerable rise in $D_{11}(\overline{H})/D_{11}^{mol}(\overline{H})$ in the red spectral region, or an increase in the aerosol concentration. For this reason, one should exercise great caution when using Volz and Goody's estimates for the thermal effect in the aerosol component.

Moreover, the behavior of the curves shown in Fig. 154, and especially the analogous curves presented by the authors for different days, by no means support their claim that aerosol layers are absent. On the contrary, the curves distinctly reveal the presence of such layers; in analyzing them we must appeal to the treatment developed above for the resolving power of the twilight-sounding method. In particular, many of the authors' curves clearly exhibit a layer of enhanced scattering power located near the 50-km level.

We may add in conclusion that Volz and Goody have found that the maximum brightness of the twilight sky occurs during the winter and spring months, and they regard an improved scattering power of the atmosphere as the explanation. It is our opinion, however, that the maximum more likely is associated with an increased transparency in the lower layers of the atmosphere, that is, a decrease in y_i, and hence in $\overline{H}$. The method of calculation adopted by Volz and Goody does not enable one to distinguish between the two effects.

Earlier in this section, we noted that the opportunity of using Eq. (V.4, 16) is restricted by the need first to remove the effects of secondary scattering.

For this reason, another procedure would seem more efficient; it rests on a small value of $\partial b_i/\partial z$, especially in the vicinity of the zenith, in accord with the considerations we have given. In this event we should use Eq. (V.4, 14) instead of Eq. (V.4, 12), whence we obtain, again neglecting the correction $\partial b_i/\partial z$,

$$\frac{\partial B_i}{\partial z} = -D_{11}(\overline{H})\,\overline{H}\,\frac{\cos\zeta}{\sin(\zeta-z)\cos z}\,.$$

$$(V.4, 17)$$

Consider two nearby points on the solar meridian having $z = z_0 \pm \delta$, where $\delta \ll 1$. Then

$$B_i(z_0 - \delta) - B_i(z_0 + \delta) = -\left(\frac{\partial B_i}{\partial z}\right)_{z=z_0} \cdot 2\delta,$$

(V.4, 18)

so that

$$D_{i1}(\overline{H})\,\overline{H} = \frac{[B_i(z_0 - \delta) - B_i(z_0 + \delta)]\cos z \sin(\zeta - z)}{2\delta \cos \zeta},$$

(V.4, 19)

or, with Eq. (V.4, 10),

$$D_{i1}(\overline{H})\overline{H} =$$

$$= \frac{\left[S_i(z_0 - \delta)\,P^{m(z_0+\delta)-m(z_0-\delta)}\dfrac{m(z_0+\delta)}{m(z_0-\delta)} - S_i(z_0+\delta)\right]}{2\delta \cos \zeta I_0 \omega_0 P^{m(z_0+\delta)}\,m(z_0+\delta)}\cos z_0 \sin(\zeta - z_0).$$

(V.4, 20)

Evidently the accuracy of the result will depend entirely on the accuracy with which P is determined, and since this is not high, the measurements will not be reliable.

The case $z_0 = 0$ is an exception, however, for the expression (V.4, 20) reduces to

$$D_{i1}(\overline{H})\,\overline{H} = \frac{[S_i(-\delta) - S_i(+\delta)]\sin \zeta}{2I_0 \omega_0 \delta \cos \zeta P};$$

(V.4, 21)

the P-dependence is now much weaker, and is no longer of decisive import-ance. In selecting a value for δ, we are restricted by the need to avoid scat-tering-diagram effects as much as possible, and the need for Eq. (V.4, 18) to be valid, with the series broken off at the first term. Hence δ should not ex-ceed about 20° But values $\delta < 10°$ would hardly be suitable because of the weak z-dependence of $\overline{H}$; and in any event we would require specialized in-strumentation such as that described in [108] and [160].

Apart from Rundle, Hunten, and Chamberlain's special study of the twi-light flash of sodium [160], Karandikar's measurements [108] are the only ones that have been made by the procedure described here. Karandikar has unfor-tunately published only one of his experimental curves (Fig. 155). The meas-urements were carried out with a special photometer that furnishes the intensity difference ΔI directly as a function of ζ, for $\delta = 6°$. The ordinate in Fig. 155 is the value of ΔI in W m^{-2} sterad^{-1} for a spectral interval of 1 μ (interfer-ence filters were employed). Unfortunately the author does not report all the

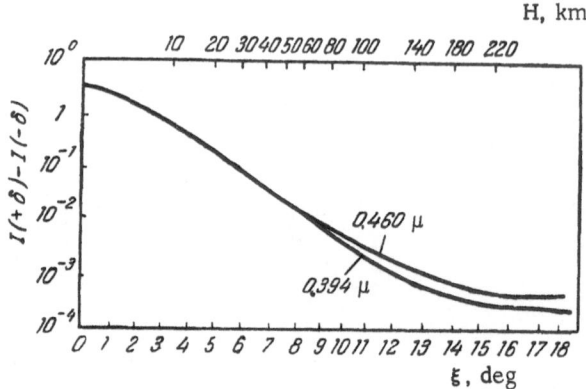

Fig. 155. Typical curves for $I(+\delta) - I(-\delta)$ as a function of ξ.

ence filters were employed). Unfortunately the author does not report all the details needed for a full computation using Eq. (V.4, 21), but for the plausible assumption $\bar{y} \approx 25$-30 km the relative behavior of $D_{i1}(\bar{H})$ agrees well with the rocket measurements shown in Fig. 146 (§2). The cleavage in the curve with respect to λ for $\xi \gtrsim 10°$ probably is caused by the night-sky glow, as it begins to emerge through the haze of scattered light. We note that Karandikar's treatment is based on ideas qualitatively similar to those leading to our Eq. (V.4, 21), expect that he regards the earth's geometric shadow as having a sharp boundary, so that $\bar{H}$ is replaced by H. One would then no longer obtain agreement with the data of Fig. 146, which may explain why Karandikar has refrained from giving an interpretation of his material.

But the principal source of uncertainty in evaluating the quantity $D_{i1}(\bar{H})\bar{H}$ remains the neglected contribution of secondary scattering, the quantity db_i/dz.

Since

$$\Delta \ln E \approx \Delta \ln b_i \approx \frac{\partial b_i}{\partial z} \cdot \frac{2\delta}{b_i} \lesssim 2.5 \cdot 10^{-2} \operatorname{tg} \delta \approx 0.1\delta,$$

we have

$$\frac{\partial b_i}{\partial z} \lesssim 5 \cdot 10^{-2} b_i.$$

On the other hand, we know that $-d(\ln \tau'_{i1})/dH \gtrsim 0.02$, corresponding to $D_{i1} \gtrsim 0.05\,\tau'_{i1}$, and since $\cos \zeta \approx -\xi$, we find from Eq. (V.4, 17):

$$\frac{\partial B_i}{\partial z} \gtrsim 5 \cdot 10^{-2} \bar{H} \xi \tau'_{i1} + \frac{\partial b_i}{\partial z}.$$

Suppose now that the quantities $\tau'_{11}(\overline{H})$ and $b_i(\zeta)$ in Eq. (V.4, 11) are of the same order of magnitude. Then

$$\frac{\partial b_i}{\partial z} \ll \frac{1}{1+\overline{H}\xi} \frac{\partial B_i}{\partial z} . \tag{V.4, 22}$$

During the total-twilight period, $\xi \approx 0.1$ and $\overline{H} > 20$ km, so that

$$\frac{\partial b_i}{\partial z} \ll \frac{1}{3} \frac{\partial B_i}{\partial z} ,$$

although this estimate is too rough in the sense of exaggerating $\partial b_i/\partial z$. The observations actually show (Chap. IV, §7) that $b_i < \tau'_{11}$ during the total-twilight period, so that $\partial b_i/\partial z$ can hardly exceed $\frac{1}{5} \partial B_i/\partial z$, and we may regard the value of $D_{11}(\overline{H})\overline{H}$ as determined with an uncertainty of about 25-30%.

If the value of $\overline{H}$ were known, the inverse problem of twilight theory would be completely solved. In fact, however, $\overline{H}$ can vary, although not over a very wide range, and the estimate of $\overline{H}$ has to be improved through actual observations of the twilight sky in each particular case. The significant uncertainty here is in the determination of $\overline{H}$ as the argument of the function $D_{11}(\overline{H})$ rather than as a factor, since D_{11} depends quite strongly on $\overline{H}$. But from general considerations of instrumental-profile theory, we cannot expect that $\overline{H}$, which depends in particular on the form of the function $D_{11}(\overline{H})$, will be known in advance to better than ± 3 km, corresponding to an uncertainty in D_{11} of a factor of ≈ 2, for a given geometrical height. And in general one can hope for no higher accuracy than this from twilight sounding of the atmosphere. Nevertheless, this estimate refers to the absolute value of $D_{11}(\overline{H})$, not to the relative variation of $D_{11}(\overline{H})$ with height, which can be ascertained to within about 25-30%, as we have seen.

Before examining in more detail the possibilities for deriving $\overline{H}$ from the data of twilight observations, let us consider how b_i can be determined from these data.

A reduction of the observational data will yield, through Eq. (V.4, 21), the quantity

$$G_i(\zeta) = D_{11}[\overline{H}(\zeta)] \overline{H}(\zeta), \tag{V.4, 23}$$

where by Eq. (V.4, 8) we have

$$\overline{H} = \frac{R(1-\sin\zeta)+\overline{y}}{\sin\zeta} \approx H + \overline{y}, \tag{V.4, 24}$$

since the denominator can be neglected in view of the uncertainty in $\bar{y}$ and the closeness of sin ζ to unity. We readily find that

$$G_i(\bar{H})\,d\ln\bar{H} = D_{i1}(\bar{H})\,d\bar{H} = d\tau'_{i1}(\bar{H}). \qquad (V.4, 25)$$

Moreover, since $(d\bar{H})_y = dH$,

$$d\ln\bar{H} = \frac{d\bar{H}}{\bar{H}} = \frac{dH}{\bar{H}}, \qquad (V.4, 26)$$

and if we regard G_i as a function of H,

$$d\tau'_{i1}[\bar{H}(H)] = G_i(H)\frac{dH}{\bar{H}} = \frac{G_i(H)}{H+\bar{y}}\,dH. \qquad (V.4, 27)$$

During total twilight, $\bar{y}$ will be of the same order as H, especially in the short-wave spectral region. Thus $\bar{y}$ may not be neglected in the denominator; but an estimate of $\bar{y}$ to within about 5 km will usually suffice. We may now integrate Eq. (V.4, 27) graphically, say, to obtain the value of $\tau'_{i1}(\bar{H})$ for given H or ζ:

$$\tau'_{i1}(H+\bar{y}) = \int_H^\infty \frac{G_i(H)}{H+\bar{y}}\,dH. \qquad (V.4, 28)$$

Substituting this value into Eq. (V.4, 11), we obtain

$$b_i(H,\lambda) = \frac{S_i(H,\lambda)}{I_0\omega_0 P^m(\lambda)m} - \int_H^\infty \frac{G_i(H)}{H+\bar{y}}\,dH \qquad (V.4, 29)$$

and the ratio of the Stokes parameters for doubly and singly scattered light at the zenith will be

$$\frac{S_i^{\,doub}(H,\lambda)}{S_i^{\,sing}(H,\lambda)} = \frac{b_i}{\tau'_{i1}} = \frac{S_i(H,\lambda)}{I_0\omega_0 P^m(\lambda)\,m\tau'_{i1}(H+\bar{y})} - 1. \qquad (V.4, 30)$$

Since the quantity $\tau'_{i1}(H+\bar{y})$ is determined here with an uncertainty of about 30-40%, the error in the estimate of the ratio S_i^{doub}/S_i^{sing} will reach 60-80%, and will be systematic in character since its main sources will be the difference of db_i/dz from zero and the inaccurate value of $\bar{y}$, the latter error declining with increasing H. But one can easily see that the uncertainty in the ratio will be no hindrance in identifying the region of heights where $b_i \lesssim \tau'_{i1}$, the region where the relation (V.4, 21) is applicable, and it will not interfere with the use of this relation there.

We may add that the relation (V.4, 21) determines not only the intensity, but in fact all the components in the first column of the scattering matrix at height $\overline{H}$. Furthermore, Eqs. (V.4, 28) and (V.4, 29) enable us to determine the character of the polarization of singly and doubly scattered light at the zenith. Specifically, by Eq. (II.2, 11),

$$I = S, \quad p = \frac{\sqrt{S_2^2 + S_3^2}}{S_1}, \quad q = \frac{S_4}{S_1}, \quad \mathrm{tg}\ 2\psi_0 = \frac{S_3}{S_1},$$

(V.4, 31)

whence we have for doubly scattered light

$$I = I_0 \omega_0 P^m m b_1,$$
$$p = \frac{\sqrt{b_2^2 + b_3^2}}{b_1}, \quad q = \frac{b_4}{b_1}, \quad \mathrm{tg}\ 2\psi_0 = \frac{b_3}{b_1},$$

(V.4, 32)

with analogous expressions for singly scattered light if b_i is replaced by τ'_{i1}.

It is important to note here that we are dealing with relative measurements, so that the error in the determination of p, q, and tg $2\psi_0$ will be much smaller than the error in I. In particular, the errors associated with the uncertainty in the height $\overline{y}$ and the imprecise knowledge of the atmospheric transparency will completely disappear.

It remains for us to explore the question of how the estimate of $\overline{y}$ might be improved. Our primary obligation will of course still be a direct measurement of the function T(y). Another way would be to use Eq. (IV.5, 10), deriving $CE_{\lambda_2}^{\lambda_1}$ and Q(H, λ) experimentally for a series of wavelength pairs, and determining the quantity $\Delta \overline{\eta}$ (and thereby $\Delta \overline{y}$) as a function of λ and $\Delta\lambda$, with subsequent extrapolation to longer wavelengths. This procedure might well result in a $\overline{y}(\lambda)$ curve having an error no greater than the limiting value of ± 3 km.

Again, if the upper atmosphere should contain extended, well-defined layers such as noctilucent clouds or the layer appearing in Fig. 153, it might be possible to detect them through measurements in different directions and in different spectral regions. Then by comparing the heights H (or zenith distances ζ) at which the twilight curves evince these features, we should have some basis for estimating the values and dispersion of the effective perigee heights $\overline{y}$.

Indeed, suppose that a cloud layer located at height h_c is observed at both $z = z_1$ and $z = z_2$ ($A = 0, \pi$). Then an extremum will be observed in the function $G_i(H)$, at different values of H in different directions, but all corresponding to the same value of $\bar{H}$, as given by the condition

$$\bar{H} = -\left(\frac{d \ln D_{i1}}{d\bar{H}}\right)^{-1}, \qquad \text{(V.4, 33)}$$

a level close to h_c, generally speaking. This means that we shall have the equation

$$\frac{[R(1 - \sin \zeta_1) + \bar{y}] \cos z_1}{\sin (\zeta_1 - z_1)} = \frac{[R(1 - \sin \zeta_2) + \bar{y}] \cos z_2}{\sin (\zeta_2 - z_2)}, \qquad \text{(V.4, 34)}$$

where ζ_1 and ζ_2 are the zenith distances joining at the extremes, observed in the directions z_1 and z_2, respectively; or the equivalent expression

$$H_1 - H_2 = \bar{y}\left(\frac{\cos z_2}{\sin (\zeta_2 - z_2)} - \frac{\cos z_1}{\sin (\zeta_1 - z_1)}\right), \qquad \text{(V.4, 35)}$$

where H_1 and H_2 are the geometric shadow heights corresponding to the extremum, as observed in the directions z_1 and z_2 respectively.

If the difference between z_1 and z_2 is large enough (say $z_1 = +70°$, $z_2 = -70°$), the quantity in parentheses, $\gamma(z_2, \zeta_2) - \gamma(z_1, \zeta_1)$, may attain values that would ensure the required accuracy in the determination of $\bar{y}$, despite the rough evaluation of the difference $H_1 - H_2$.

But actually we do not need to have a cloud layer at all. Suppose that measurements carried out at various z have enabled us to determine the function $G_i(H)$. Then the same heights H will correspond to the same values of $G_i(H)$, so that Eqs. (V.4, 34) and (V.4, 35) will again be valid. This situation enables us to obtain, during any twilight, a series of independent values of $\bar{y}$ for various z, and because of the weak $\bar{y}(\bar{H})$ dependence, to raise the accuracy in the determination of $\bar{y}$ by averaging the values. To do so, however, we must ensure adequate precision in measuring the atmospheric transparency, as this quantity forms the main source of error when $G_i(H)$ is measured in directions differing from the zenith direction. Moreover, the method ignores scattering-diagram effects, which also might be responsible for significant error.

Let us suppose, finally, that a change has occurred in $\bar{y}$ at some moment because of changing conditions in the vicinity of the terminator. This effect will appear, as we have seen, as a simultaneous alteration in the behavior of the $G_i(H)$ curves at all points of the solar meridian, beginning at different values of H depending on the zenith distance z. Thus it is more convenient here to regard G_i as a function of ζ. The slope of the $G_i(\zeta)$ curves will evidently be given by the relation

$$\frac{dG_i(\zeta)}{d\zeta} = \frac{dG_i(\bar{H})}{d\bar{H}} \left[\frac{dH}{d\zeta} + \gamma \frac{d\bar{y}}{d\zeta} \right], \qquad (V.4, 36)$$

where $dH/d\zeta \approx -R\cos\zeta$ and the quantity $\bar{y}\, d\gamma/d\zeta$ is small enough to be neglected. If we write Eq. (V.4, 36) in the form of finite differences, then

$$\Delta G_i = \frac{dG_i(\bar{H})}{d\bar{H}} [-R\cos\zeta\Delta\zeta + \gamma\Delta\bar{y}] \qquad (V.4, 37)$$

or, on a logarithmic scale,

$$\Delta\ln G_i = \frac{d\ln G_i(H)}{d\bar{H}} [-R\cos\zeta\Delta\zeta + \gamma\Delta\bar{y}]. \qquad (V.4, 38)$$

If the variation in $\bar{y}$ is of a short-term character, the derivatives $dG_i(\bar{H})/d\bar{H}$ or $d[\ln G_i(H)]/d\bar{H}$ can be evaluated for the "normal" behavior of the curves before and after the constancy of $\bar{y}$ is violated. In this way we can determine the value of $\Delta\bar{y}$ directly. Since under average conditions $d(\ln G_j)/d\bar{H} \approx 0.02-0.10$ km^{-1} (see Fig. 149) and $\gamma \approx 1$, the departure from the normal relation $\ln G_i(\zeta)\delta(\Delta\ln G_i) = [d(\ln G_i)/d\bar{H}]\gamma\Delta\bar{y}$ due to the change in $\bar{y}$ will be approximately $\delta(\Delta\ln G_i) \approx (0.02-0.10)\Delta\bar{y}$, and a variation $\Delta\bar{y}$ of 1-2 km will already be easily perceptible in the $\ln G_i(\zeta)$ curves. If we should also be able to determine the λ-dependence of $\Delta\bar{y}$, we would be able to estimate both the dispersion in $\bar{y}$ and the height of the barrier that has arisen in the vicinity of the terminator.

We conclude this section by noting that if the contribution of secondary scattering is small or of known amount, we can use Eqs. (V.4, 16) and (V.4, 17) to evaluate $\bar{H}$. Dividing the second of these by the first, we obtain

$$\bar{H} = -R\cos^2 z \frac{\dfrac{\partial B_i}{\partial z}}{\dfrac{\partial B_i}{\partial \zeta}}, \qquad (V.4, 39)$$

and the relative error in the determination of $\overline{H}$ caused by neglecting secondary scattering will evidently be

$$\frac{\delta \overline{H}}{\overline{H}} \approx \frac{\dfrac{\partial b_i}{\partial z}}{\dfrac{\partial B_i}{\partial z}} \,.$$

(V.4, 40)

CONCLUDING REMARKS

We have been compelled to invite the reader to accompany us along a lengthy and at times laborious route — a detailed analysis of the manifold phenomena that occur when sunlight is propagated and scattered in the earth's atmosphere under twilight conditions. Only in this way have we been able to treat the phenomena of twilight quantitatively without losing sight of their physical meaning. Many of them have emerged in a new and unexpected setting. Thus we have often had to reconsider accepted explanations resting mainly on intuitive ideas. For instance, we have found an entirely new interpretation for the anomalous-transparency effect, the coloration of the twilight sky, the purple light, the maximum in the $d(\ln I)/d\zeta$ curves as a function of ζ for small depressions of the sun, the color of doubly scattered light, and many other phenomena. A new division of twilight into phases has been achieved. But in the author's opinion the main result of this survey has been the opportunity of approaching the analysis of twilight from a unified quantitative standpoint which decisively rejects the speculative developments that still persist.

No less important, from the author's point of view, is the finding that while doubly scattered light does make a significant contribution during the semitwilight and total-twilight periods, it nevertheless remains a subsidiary factor, with the principal twilight phenomena controlled by single scattering of sunlight. Finally, it has been the author's aim to rehabilitate the twilight method of sounding the upper atmosphere; in particular, he has been encouraged to formulate certain new techniques for organizing and reducing twilight observations. Most of the results obtained by the author are here published for the first time. It is these circumstances that have governed the entire character of the book now commended to the attention of the reader. Much more space has been allotted to original developments than to the work of other investigators, although the author has tried to give a reasonably adequate survey of the literature and to provide the reader with a convenient key to it.

Many of the developments on which this book is based were worked out some fifteen years ago, and are partially represented in papers by the author dating from that time. However, the level of scientific achievement and the fund of observational material available in the mid-1940's did not allow the treatment to be carried through to a definitive form. By 1956, when the author was able to return to experimental research in the field of atmospheric optics and once again to take up the problem of the twilight sky, the situation had changed radically. Of particular value for the author was an extended stay in 1959 at the Abastumani Astrophysical Observatory, at the kind invitation of its Director, E. K. Kharadze. This opportunity to become familiar in detail with the extensive materials obtained by T. G. Megrelishvili over many years of observation has not only helped to refine the theoretical treatment but has also clarified a number of matters that had previously escaped the author's attention. The completion of this book therefore owes a great deal to E. K. Kharadze and T. G. Megrelishvili, and the author would like to express his sincere appreciation to them.

The author feels it would be well to point out in closing that the analysis presented here suggests a variety of new observational and instrumental problems. First among them is the need to set up a network of regularly operating stations, widely dispersed geographically, to carry out upper-atmosphere research by the twilight method. If properly organized, such a network would be able to secure information not only on the "climate," but also on the "weather" in the high layers of the atmosphere, including such parameters as the pressure, density, temperature, aerosol content, and possibly the wind. And the accuracy of the data might be as high as that attainable by current rocket-sounding techniques. In developing such a network, one must provide above all for the solution of a number of procedural problems, including the planning of standardized apparatus and methods for obtaining and reducing the observational data. In this regard, one should remember that the immense number of twilight measurements acquired in the past are of little value from the standpoint of interpretation, simply because the bulk and character of these data afford no opportunity of their being reliably interpreted. Furthermore, one of the primary tasks ahead will be to measure the function $T(y)$ directly, and to make a direct comparison of the twilight and rocket methods for sounding the atmosphere. If the publication of this book contributes toward the formulation and solution of these problems, its objective will have been achieved.

APPENDIX

Research on twilight phenomena has progressed in several directions since the original publication of this book and its condensed version [299]; the following are among the pertinent developments.

Divari [300-304] has further analyzed the Alma-Ata photoelectric photometry of the twilight sky along the solar vertical [298]. For $\xi \lesssim 6°$, approximate proportionality obtains (outside the shadow) between the brightness of primary twilight (singly scattered sunlight, described by the twilight ray) and secondary twilight (tropospheric scattering of the twilight ray) [300]. Application of Fesenkov's method (p. 265) yields information on the height distribution of atmospheric dust [301]. Polarimetry confirms that aerosol scattering sets in at h = 80 km, strengthening steadily (for $6° \leq \xi \leq 14°$) to 50-fold predominance at 130 km. The scale height $H_{dust} = 18$ km at h = 100 km, increasing with h; $H_{dust} = 1.2$-2.5 H_{mol}. An enhanced dust concentration from late August onward implicates the Perseid meteor stream.

The aerosol and molecular components of the primary- and secondary-twilight brightness have been computed [304] as functions of λ, ξ, z, and azimuth A for a model atmosphere utilizing twilight determinations of the dust profile [303].

Further twilight evidence for a turbid meteoritic dust layer has come from Dietze's zenith photometry and polarimetry [305, 123]. Computed polarization anomalies for $1° \leq \xi \leq 15°$ agree with Link's 2-μ particle layer, maintained semipermanently at h = 80-120 km.

In the Crimea, Bondarenko [306] has measured the λ-dependence of the zenith polarization p. With increasing ξ, p declines more rapidly than its Rayleigh value, especially in the morning and toward red wavelengths, and $p(\lambda)$ develops two maxima.

The Alma-Ata scattering and polarization conference (see [275, 298]) comprehensively surveyed Soviet programs. Yudalevich [307] reported preliminary results of his twilight computations (p. 255), while Darchiya [308] reviewed spectrophotometric evidence for varying ozone content (p. 43).

With the twilight-layer concept, Morozov [309] has obtained formulas expressing the brightness along the solar vertical due to molecular scattering for varying ξ; the treatment is consistent with Divari's measurements. The utility of the reversal effect (p. 165) for atmospheric sounding, in particular the ozone distribution, has been examined critically by Mateer [310]. Goody [311] has given a convenient approximation for the height h of a refracted ray in terms of y and ξ; he stresses the importance of the corresponding scale height H.

Twilight photometry has now been applied to study the lunar tide in the upper atmosphere. Although only 6% as strong as the solar tide, it significantly alters the distribution of scattering particles. Photoelectric and photographic Na airglow measurements by Barber [312] yield brightness maxima (constant ξ) at lunar ages of ≈ 13 and 23^d; the Na emission fluctuates up to $\pm 30\%$. Barber's harmonic analysis of Ljunghall's zenith photometry [67] indicates a 900-m semidiurnal displacement at $h \approx 110$ km, and an even more delicate true 29.5-day oscillation; both components at $h \approx 32$ km. Divari's observations confirm the lunar tide [302] but imply a stronger amplitude at his lower latidute, rising to 11 km at $h \approx 160$ km.

Of value as background, an extended review of molecular and aerosol scattering processes in the atmosphere has been contributed by Bullrich [313]. Recent Soviet work includes determination of scattering functions from photoelectric sky photometry at Alma-Ata [314-316], and a treatment of integral equations describing scattering in a spherical atmosphere [317], a generalization requisite for twilight analysis (p. 93).

Sodium layers at $h \approx 64$ and 96 km have been disclosed by interferometry of twilight Na emission, consistent with resonance scattering [318], while a twilight value $D_2/D_1 \approx 1.6$ for the Na doublet ratio has yielded information on the Na distribution [319]. In New Zealand, Gasden [320] has observed several hundred twilights with a photoelectric scanning spectrometer, finding seasonal and evening/morning effects in the Na ionization.

Twilight N_2^+ emission has been attributed to solar far-ultraviolet irradiation [321], and AlO bands have been observed [322]. Sullivan and Hunten [323] have derived atmospheric profiles for Na, K, and Li from twilight-glow emissions; sea water may furnish the Na and K, while the Li variation correlates strikingly with meteoritic influx and high-altitude thermonuclear detonations. Many observers have in fact reported intensities of twilight emissions, but discussions of the corresponding ionization—excitation mechanisms and photochemical processes fall beyond the scope of this book (see [255, 324]).

Cole [325] suggests that photoelectrons backscattered from the sunlit magnetically conjugate ionosphere, especially under an active winter sun, can raise the F-region electron temperature to yield the observed λ 6300 [O I] twilight enhancement. The 1963 volcanic eruption on Bali generated a scattering layer, explored through its brilliant color effects at twilight[326].

The analogous twilight phenomenon on Venus has been exploited to probe its atmosphere. Edson [327] recounts how measurement and photometry of the prolonged cusps of the planet's illuminated crescent yield accurate heights, typically a few kilometers, for the top of a thin gaseous haze layer above various parts of the opaque cloud deck, and offer a theoretical approach to the atmospheric heat-balance problem.

R. B. R.

PRINCIPAL NOTATION

A	Azimuth relative to solar vertical
A	Albedo of underlying surface
a	Term in approximation to y_i [Eq. (IV.1, 5)]
B_i	Functions (V.4, 10)
b	Term in approximation to y_i [Eq. (IV.1, 5)]
b_i	Directional factors in multiple-scattering function (V.4, 2)
$C^{\lambda_1}_{\lambda_2}$	Constant term in color index [Eq. (IV.5, 11)]
$CE^{\lambda_1}_{\lambda_2}$	Color index [Eq. (I.3, 1)]
D	Optical density (§II.2)
D_{ij}	Scattering matrix [Eq. (II.2, 16)]
D_{i1}	Scattering components for depolarized incident light
D_{11}	Differential scattering cross section
Dis(φ)	Scattering dissymmetry [Eq. (II.2, 20)]
E	Illuminance, lux
$f_{ij}(\varphi)$	Normalized scattering matrix [Eq. (II.2, 18)]
$\tilde{f}_{ij}(\varphi)$	Reduced scattering matrix [Eq. (II.2, 22)]
$f_{11}(\varphi)$	Scattering function
G(λ)	Spectral sensitivity, instrumental
G_i	$D_{i1} \cdot \bar{H}$ [Eq. (V.4, 23)]
g	Acceleration of gravity
$g(\vartheta, \psi)$	Angular distribution of scattered light
H	Scale height
H	Height of earth's geometric shadow [Eq. (III.1, 28); Figs. 84, 86]
$\bar{H}$	Effective shadow height (§IV.3)
h	Height above earth's surface
I	Intensity (brightness)

I_0	Intensity of sunlight outside atmosphere
I_1, I_2	Intensity of singly, doubly scattered light
I_λ	Monochromatic intensity
$I_\parallel$, $I_\perp$	Intensities of orthogonally polarized components
K	Elongation of scattering diagram (Fig. 74)
K	Logarithmic gradient of attenuation coefficient [Eq. (III.5, 4)]
$K(\rho)$	Attenuation efficiency factor [Eqs. (II.2, 27)]
k	Attenuation (extinction) coefficient (§II.2)
L	Height of observer
l	Path length; distance to observed point
M	Molecular weight of air
m	Air mass [Eq. (II.3, 3)]
m_{eff}	Effective air mass [Eq. (III.5, 1)]
m_t	Air mass at bend in log I(m) curve (Fig. 102)
m_z	Air mass along line of sight
m_ζ, m_ξ	Air mass for given solar zenith distance or depression at observed point
m_0	Air mass for ozone layer
N, n	Particle concentration
N	Point representing observer
n	Index of refraction
P	Observed point in atmosphere
$P(\lambda)$	Vertical transparency of atmosphere
p	Degree of polarization
p	Air pressure
p	Factor $\gamma m\, d\alpha/dy$ [Eq. (III.5, 27)]
Q	Point representing perigee
$Q(h, \lambda)$	Logarithmic gradient of optical thickness [Eq. (IV.5, 5)]
q	Specific humidity, g/kg
q	Polarization ellipticity [Eq. (II.2, 11)]
q	Logarithmic gradient of sky brightness [Eq. (IV.4, 2)]
R	Universal gas constant
R	Radius of earth
r	Radius of particle; distance
r	Amount of polarization (§II.2)
S	Surface area
S	Meteorological visibility
S	Stokes vector-parameter (§II.2); $\mathbf{S}^0$, for incident beam
S_i	Stokes parameters [Eqs. (II.2, 10)]
S'_i	Multiple-scattering terms [Eqs. (V.1, 1), (V.4, 2)]
$s(\tau_0)$	Function (III.4, 4), Fig. 100
T	Idealized transparency [Eq. (II.2, 42)]

T	Temperature
T'	Gradient of transparency with respect to perigee height
T^{-1}	Opacity (§II.2)
t	Instrumental transparency [Eq. (II.2, 51)]
t	Time
t_d	Instrumental transmissivity [Eq. (II.2, 48)]
u	Dimensionless distance to point on ray [Eqs. (III.1, 8)]
V	Volume
v	Velocity
v	$1 - \sin \zeta + x$ [Eqs. (III.1, 8)]
$v(\lambda)$	Visibility function (§II.3)
w	Dimensionless height of point on ray [Eqs. (III.1, 8)]
x	Dimensionless perigee height of ray [Eqs. (III.1, 8)]
y	Perigee height of solar ray
y_i	Perigee at inflection on transparency curve (Figs. 103, 104)
z	Zenith distance of line of sight at observer
z'	Zenith distance of line of sight at observed point
α	Absorption coefficient
α	Altitude of sun [Eq. (III.5, 16)]
$-\alpha$	Exponent in particle distribution by size [Eq. (II.1, 2)]
β	Specific absorption coefficient [Eq. (II.2, 41)]
γ	$\cos z/\cos \psi$ [Eq. (III.1, 26), Figs. 85]
Δ	Depolarization (§II.2)
$\dot\Delta$	$y - a$ (§IV.3)
δ_{ij}	Kronecker symbol
ε	$\sin^2 z \sin^2 A$ [Eqs. (III.1, 8)]
ζ	Zenith distance of sun at observer
ζ'	Zenith distance of sun at observed point
η	Height of observed point above shadow [Eq. (III.1, 29)]
$\bar\eta$	Effective shadow boundary (§IV.3)
Θ	Total transmissivity [Eq. (II.2, 43)]
Θ_d	Diffuse transmissivity [Eq. (II.2, 45)]
ϑ	Angle of incidence of ray
ϑ	Geocentric angle between observer and observed point (Figs. 82)
$\varkappa_{ij}$	Extinction matrix [Eq. (II.2, 14)]
λ	Wavelength of light
ν	Angle of refraction
ν_z, ν_ζ	Refraction angle at observer, at perigee of ray
ξ	Depression of sun below horizon $= \zeta - \pi/2$
ρ	Air density
ρ	Radial parameter $2 \pi r/\lambda$ [Eq. (II.2, 26)]

$\Sigma(\rho)$ Scattering efficiency factor [Eqs. (II.2, 27)]

σ Isotropic scattering coefficient [Eq. (II.2, 17)]

τ Optical thickness below a given level [Eq. (II.2, 6)]

τ' Optical thickness above a given level

τ^* Optical thickness of entire atmosphere

τ'_{ik} Optical thickness for scattered light [Eq. (IV.1, 11)]

Φ Luminous-flux density [Eq. (II.2, 44)]

φ Scattering angle (§I.1)

χ Rotation angle of analyzer (§II.2)

χ Geocentric angle between observed point and perigee of ray (Figs. 82)

ψ Rotation angle of quarter-wave compensator (§II.2)

ψ Azimuthal angle of ray

ψ Angle to normal to ray through zenith of perigee y (Figs. 82)

ψ_0 Direction of maximum polarization

ω Solid angle

ω_0 Solid angle of solar disk

REFERENCES

Abbreviations generally follow the conventions
of the World List of Scientific Periodicals

1. Ferdinand Rosenberger, Die Geschichte der Physik in Grundzügen,
   1 (Braunschweig, Vieweg, 1882); Russian translation (Moscow,
   United Sci. Techn. Presses, 1933).
2. Johann Kepler, Ad Vitellionem paralipomena, quibus Astronomiae
   pars optica traditur (Frankfurt, 1604).
3. W. von Bezold, "Beobachtungen über die Dämmerung," Annln.
   Phys., (2) **123**, 240-276 (1864).
4. G. V. Rozenberg, "Light scattering in the earth's atmosphere," Usp.
   fiz. Nauk, **71**, 173-213 (1960) [Soviet Phys. Usp., **3**, 346-371(1960)].
5. V. G. Fesenkov, "The structure of the atmosphere" [in Russian],
   Trudy glav. ross. astrofiz. Obs., **2**, 7 (1923).
6. C. Jensen, "Die Himmelsstrahlung," Handb. Phys., **19**, 70-152(1928).
7. C. Dorno, "Himmelshelligkeit, Himmelspolarisation und Sonnen-
   intensität in Davos 1911 bis 1918," Veröff. K. preuss. met. Inst., **6**,
   No. 303 (1919).
8. Carl W. M. Dorno, Physik der Sonnen- und Himmelsstrahlung (Braunschweig, 1919).
9. E. V. Pyaskovskaya-Fesenkova, Investigation of the Scattering of
   Light in the Earth's Atmosphere [in Russian] (Moscow, USSR Acad.
   Sci. Press, 1957), 217 pp.
10. S. Chandrasekhar, Radiative Transfer (Oxford Univ. Press, 1950),
    393 pp.; Russian translation (Moscow, Foreign Lit. Press, 1953).
11. Viktor V. Sobolev, Transfer of Radiant Energy in Stellar and Plan-
    etary Atmospheres [in Russian] (Moscow, Techn. Theor. Press, 1956),
    391 pp.; English translation by S. I. Gaposchkin, A Treatise on Ra-
    diative Transfer (Princeton, Van Nostrand, 1963), 319 pp.

12.　　　　E. S. Kuznetsov, "Application of formulas from the theory of non-horizontal visibility to a calculation of the sky brightness and the range of visibility for the simplest types of scattering diagram" [in Russian], Izv. Akad. Nauk SSSR, Ser. geogr. geofiz., **9**, 204-229 (1945).

13.　　　　E. M. Feigel'son, M. S. Malkevich, S. Ya. Kogan, T. D. Koronova, K. S. Glazova, and M. A. Kuznetsova, Calculation of the Brightness of Light in the Case of Anisotropic Scattering, Part **1** – Trudy Inst. Fiz. Atmos. Akad. Nauk SSSR, No. 1 (Moscow, USSR Acad. Sci. Press, 1958); English translation (New York, Consultants Bureau, 1960), 104 pp. V. S. Atroshenko, E. M. Feigel'son, K. S. Glazova, and M. S. Malkevich, idem, Part 2 – Trudy Inst. Fiz. Atmos. Akad. Nauk SSSR, No. 3 (Moscow, USSR Acad. Sci. Press, 1962); English translation by R. B. Rodman (New York, Consultants Bureau, 1963), 226 pp.

14.　　　　K. S. Shifrin and N. P. Pyatovskaya, Tables of the Oblique Range of Visibility and Daytime Sky Brightness [in Russian] (Leningrad, Hydrol. Met. Press, 1959).

15.　　　　S. Chandrasekhar and D. D. Elbert, "The illumination and polarization of the sunlit sky on Rayleigh scattering," Trans. Am. phil. Soc., (2) **44**, 643-728 (1954).

16.　　　　Kinsell L. Coulson, J. V. Dave, and Z. Sekera, Tables Related to Radiation Emerging from a Planetary Atmosphere with Rayleigh Scattering (Berkeley, Univ. California Press, 1960), 548 pp.

17.　　　　K. S. Shifrin, Scattering of Light in a Turbid Medium [in Russian] (Moscow, Techn. Theor. Press, 1951), 288 pp.

18.　　　　Hendrik C. van de Hulst, Light Scattering by Small Particles (New York, Wiley, 1957), 470 pp.

19.　　　　Yu. S. Georgievskii, A. Ya. Driving, N. V. Zolotavina, G. V. Rozenberg, E. M. Feigel'son, and V. S. Khazanov (general editor, Prof. G. V. Rozenberg), Searchlights in the Atmosphere [in Russian] (Moscow, USSR Acad. Sci. Press, 1960), 245 pp.

20.　　　　O. D. Barteneva, "Scattering functions of light in the atmospheric boundary layer," Izv. Akad. Nauk SSSR, Ser. geofiz.,(1960) 1852-65 [Bull. Acad. Sci. USSR, Geophys. Ser.,(1960) 1237-44].

21.　　　　G. Sh. Livshits, "Brightness of the cloudless sky above snow cover" [in Russian], Izv. astrofiz. Inst. Akad. Nauk kazakh. SSR, **4**, 161-218 (1957).

22.　　　　E. L. Krinov, "The spectral brightness of the sky" [in Russian], Izv. Akad. Nauk SSSR, Ser. Geogr. Geofiz., **6**, No. 4 (1942).

23.　　　　P. Hess, "Die spektrale Energieverteilung der Himmelsstrahlung," Beitr. Geophys., **55**, 204-220 (1939).

24. F. W. P. Götz and E. Schönmann, "Die spektrale Energieverteilung von Himmels- und Sonnenstrahlung," Helv. phys. Acta, **21**, 151-168 (1948).

25. François Arago, Oeuvres complètes, **7** (Paris, Gide, 1858).

26. D. Brewster, "Additional observations on the polarisation of the atmosphere," Phil. Mag., (4) **33**, 290-304 (1867).

27. Robert Rubenson, Mémoire sur la polarisation de la lumière atmosphérique (Uppsala, Leffler, 1864), 145 pp. + appendix.

28. Z. Sekera, "Recent developments in the study of the polarization of sky light," Adv. Geophys., **3**, 43-104 (1956).

29. D. G. Stamov, "The possibility of a polarimetric determination of atmospheric turbidity in various directions" [in Russian], Actinometry and Atmospheric Optics, Proc. interdept. Conf. **2**, 1959 (Leningrad, Hydrol. Met. Press, 1961), 115-124.

30. E. V. Pyaskovskaya-Fesenkova, "Some optical properties of the atmosphere above the Lybian desert" [in Russian], Dokl. Akad. Nauk SSSR, **123**, 269-271 (1958).

31. I. N. Yaroslavtsev, Zh. Geofiz., **5**, 499 (1935).

32. D. G. Stamov, Izv. krym. pedag. Inst., Simferopol, **29**, 283 (1957).

33. G. Dietze, "Einige Ergebnisse einer homogenen siebeneinhalbjährigen Messreihe der Maximalpolarisation," Z. Met., **11**, 211-219 (1957).

34. S. I. Sivkov, "Qualitative characteristics of light depolarization by atmospheric aerosol particles" [in Russian], Actinometry and Atmospheric Optics, Proc. interdept. Conf. **2**, 1959 (Leningrad, Hydrol. Met. Press, 1961), 124-132.

35. E. V. Pyaskovskaya-Fesenkova, summary in Conference on Problems of Actinometry, Atmospheric Optics, and Nuclear Meteorology [in Russian] (Vilnius, Inst. Geol. Geogr., Lithuanian Acad. Sci., 1960), 128 pp.

36. I. I. Tikhanovskii, Zh. russk. fiz.-khim. Obshch., fiz. Otd., **56**, 664 (1928).

37. Z. Sekera, "Lichtelektrische Registrierung der Himmelspolarisation," Beitr. Geophys., **44**, 157-175 (1935).

38. V. G. Fesenkov, "The polarization of night-sky emission lines" [in Russian], Astr. Żh., **37**, 794-798 (1960).

39. G. V. Rozenberg, "Polarization of doubly scattered light in the case of molecular scattering" [in Russian], Izv. Akad. Nauk SSSR, Ser. geogr. geofiz., **13**, 154-161 (1949).

40. N. Pil'chikov, "Sur la polarisation spectrale du ciel," C.r. Acad. Sci., Paris, **115**, 555-558 (1892).

324 REFERENCES

41. J. M. Pernter and F. M. Exner, Meteorologische Optik, 2. Aufl. (Wien and Leipzig, Braumüller, 1922), 907 pp.

42. K. Bullrich, "Neue Messungen und Berechnungen der Himmelslichtpolarisation," Ber. dt. Wetterd., Offenbach, 7 (No. 51), 74-78 (1959).

43. K. B. Panshin, I. A. Khvostikov, and V. I. Chernyaev, Proc. Elbrus Expedition [in Russian] (1934-35).

44. M. A. Schirmann, "Dispersion und Polychroïsmus des polarisierten Lichtes, das von Einzelteilchen von der Grössenordnung der Wellenlänge des Lichtes abgebeugt wird," Annln. Phys., (4) 59, 493-537 (1919).

45. J. W. Strutt (Lord Rayleigh), "On the light from the sky, its polarization and colour," Phil. Mag., (4) 41, 107-120 (1871); "On the scattering of light by small particles," Phil. Mag., (4) 41, 447-454 (1871).

46. P. I. Brounov, Atmospheric Optics [in Russian] (Moscow, Techn. Press, 1924), 220 pp.

47. F. Roggenkamp, "Die neutralen Punkte von Arago und Babinet in den Jahren 1910 bis 1925," Met. Z., 50, 111-113 (1933).

48. W. Smosarski, "Polarisation des Himmelslichtes im Weltpol und andere Beobachtungen," Beitr. Geophys., 48, 213-224 (1936).

49. I. I. Tikhanovskii, "The scattering of light in the atmosphere" [in Russian], Usp. fiz. Nauk, 6, 291-309 (1926).

50. O. Knopf, "Beobachtungen des Aragoschen neutralen Punktes der atmosphärischen Polarisation," Beitr. Phys. frei. Atmos., 8, 57-72 (1919).

51. J. L. Soret, "Sur la polarisation atmosphérique," Annls. Chim. Phys., (6) 14, 503-541 (1888); Archs. Sci. phys. nat., (3) 20, 439 (1888).

52. F. Ahlgrimm, Jb. hamb. wiss. Anst., 32 (1914).

53. I. I. Tikhanovskii, "Die Bestimmung des optischen Anisotropiekoeffizienten der Luftmoleküle durch Messungen der Himmelspolarisation," Phys. Z., 28, 252-260 (1927).

54. M. A. Schirmann, "Versuch einer einheitlichen Erklärung der Erscheinungen der atmosphärischen Polarisation," Annln. Phys., (4) 61, 195-200 (1920), "Neue theoretische Untersuchungen über die Polarisation des Lichtes an trüben Medien und deren Konsequenzen für die Probleme der atmosphärischen Polarisation," Met. Z., 37, 12-22 (1920).

55. W. Milch, "Über den Einfluss grösserer Teilchen in der Atmosphäre auf das Polarisationsverhältnis des Himmelslichtes," Z. Geophys., 1, 109-117 (1925).

56. P. P. Feofilov, "Night illumination and energy distribution in the spectrum of the night sky," Dokl. Akad. Nauk SSSR, 34, 228-232 (1942); Izv. Akad. Nauk SSSR, Ser. geogr. geofiz., 6, 290 (1942).

57. N. B. Divari, "The brightness of the sky at the beginning and end of night" [in Russian], Izv. Akad. Nauk SSSR, Ser. geogr. geofiz., **13**, 242-246 (1949).

58. W. Brunner, "Beiträge zur Photometrie des Nachthimmels, unter besonderer Berücksichtigung des Zodiakallichtes und der Dämmerungserscheinungen," Publnen. eidg. Sternw. Zürich, **6** (1935).

59. V. V. Sharonov, Tables for Calculating Natural Illumination and Visibility [in Russian] (Moscow, USSR Acad. Sci. Press, 1945).

60. Pentti Kalaja, "Die Zeiten von Sonnenschein, Dämmerung und Dunkelheit in verschiedenen Breiten," Suom. geod. Lait. Julk., No. 49 (Helsinki, 1958), 63 pp.

61. V. V. Sharonov, "Illumination of a horizontal plane at twilight and at night," Dokl. Akad. Nauk SSSR, **42**, 296-300 (1944); Trudy yubileinoi nauch. Sess. leningr. gos. Univ., (1948) 47.

62. H. H. Kimball, "Photometric measurements of daylight illumination on a horizontal surface at Mount Weather, Va.," Mon. Weath. Rev. U.S. Dept. Agric., **42**, 650-653 (1914); "Duration of twilight," Mon. Weath. Rev., **44**, 12-13 (1916); "The duration and intensity of twilight," Mon. Weath. Rev., **44**, 614-620 (1916); "Daylight illumination, and the intensity and duration of twilight," Trans. illum. Engng. Soc., N.Y., **11**, 399-413 (1916).

63. F. Schmid, "Die Beziehungen der Abend- und Morgendämmerung zum Zodiallichte," Sirius, Lpz., **55**, 95-100 (1922); "Die Nachtdämmerung," Sirius,Lpz., **56**, 1-15 (1923).

64. K. Kähler, "Über die Helligkeit nach Sonnenuntergang," Met. Z., **44**, 212-217 (1927).

65. N. N. Kalitin, "Bemerkungen über die spektrale Zusammensetzung des diffusen Lichtes während der Dämmerung," Beitr. Geophys., **25**, 348-359 (1930).

66. I. N. Yaroslavtsev, "Dämmerungshelligkeit," Beitr. Geophys., **29**, 161-167 (1931); Zh. Geofiz., **3**, 222 (1933).

67. A. Ljunghall, "The intensity of twilight and its connection with the density of the atmosphere," Meddn. Lunds astr. Obs., (2) **13**, No. 125, 3-171 (1949).

68. N. N. Kalitin, "Twilight illumination from the zenith of the cloudless sky" [in Russian], Meteorologiya Gidrol., **6** (No. 11), 93-95 (1940).

69. I. N. Yaroslavtsev, Zh. Geofiz. Met., **5**, 1 (1928); Zh. Geofiz. Met., **6**, 143 (1929).

70. F. Schembor, "Ergebnisse von Helligkeitsmessungen mit der Kaliumzelle in der Dämmerung," Beitr. Geophys., **28**, 279-292 (1930).

71. Paul Gruner and H. Kleinert, Die Dämmerungserscheinungen —
     Probl. kosm. Phys., No. 10 (Hamburg, Grand, 1927), 124 pp.

72. N. I. Kucherov, "Twilight observations as a method for sounding
     the lower atmosphere" [in Russian], Izv. Akad. Nauk SSSR, Ser.
     geogr. geofiz., 11, 465-487 (1947).

73. M. G. J. Minnaert, Licht en Kleur in het Landschap (Zutphen,
     Thieme, 1937), 353 pp.; English translation, Light and Colour in
     the Open Air (London, Bell, 1940; New York, Dover, 1954), 362 pp.;
     Russian translation (Moscow, Phys. Math. Press, 1958).

74. E. Bauer, A. Danjon, and J. Langevin, "Les phénomènes crépusculaires
     au mont Blanc," C.r. Acad. Sci., Paris, 178, 2115-17 (1924).

75. J. Dufay, "Recherches sur la lumière du ciel nocturne," Bull. Obs.
     Lyon, 10, No. 9, 1-188 (1928).

76. Lord Rayleigh, "A photoelectric method of measuring the light of
     the night sky with studies of the course of variation through the
     night," Proc. R. Soc., A124, 395-408 (1929).

77. V. G. Fesenkov, "Investigation of the structure of the atmosphere
     by means of twilight phenomena" [in Russian], Astr. Zh., 7, 100-
     107 (1930).

78. F. Schembor, "Photographische Photometrie der Dämmerung,"
     Beitr. Geophys., 29, 69-94 (1931).

79. K. Graff, "Prüfung der winterlichen Himmelsklarheit während der
     Dämmerung auf der Insel Mallorca," Sber. Akad. Wiss. Wien, Abt.
     IIa, 140, 513-518 (1931).

80. K. Graff, "Beobachtung einiger meteorologisch-optischer Dammer-
     ungserscheinungen auf der Insel Mallorca," Sber. Akad. Wiss. Wien,
     Abt. IIa, 141, 173-185 (1932).

81. K. Graff, "Messungen der Dämmerungshelligkeit auf Mallorca bei
     Sonnenhöhen zwischen −9° und −13°," Sber. Akad. Wiss. Wien, Abt.
     IIa, 141, 509-513 (1932).

82. W. M. Smart, "Photometric observations of twilight," Mon. Not. R.
     astr. Soc., 93, 441-443 (1933).

83. V. G. Fesenkov, "Stratospheric research by photometric analysis of
     twilight" [in Russian], Izv. Akad. Nauk SSSR, Otd. mat. estest. Nauk,
     (7) 7, 1501-15 (1934).

84. F. Link, "Sondages de la haute atmosphère à l'aide des phénomènes
     crépusculaires," J. Obsrs. Obs. Marseille, 17, 161-169 (1934).

85. N. M. Shtaude, "Photometric observations of twilight as a method
     of studying the upper stratosphere" [in Russian], Trudy Kom. Izuch.
     Stratos., 1 (1936).

86. V. I. Chernyaev and M. F. Vuks, "The spectrum of the twilight sky,"
     Dokl. Akad. Nauk SSSR, 14, 77-80 (1937).

87.   C. Wirtz, "Sonnenstrahlung, atmosphärische Extinktion und Strahlungs-
filter," Annln. Hydrogr., Berl., **63**, 66-83 (1935); "Zur Helligkeit und
Farbe des klaren Himmels an der Nordseeküste Schleswig-Holsteins,"
Annln. Hydrogr., **63**, 170-173 (1935); "Über die blaue Farbe des
klaren Himmels ...," Annln. Hydrogr., **63**, 482-487 (1935); "Zur
objectiven Messung der blauen Farbe des klaren Himmels," Annln.
Hydrogr., **64**, 33-34 (1936); "Sonnenstrahlung und Transparenz in
stauberfüllten Atmosphäre," Annln. Hydrogr., **64**, 473-479 (1936);
"Zur photometrischen Messung der Himmelshelligkeit und des
Farbenindex," Annln. Hydrogr., **65**, 130-131 (1937); "Die Selen-
Sperrschicht-Photozelle als Aktinometer," Annln. Hydrogr., **65**, 269-
276 (1937).

88.   W. Smosarski, "Dämmerungsfarben-Intensität in den Jahren 1913-
1936," Beitr. Geophys., **50**, 252-263 (1937).

89.   F. Schmid, "Symmetrien und Asymmetrien des Purpurlichtes," Met.
Z., **54**, 10-15 (1937).

90.   R. Grandmontagne, "Décroissance des lumières bleue et rouge à la
fin du crépuscule," C.r. Acad. Sci., Paris, **207**, 1436-38 (1938).

91.   E. O. Hulburt, "Brightness of the twilight sky and the density and
temperature of the atmosphere," J. Opt. Soc. Am., **28**, 227-236
(1938).

92.   A. V. Mironov, "Photoelectric photometry of twilight" [in Russian],
Izv. Akad. Nauk SSSR, Ser. geogr. geofiz., **4**, 843-848 (1940).

93.   I. A. Khvostikov, E. N. Magid, and A. A. Shubin, "A study of the
spectral composition of the twilight sky" [in Russian], Izv. Akad.
Nauk SSSR, Ser. geogr. geofiz., **4**, 675-684 (1940).

94.   R. Grandmontagne, "Études photoélectriques sur la lumière du ciel
nocturne," Annls. Phys., (11) **16**, 253-305 (1941).

95.   G. V. Rozenberg, "On a new phenomenon in the scattered light of
a twilight sky," Dokl. Akad. Nauk SSSR, **36**, 270-274 (1942).

96.   G. V. Rozenberg, "Characteristics of the polarization of light scattered by the atmosphere under conditions of twilight illumination"
[in Russian], Dissertation (Moscow, 1946).

97.   T. G. Megrelishvili, "Electrocolorimetry of the twilight sky," Dokl.
Akad. Nauk SSSR, **53**, 123-126 (1946).

98.   N. M. Shtaude, Izv. Akad. Nauk kazakh. SSR, Ser. astr. fiz., No. 2,
108 (1946).

99.   N. M. Shtaude, "The twilight phenomenon in relation to solar activity," Dokl. Akad. Nauk SSSR, **55**, 27-30 (1947).

100.   N. M. Kopylov, "On the photometric analysis of twilight" [in Rus-
sian], Meteorologiya Gidrol., No. 2, 3-14 (1947).

101.   T. G. Megrelishvili, Byull. abastuman. astrofiz. Obs., No. 9, 1-142
(1948).

102.   G. de Vaucouleurs, "Observations des discontinuités crépusculaires,"
C.r. Acad. Sci., Paris, **232**, 342-344 (1951).

103.   E. V. Ashburn, "The density of the upper atmosphere and the brightness of the twilight sky," J. geophys. Res., **57**, 85-93 (1952).

104.   M. J. Koomen, C. Lock, D. M. Packer, R. Scolnik, R. Tousey, and
E. O. Hulburt, "Measurements of the brightness of the twilight sky,"
J. Opt. Soc. Am., **42**, 353-356 (1952).

105.   N. B. Divari, "Deep twilight near the horizon" [in Russian], Izv.
Akad. Nauk SSSR, Ser. geofiz., (1952) 79-84.

106.   M. W. Chiplonkar and J. D. Ranade, "The brightness of the zenith
sky during twilight. II," Proc. Indian Acad. Sci., **A18**, 121-125(1943).

107.   E. D. Sholokhova and M. S. Frish, "The luminosity of the twilight
sky in the 1-μ region" [in Russian], Dokl. Akad. Nauk SSSR, **105**,
1218-20 (1955).

108.   R. V. Karandikar, "Search for discontinuities in the brightness of the
twilight sky," J. Opt. Soc. Am., **45**, 389-392 (1955).

109.   T. G. Megrelishvili, "The range of applicability of the twilight
method of atmospheric research" [in Russian], Izv. Akad. Nauk SSSR,
Ser. geofiz., (1956) 976-983.

110.   J. V. Dave and K. R. Ramanathan, "On the intensity and polarization of the light from the sky during twilight," Proc. Indian Acad.
Sci., **A43**, 67-78 (1956).

111.   E. K. Bigg, "Detection of atmospheric dust and temperature inversions by twilight scattering," Nature, Lond., **177**, 77-79 (1956);
idem, J. Met., **13**, 262-268 (1956).

112.   T. G. Megrelishvili, "The luminescence of the twilight sky in the
infrared spectral region" [in Russian], Dokl. Akad. Nauk SSSR, **116**,
766-768 (1957).

113.   N. B. Divari, "The polarization of the light at the zenith of the twilight sky" [in Russian], Dokl. Akad. Nauk SSSR, **112**, 217-220 (1957),

114.   M. Gadsden, "The colour of the zenith twilight sky: absorption due
to ozone," J. atmos. terr. Phys., **10**, 176-180 (1957).

115.   T. G. Megrelishvili, "The possibility of investigating aerosol layers
by the twilight method," Izv. Akad. Nauk SSSR, Ser. geofiz.,(1958)
560-563 [Bull. Acad. Sci. USSR, Geophys. Ser.,(1958) 315-317];
idem, Indian J. Met. Geophys., **9**, 271-276 (1958).

116.   N. B. Divari, "Variations in the color of the twilight sky" [in Rus-
sian], Dokl. Akad. Nauk SSSR, **122**, 795-798 (1958).

117.   F. Link, L. Neužil, and I. Zacharov, Publs. astr. Inst. Czech. Acad.
Sci., No. 38 (1958).

118.   O. D. Barteneva and A. N. Boyarova, "Brightness of twilight and of
        the night sky" [in Russian], Trudy glav. geofiz. Obs. Voeikova, No.
        100, 133-140 (1960).

119.   K. Ya. Kondrat'ev and O. A. Zigel', "Thermal sounding of the at-
        mosphere by twilight photometry" [in Russian], Vest. leningr. gos.
        Univ., (1960) No. 10 (Ser. Fiz. Khim., No. 2), 45-48.

120.   A. Kh. Darchiya, "A spectrophotometric study of twilight-glow phe-
        nomena" [in Russian], Izv. glav. astr. Obs. Pulkove, 21, No. 6 (No.
        165), 114-151 (1960).

121.   T. G. Megrelishvili, "A study of the optical properties of the earth's
        atmosphere by the twilight method" [summary, in Russian], Actin-
        ometry and Atmospheric Optics, Proc. interdept. Conf. 2, 1959
        (Leningrad, Hydrol. Met. Press, 1961), 105.

122.   A. Ya. Driving, G. V. Rozenberg, and N. K. Turikova, "Some re-
        sults of spectral polarization studies of the daytime and twilight
        sky" [summary, in Russian], Actinometry and Atmospheric Optics,
        Proc. interdept. Conf. 2, 1959 (Leningrad, Hydrol. Met. Press, 1961),
        104.

123.   G. Dietze, Report to All-Union Meteorological Conference, Lenin-
        grad, June 1961.

124.   F. Volz and R. M. Goody, "Twilight intensity at 20° elevation,"
        Blue Hill Met. Obs., Harvard Univ., Contract AF 19(604)-4546, Sci.
        Rep. No. 1, AFCRC-TN-60-284 (1960), 46 pp.

125.   I. A. Khvostikov, The Night Airglow [in Russian], 2nd edn. (Moscow,
        USSR Acad. Sci. Press, 1948), 495 pp.

126.   Lord Rayleigh, "Absolute intensity of the aurora line in the night
        sky, and the number of atmospheric transitions required to maintain
        it," Proc. R. Soc., A129, 458-467 (1930).

127.   J. Cabannes and J. Dufay, "Variation annuelle de l'intensité des
        raies brillantes du ciel nocturne," C.r. Acad. Sci., Paris, 200, 878-
        880 (1935).

128.   N. M. Shtaude, "The twilight method of stratospheric research" [in
        Russian], Izv. Akad. Nauk SSSR, Ser. geogr. geofiz., 13, 307-319
        (1949).

129.   O. B. Vasil'ev, Proc. Conf. Noctilucent Clouds 6 [in Russian] (Riga,
        1961), 35.

130.   G. V. Rozenberg, "The anatomy of the twilight glow" [summary,
        in Russian], Actinometry and Atmospheric Optics, Proc. interdept.
        Conf. 2, 1959 (Leningrad, Hydrol. Met. Press, 1961), 105-107.

131.   P. Hautefeuille and J. Chappuis, "Sur la liquéfaction de l'ozone et
        sur la couleur à l'état gazeux," C.r. Acad. Sci., Paris, 91, 522-525
        (1880).

132.   J. Gauzit, "Étude de l'ozone atmosphérique par spectroscopie visu-
        elle," Annls. Phys., (11) **4**, 450-532 (1935).

133.   E. O. Hulburt, "Explanation of the brightness and color of the sky,
        particularly the twilight sky," J. Opt. Soc. Am., **43**, 113-118 (1953).

134.   J. Dufay and J. Gauzit, "Mesures spectrophotométriques de la bril-
        lance du ciel au crépuscule," Annls. Astrophys., **9**, 135-142 (1946).

135.   K. Ya. Kondrat'ev, The Radiant Energy of the Sun [in Russian]
        (Leningrad, Hydrol. Met. Press, 1954), 600 pp.

136.   V. M. Slipher, "Spectrographic studies of the planets," Mon. Not.
        R. astr. Soc., **93**, 657-668 (1933), p. 666.

137.   H. Garrigue, "Nouveaux résultats sur la lumière du ciel nocturne,"
        C.r. Acad. Sci., Paris, **202**, 1807 -09 (1936).

138.   G. Courtès, "La raie ^{4}S − ^{2}D de l'azote observée au crépuscule,"
        C.r. Acad. Sci., Paris, **231**, 62-63 (1950). M. Dufay, "La raie
        interdite ^{4}S − ^{2}D de l'atome neutre d'azote dans les spectres du ciel
        nocturne et du crépuscule," C.r. Acad. Sci., Paris, **233**, 419-420
        (1951); "Étude de l'émission de la molécule d'azote ionisée et de
        l'atome neutre d'azote au crépuscule," Annls. Phys., (12) **8**, 813-
        862 (1953).

139.   A. B. Meinel, "The spectrum of the airglow and the aurora," Rep.
        Prog. Phys., **14**, 121-146 (1951).

140.   P. Berthier, "Sur l'émission crépusculaire et nocturne des bandes de
        OH et O_2 dans le proche infrarouge," C.r. Acad. Sci., Paris, **236**,
        1808-10 (1953).

141.   A. V. Jones, "Ca II emission lines in the twilight spectrum," Nature,
        Lond., **178**, 276-277 (1956).

142.   J. Delannoy and G. Weill, "Observation d'une nouvelle raie d'émis-
        sion crépusculaire atmosphérique," C.r. Acad. Sci., Paris, **247**, 806-
        807 (1958).

143.   V. I. Krasovskii, "Some results of IGY and IGC research on the
        aurora and night glow," Usp. fiz. Nauk, **75**, 501-525 (1961) [Soviet
        Phys. Usp., **4**, 904 918 (1962)].

144.   J. W. Chamberlain, "Interpretation of twilight spectra," Annls.
        Géophys., **14**, 196-203 (1958).

145.   J. Bricard and A. Kastler, "Recherches sur la radiation D du sodium
        dans la lumière du ciel crépusculaire et nocturne," Annls. Géophys.,
        1, 53-91 (1944).

146.   J.-É. Blamont and A. Kastler, "Réalisation d'un photomètre photo-
        électrique pour l'étude de l'émission crépusculaire de la raie D du
        sodium dans la haute atmosphère," Annls. Géophys., **7**, 73-89 (1951).

147.   D. M. Hunten, "A study of sodium in twilight. I. Theory," J. atmos.
        terr. Phys., **5**, 44-56 (1954).

148. D. M. Hunten and G. G. Shepherd, "A study of sodium in twilight. II. Observations on the distribution," J. atmos. terr. Phys., 5, 57-62 (1954).

149. T. M. Donahue and R. Resnick, "Resonance absorption of sunlight in twilight layers," Phys. Rev., 98, 1622-25 (1955).

150. T. M. Donahue and A. Foderaro, "The effect of resonance absorption on the determination of the height of airglow layers," J. geophys. Res., 60, 75-86 (1955).

151. J.-É. Blamont, T. M. Donahue, and V. R. Stull, "The sodium twilight airglow 1955-1957. I," Annls. Géophys., 14, 253-281 (1958).

152. J.-É. Blamont, T. M. Donahue, and W. Weber, "The sodium twilight airglow 1955-1957. II," Annls. Géophys., 14, 282-304 (1958).

153. J. W. Chamberlain, "Resonance scattering by atmospheric sodium. I. Theory of the intensity plateau in the twilight airglow," J. atmos. terr. Phys., 9, 73-89 (1956).

154. J. W Chamberlain and B. J. Negaard, idem, "II. Nightglow theory," J. atmos. terr. Phys., 9, 169-178 (1956).

155. D. M. Hunten, idem, "III. Supplementary considerations," J. atmos. terr. Phys., 9, 179-183 (1956).

156. J. W. Chamberlain, D. M. Hunten, and J. E. Mack, idem, "IV. Abundance of sodium in twilight," J. atmos. terr. Phys., 12, 153-165 (1958).

157. J. C. Brandt and J. W. Chamberlain, idem, "V. Theory of the day airglow," J. atmos. terr. Phys., 13, 90-99 (1958).

158. T. M. Donahue and D. M. Hunten, "The correction of sodium twilight glow observations," J. atmos. terr. Phys., 13, 165-166 (1958).

159. E. A. Lytle and D. M. Hunten, "The ratio of sodium to potassium in the upper atmosphere," J. atmos. terr. Phys., 16, 236-245 (1959).

160. H. N. Rundle, D. M. Hunten, and J. W. Chamberlain, "Resonance scattering by atmospheric sodium. VII. Measurement of the vertical distribution in twilight," J. atmos. terr. Phys., 17, 205-219 (1960).

161. D. M. Hunten, idem, "VIII. An improved method of deducing the vertical distribution," J. atmos. terr. Phys., 17, 295-301 (1960).

162. T. M. Donahue and J.-É. Blamont, Report to Symposium on Aeronomy, Copenhagen, July 1960.

163. A. Wegener, "Optik der Atmosphäre," Müller-Pouillets Lehrbuch der Physik, 11. Aufl., 5 (Braunschweig, Vieweg, 1928), 1. Hälfte, 199-289.

164. C. Jensen, "Atmosphärisch-optische Messungen in Ilmenau," Beitr. Geophys., 35, 166-188 (1932).

332 REFERENCES

165. C. Jensen, "Die Verfolgung der neutralen Punkte der atmosphärischen Polarisation in Arnsberg i. W. während eines Zeitraums von 19 Jahren," Met. Z., **54**, 90-97 (1937).

166. Friedrich Busch and C. Jensen, Tatsachen und Theorien der atmosphärischen Polarisation (Hamburg, Lucas Gräfe, 1911), 532 pp.

167. I. A. Khvostikov and A. N. Sevchenko, "Applications de la méthode polarimétrique à l'étude de la structure des couches supérieures de l'atmosphère," Dokl. Akad. Nauk SSSR, **13**, 359-363 (1936).

168. I. A. Khvostikov, "An outline of the physics of the earth's atmosphere" [in Russian], Usp. fiz. Nauk, **19**, 49 (1938).

169. V. M. Bovsheverov, A. V. Mironov, I. M. Mikhailin, V. M. Morozov, Z. L. Ponizovskii, S. P. Sokolov, and I. A. Khvostikov, "A complex study of the ionosphere by the optical method and the method of radio-wave reflection" [in Russian], Izv. Akad. Nauk SSSR, Ser. geogr. geofiz., **4**, 657-674 (1940).

170. Z. L. Ponizovskii and G. V. Rozenberg, "On polarization anomalies in scattered light of a twilight sky as connected with the condition of ionosphere," Dokl. Akad. Nauk SSSR, **37**, 218-220 (1942).

171. G. V. Rozenberg, Trudy geofiz. Inst., No. 12 (139), 35 (1948).

172. R. Robley, "La diffusion multiple dans l'atmosphère déduite des observations crépusculaires. I. Résultats expérimentaux," Annls. Géophys., **6**, 191-211 (1950).

173. W. M. Cohn, "Some observations of the sky polarization during the total solar eclipses of August 31, 1932 and February 14, 1934," Phys. Rev., **45**, 848 (1934).

174. V. G. Fesenkov, "Some results of observations during the total solar eclipse of Sept. 21, 1941" [in Russian], Astr. Zh., **22**, 45-51 (1945).

175. Sisir K. Mitra, The Upper Atmosphere, 2nd edn. (Calcutta, Asiatic Society, 1952), 713 pp.; Russian translation (Moscow, Foreign Lit. Press, 1955).

176. B. A. Mirtov, "Rocket studies of the composition of the atmosphere at high elevations" [in Russian], Usp. fiz. Nauk, **63**, 181-196 (1957).

177. R. M. Goody, The Physics of the Stratosphere (Cambridge Univ. Press, 1954), 187 pp.; Russian translation (Hydrol. Met. Press, 1958).

178. B. A. Mirtov, Report to All-Union Meteorological Conference, Leningrad, June 1961.

179. V. G. Istomin, "Investigations of the ion composition of the earth's atmosphere by rockets and satellites," Iskusst. Sputn. Zemli, **2**, 32-35 (1958) [Artif. Earth Satell., **2**, 40-44 (1960)].

180. C. Y. Johnson, E. B. Meadows, and J. C. Holmes, "Ion composition of the Arctic ionosphere," J. geophys. Res., **63**, 443-444 (1958).

181.   I. A. Prokof'eva, Atmospheric Ozone [in Russian], in series Results and Problems of Modern Science (Moscow, USSR Acad. Sci. Press, 1951), 231 pp.

182.   I. A. Khvostikov, "Ozone in the stratosphere" [in Russian], Usp. fiz. Nauk, **59**, 228-323 (1956).

183.   F. S. Johnson, J. D. Purcell, R. Tousey, and K. Watanabe, "Direct measurements of the vertical distribution of atmospheric ozone to 70 kilometers altitude," J. geophys. Res., **57**, 157-176 (1952).

184.   S. V. Venkateswaran, J. G. Moore, and A. J. Krueger, "Determination of the vertical distribution of ozone by satellite photometry," J. geophys. Res., **66**, 1751-71 (1961).

185.   K. Ya. Kondrat'ev and O. P. Filipovich, Thermal regime in the upper atmospheric layers [in Russian] (Leningrad, Hydrol. Met. Press, 1960), 355 pp.

186.   E. G. Shvidkovskii, "Meteorological measurements by rockets" [in Russian], Trudy tsent. aérol. Obs., No. 29, 5-50 (1960).

187.   A. I. Repnev, "Properties of the upper atmosphere and artificial earth satellites" [in Russian], Trudy tsent. aérol. Obs., No. 25, 5-62 (1959).

188.   E. G. Shvidkovskii, "Some results of measurements of thermodynamic parameters of the stratosphere by meteorological rockets," Iskusst. Sputn. Zemli, **2**, 10-16 (1958) [Artif. Earth Satell., **2**, 12-19 (1960)].

189.   W. G. Stroud, W. R. Bandeen, W. Nordberg, et al., "Temperatures and winds in the Arctic as obtained by the rocket grenade experiments," IGY Rocket Rep. Ser., No. 1, 58-79 (1958).

190.   L. E. Miller, " 'Molecular weight' of air at high altitudes," J. geophys. Res., **62**, 351-365 (1957).

191.   V. V. Mikhnevich, "Measurement of the pressure in the upper atmosphere" [in Russian], Usp. fiz. Nauk, **63**, 197-204 (1957).

192.   I. A. Khvostikov, Proc. Conf. Noctilucent Clouds 6 [in Russian] (Riga, 1961).

193.   C. Junge, "The size distribution and aging of natural aerosols as determined from electrical and optical data on the atmosphere," J. Met., **12**, 13-25 (1955).

194.   C. Junge, "Remarks about the size distribution of natural aerosols," in Artificial Stimulation of Rain, Proc. Conf. Phys. Cloud and Precip. Particles **1** (Pergamon Press, 1957), 3-17.

195.   A. G. Laktionov, "Results of investigations of natural aerosols over the various regions of the USSR," Izv. Akad. Nauk SSSR, Ser. geofiz., (1960) 566-574 [Bull. Acad. Sci. USSR, Geophys. Ser., (1960) 373-378]; "Distribution by height of concentration of aerosol particles and measurement of coefficients of vertical displacement in a

free atmosphere," Izv. Akad. Nauk SSSR, Ser. geofiz., (1960) 1397-1406 [Bull. Acad. Sci. USSR, Geophys. Ser., (1960) 931-936]; "Determination of the vertical-mixing coefficients of particles in the free atmosphere" [in Russian], Dokl. Akad. Nauk SSSR, **133**, 838-840 (1960).

196. G. P. Faraponova, "Measurement of the index of attenuation of sunlight in the atmosphere" [in Russian], Trudy tsent. aérol. Obs., No. 23, 52-62 (1957); "Measurement of the attenuation of sunlight in the free atmosphere" [in Russian], Trudy tsent. aérol. Obs., No. 32, 3-16 (1959).

197. V. F. Belov, "Investigation of scattering functions in the troposphere and lower stratosphere" [in Russian], Trudy tsent. aérol. Obs., No. 23, 63-77 (1957); Report to All-Union Meteorological Conference, Leningrad, June 1961. B. A. Chayanov, "Scattering functions in the free atmosphere" [in Russian], Trudy tsent. aérol. Obs., No. 32, 17-27 (1959).

198. F. H. Ludlam, "Noctilucent clouds," Tellus, **9**, 341-364 (1957) [Usp. fiz. Nauk, **65**, 407-440 (1958)].

199. Sergei I. Vavilov, Collected Works [in Russian], **2** (Moscow, USSR Acad. Sci. Press, 1952), 431.

200. K. Ya. Kondrat'ev, Radiant Heat Exchange in the Atmosphere [in Russian] (Leningrad, Hydrol. Met. Press, 1956), 419 pp.

201. G. V. Rozenberg, "The Stokes vector-parameter (matrix methods for calculating the polarization of radiation in the geometrical optics approximation)" [in Russian], Usp. fiz. Nauk, **56**, 77-110 (1955).

202. G. V. Rozenberg, "Absorption spectroscopy of dispersed materials," Usp. Fiz. Nauk, **69**, 57-104 (1959) [Soviet Phys. Usp., **2**, 666-698 (1959)].

203. G. V. Rozenberg, "Some topics in the propagation of electromagnetic waves through turbid media" [in Russian], Dissertation (Moscow, 1954).

204. V. S. Malkova, Report to Conference on Spectroscopy of Dispersed Media, Minsk, 1961.

205. F. Möller, "Strahlung in der unteren Atmosphäre," Handb. Phys., **48**, 155-253 (1957).

206. F. Volz, "Optik der Tropfen. I. Optik des Dunstes," Handb. Geophys., **8**, 822-897 (1956).

207. V. G. Kastrov, "Asymmetry of light scattering in the atmosphere" [in Russian], Trudy tsent. aérol. Obs., No. 32, 84-86 (1959).

208. I. N. Minin, Report to All-Union Meteorological Conference, Leningrad, June 1961.

209.   M. M. Gurevich and L. Chokhrov, Trudy gos. opt. Inst., 11, No. 17
        (No. 90) (1936).

210.   V. A. Timofeeva, Trudy morsk. gidrofiz. Inst., 3 (1953).

211.   L. M. Romanova, "The solution of the radiation-transfer equation
        for the case when the indicatrix of scattering greatly differs from
        the spherical one. I," Optika Spektrosk., 13, 429-435 (1962) [Op-
        tics Spectrosc., N.Y., 13, 238-241 (1962)]; "Small-angle approxi-
        mation of a solution of the equation for radiation transfer and its
        refinement," Izv. Akad. Nauk SSSR, Ser. geofiz., (1962) 1108-12
        [Bull. Acad. Sci. USSR, Geophys. Ser., (1962) 709-711].

212.   G. V. Rozenberg, Report to All-Union Meteorological Conference,
        Leningrad, June 1961; "Light characteristics of thick layers of a
        scattering medium with low specific absorption," Dokl. Akad. Nauk
        SSSR, 145, 775-777 (1962) [Soviet Phys. Dokl., 7, 706-708 (1963)];
        in Spectroscopy of Diffusing Media [in Russian] (Minsk, Belorussian
        Acad. Sci. Press, 1963).

213.   R. Penndorf, "Tables of the refractive index for standard air and the
        Rayleigh scattering coefficient ...," J. Opt. Soc. Am., 47, 176-182
        (1957).

214.   See, for example: Gerhard Herzberg, Molecular Spectra and Mo-
        lecular Structure, 2nd edn. (Princeton, Van Nostrand, 1950), 2 vol.;
        Russian translation, Spectra and Structure of Diatomic Molecules
        (Moscow, Foreign Lit. Press, 1949). A. K. Suslov, "Spectrum of
        oxygen molecules" [in Russian], Trudy Sekt. Astrobot., Alma-Ata,
        6, 65-76 (1958); "Procedure and some results of spectrophotometry
        of telluric lines of oxygen" [in Russian], Trudy Sekt. Astrobot., 7,
        226-276 (1959).

215.   P. P. Toropova, "The role of various factors in attenuating the light
        of the earth's atmosphere" [in Russian], Izv. astrofiz. Inst. Akad.
        Nauk kazakh. SSR, 6, 3-72 (1958).

216.   F. Link et al., Handb. Geophys., 8 (Berlin, Borntraeger, 1942-43).

217.   Smithsonian Meteorological Tables, 6th edn., ed. J. List (Washing-
        ton, Smithsonian Inst., 1951), 527 pp.

218.   B. S. Pritchard and W. G. Elliott, "Two instruments for atmospheric
        optics measurements," J. Opt. Soc. Am., 50, 191-202 (1960).

219.   V. G. Kastrov, "Investigation of certain errors in the determination
        of solar radiation absorption in the atmosphere with pyranometers"
        [in Russian], Trudy tsent. aérol. Obs., No. 32, 46-65 (1959); "Ab-
        sorption of solar radiation in the lower troposphere" [in Russian],
        Trudy tsent. aérol. Obs., No. 32, 73-83 (1959).

220.   Ya. I. Kogan and Z. A. Burnasheva, "Growth and measurement of
        condensation nuclei in a continuous stream" [in Russian], Zh. fiz.
        Khim., 34, 2630-39 (1960).

221. G. Sh. Livshits, "The scattering diagram of light in the atmosphere in the near infrared region" [in Russian], Izv. astrofiz. Inst. Akad. Nauk kazakh. SSR, **7**, 65-73 (1958).

222. K. Bullrich, "Streulichtmessungen in Dunst und Nebel," Met. Rdsch., **13**, 21-29 (1960); T. P. Toropova, "The wavelength dependence of the degree of polarization of light scattered in the ground layer of the atmosphere," [in Russian], Izv. astrofiz. Inst. Akad. Nauk kazakh. SSR, **11**, 105-110 (1960).

223. G. V. Rozenberg and I. M. Mikhailin, "Ellipticity of the polarization of scattered light" [in Russian], Optika Spektrosk., **5**, 671-681 (1958); "Experimental detection of ellipticity in the polarization of scattered light" [in Russian], Dokl. Akad. Nauk SSSR, **122**, 62-64 (1958).

224. P. Gruner, "Über die Gesetze der Beleuchtung der irdischen Atmosphäre durch das Sonnenlicht," Beitr. Phys. frei. Atmos., **8**, 120-156 (1919).

225. F. Link, "Théorie photométrique des éclipses de Lune," C.r. Acad. Sci., Paris, **196**, 251-253 (1933).

226. F. Link, "L'éclairement de la haute atmosphère et les Tables crépusculaires de M. Jean Lugeon," C.r. Acad. Sci., Paris, **199**, 303-305 (1934).

227. F. Link, J. Obsrs. Obs. Marseille, **17**, 41 (1934).

228. F. Link, "Tables d'éclairements crépusculaires de la haute atmosphère," Mém. Inst. natn. Mét. Pol., No. 5, 55 (1935).

229. J. Sweer, "Path of a ray of light tangent to the surface of the earth," J. Opt. Soc. Am., **28**, 327-329 (1938).

230. F. Link and Z. Sekera, "Dioptrische Tafeln der Erdatmosphäre," Publikace Praž. st. Hvězd., No. 14, 1-28 (1940). F. Link, "Beleuchtungstafeln der Erdatmosphäre," Mitt. Beob. tschech. astr. Ges., No. 6, 1 (1941); "Extension des tables dioptriques de l'atmosphère terrestre," Publikace Praž. st. Hvězd., No. 18, 1-15 (1947).

231. V. G. Fesenkov, "Theoretical brightness of the daytime sky for a spherical earth" [in Russian], Astr. Zh., **32**, 265-281 (1955).

232. V. G. Fesenkov, "Brightness of the cloudless daytime sky for a spherical earth" [in Russian], Dokl. Akad. Nauk SSSR, **101**, 845-847 (1955).

233. V. G. Fesenkov, "On the optical state of the atmosphere illuminated by twilight," Astr. Zh., **36**, 201-207 (1959) [Soviet Astr., **3**, 207-213 (1959)].

234. Jean Lugeon, Tables Crépusculaires, 2nd edn. (Warsaw, Acad. Pol. Sci., 1957), 438 pp.

235. G. V. Rozenberg, "Measurement of atmospheric refraction" [in Russian], Izv. Akad. Nauk SSSR, Ser. geogr. geofiz., **13**, 383-387 (1949).

236. C. W. Allen, Astrophysical Quantities (Univ. London, 1955); Russian translation (Moscow, Foreign Lit. Press, 1960).

237. V. G. Fesenkov, Proc. All-Union Conf. Stratospheric Research [in Russian], 1 (USSR Acad. Sci., 1936).

238. N. M. Shtaude, Proc. All-Union Conf. Stratospheric Research [in Russian], 1 (USSR Acad. Sci., 1936).

239. N. M. Shtaude, Izv. Akad. Nauk kazakh. SSR, Ser. astr. fiz., No. 2, 97 (1946).

240. N. M. Shtaude, "Principles of a simplified theory for twilight phenomena" [in Russian], Izv. Akad. Nauk SSSR, Ser. geogr. geofiz., 11, 349-370 (1947).

241. N. M. Shtaude, Izv. Akad. Nauk kazakh. SSR, Ser. astr. fiz., No. 2, 22 (1946).

242. N. M. Shtaude, "On the determination of the transparency coefficient of the earth's atmosphere" [in Russian], Izv. nauch. Inst. Lesgafta, 15, No. 1-2, 57-74 (1929).

243. D. J. K. O'Connell, "The green flash and kindred phenomena," Endeavour, 20, 131-137 (1961); D. J. K. O'Connell and C. Treusch, The Green Flash and Other Low Sun Phenomena (Vatican Obs., 1958), 192 pp.

244. F. W. Götz, "Zum Strahlungsklima des Spitzbergensommers. Strahlungs- und Ozonmessungen in der Königsbucht 1929," Beitr. Geophys., 31, 119-154 (1931).

245. S. F. Rodionov, E. N. Pavlova, and N. N. Stupnikov, "On a new anomalous effect in the short-wave end of the solar spectrum. I," Dokl. Akad. Nauk SSSR, 19, 55-57 (1938).

246. S. F. Rodionov and E. N. Pavlova, "Contribution to the problem of the Umkehr-Effekt. II," Dokl. Akad. Nauk SSSR, 19, 59-60 (1938).

247. G. V. Rozenberg, "Limits of applicability of the Bouguer law, and the inversion, anomalous-, and 'selective'-transparency effects in the atmosphere" [in Russian], Dokl. Akad. Nauk SSSR, 145, 1269-70 (1962).

248. S. F. Rodionov, E. N. Pavlova, E. V. Rdultovskaya, and N. T. Reinov, Izv. Akad. Nauk SSSR, Ser. geogr. geofiz., 6, No. 4, 135 (1942).

249. S. F. Rodionov, "Transparency of the atmosphere in the ultraviolet spectral region" [in Russian], Izv. Akad. Nauk SSSR, Ser. geogr. geofiz., 14, 334-358 (1950).

250. C. L. Pekeris, Avh. norske VidenskAkad. Oslo, math.-naturw. Kl., No. 8 (1933).

251. E. A. Polyakova, "Spectrographic observation of atmospheric transparency for solar ultraviolet radiation" [in Russian], Trudy glav. geofiz. Obs. Voeikova, 19 (No. 81), 185-192 (1950).

252.   J. Dufay, "Altitude de l'émission des raies D du sodium au crépus-
       cule et rôle de l'absorption par l'ozone," C.r. Acad. Sci., Paris,
       225, 690-692 (1947).

253.   E. A. Polyakova, "A calculation of twilight colors" [in Russian],
       Izv. Akad. Nauk SSSR, Ser. geogr. geofiz., **13**, 247-255 (1949).

254.   Yu. I. Gal'perin, "Intensity ratio of the components of the yellow
       sodium doublet in the twilight spectrum," Astr. Zh., **33**, 173-181
       (1956); idem, The Airglow and the Aurorae (Belfast sympos., Sept.
       1955) — J. atmos. terr. Phys. spec. Suppl. **5** (London, Pergamon
       Press, 1956), 91-94.

255.   Joseph W. Chamberlain, Physics of the Aurora and Airglow, Inter-
       natl. Geophys. Ser. **2** (New York, Academic Press, 1961), 704 pp.

256.   P. Gruner, "Photometrie der Dämmerungsfarben, insbesondere des
       Purpurlichtes," Beitr. Geophys., **46**, 202-207 (1935); "Photometrie
       des Purpurlichtes," Beitr. Geophys., **50**, 143-149 (1937); idem, "II,"
       Beitr. Geophys., **51**, 174-194 (1937); "Neueste Dämmerungsforschun-
       gen," Ergebn. kosm. Phys., **3** (Beitr. Geophys. Suppl., **3**), 113-154
       (1938).

257.   Johann H. Lambert, Photometria, sive De mensura et gradibus
       luminis, colorum et umbrae (1760); German translation by E. An-
       ding, Lamberts Photometrie, 2. Heft — Ostwald's Klassiker exakt.
       Wiss., **32** (Leipzig, Engelmann, 1892), Teil 3-5, pp. 96-112.

258.   Chandrasekhar V. Raman, Molecular Diffraction of Light (Univ.
       Calcutta, 1922), 103 pp.

259.   V. L. Ginzburg, "Polarization of the lines in the night sky lumi-
       nescence spectrum and in the spectrum of northern lights," Dokl.
       Akad. Nauk SSSR, **38**, 237-240 (1943).

260.   V. L. Ginzburg and N. N. Sobolev, "On secondary light scattering
       in the atmosphere and on polarization anomalies during twilight,"
       Dokl. Akad. Nauk SSSR, **40**, 223-225 (1943).

261.   G. V. Rozenberg, I. A. Khvostikov, and F. F. Yudalevich, "The
       effect of multiple scattering on the brightness of the twilight sky"
       [in Russian], Dokl. Akad. Nauk SSSR, 59, 1277-79 (1948).

262.   William J. Humphreys, Physics of the Air, 3rd edn. (McGraw-Hill,
       1940), 676 pp.; Russian translation (1936).

263.   N. M. Shtaude, "Second-order scattering during twilight" [in Rus-
       sian], Izv. Akad. Nauk SSSR, Ser. geogr. geofiz., **12**, 387-393 (1948);
       "The effect of second-order scattering on the brightness of twilight
       at the zenith" [in Russian], Dokl. Akad. Nauk SSSR, **59**, 1281-82
       (1948); "Secondary scattering during twilight for various structures
       of the atmosphere" [in Russian], Dokl. Akad. Nauk SSSR, **64**, 819-
       822 (1949).

264.   F. L. Whipple, "Meteors and the earth's upper atmosphere," Rev.
       mod. Phys., **15**, 246-264 (1943).

265.   L. Harang, "The auroral luminosity-curve," Terr. Magn. atmos.
       Elect., **51**, 381-400 (1946).

266.   F. Link, "Die Dämmerungshelligkeit im Zenit und die Luftdichte
       in der Ionosphäre," Met. Z., **59**, 7-12 (1942); "Situation actuelle
       des recherches crépusculaires," Bull. astr. Insts. Czech., **1**, 135-140
       (1949).

267.   F. F. Yudalevich, "On the formulation of a theory for twilight phe-
       nomena with allowance for secondary scattering" [in Russian], Dokl.
       Akad. Nauk SSSR, **75**, 799-802 (1950); "Principles of a theory for
       twilight phenomena with allowance for the effect of secondary scat-
       tering in the atmosphere" [in Russian], Izv. Akad. Nauk SSSR, Ser.
       geogr. geofiz., **14**, 562-570 (1950).

268.   F. F. Yudalevich, "On the role of secondary scattering in the twi-
       light phenomenon" [in Russian], Dokl. Akad. Nauk SSSR, **55**, 717-
       720 (1947).

269.   R. Robley, "La diffusion multiple dans l'atmosphère déduite des ob-
       servations crépusculaires. II. Considérations théoriques," Annls.
       Géophys., **8**, 1-20 (1952).

270.   J. V. Dave, "On the intensity and polarisation of the light from the
       sky during twilight. Part II," Proc. Indian Acad. Sci., A43, 336-
       358 (1956).

271.   T. Sato, "Studies on the twilight by the scattering," Sci. Rep.
       Tôhoku Univ., Ser. 5 (Geophys.), **3**, 114-129 (1951); idem [in Jap-
       anese], J. met. Soc. Japan, **30**, 1-22 (1952); "The contribution of
       the intensity in scattered light of each wavelength to the sky light
       at sunrise and sunset," J. met. Soc. Japan, **39**, 116-133 (1961).

272.   F. F. Yudalevich, "On calculating the distribution of intensity and
       degree of polarization in the twilight sky, allowing for the earth's
       sphericity," Izv. Akad. Nauk SSSR, Ser. geofiz., (1961) 1199-1208
       [Bull. Acad. Sci. USSR, Geophys. Ser., (1961) 790-795].

273.   V. G. Fesenkov, "On the atmospheric component of zodiacal light,"
       Astr. Zh., **33**, 708-714 (1956).

274.   F. F. Yudalevich, "Multiple scattering in the infrared spectral re-
       gion" [in Russian], Izv. Akad. Nauk SSSR, Ser. geofiz., (1956) 862-
       864.

275.   V. G. Fesenkov, "The twilight method for investigating the optical
       properties of the atmosphere" [in Russian], Scattering and Polariza-
       tion of Light in the Earth's Atmosphere (All-Union conf., Alma-
       Ata, Oct. 1961) — Trudy astrofiz. Inst., Alma-Ata, **3**, 214-235 (1962).

276.   T. G. Megrelishvili and I. A. Khvostikov, "The structure of the upper atmosphere from twilight observations" [in Russian], Dokl. Akad.
Nauk SSSR, **59**, 1283-86 (1948).

277.   H. K. Kallmann, Problemy sovrem. Fiz., No. 4 (1954); Rocket Panel,
"Pressures, densities, and temperatures in the upper atmosphere,"
Phys. Rev., **88**, 1027-32 (1952).

278.   R. N. Bracewell, "Simple graphical method of correcting for instru-
mental broadening," J. Opt. Soc. Am., **45**, 873-876(1955); R. N.
Bracewell and J. A. Roberts, "Aerial smoothing in radio astronomy,"
Aust. J. Phys., **7**, 615-640 (1954).

279.   F. Link, "Die Dämmerungshelligkeit im infraroten Lichte und die
Luftdichte der Ionosphäre," Met. Z., **61**, 87-90 (1944).

280.   F. Link, "Mesures de la brillance du ciel crépusculaire dans l'in-
frarouge et densité de l'ionosphère," C.r. Acad. Sci., Paris, **222**,
333-334 (1946).

281.   I. Zacharov, "Dämmerungsmessungen im Laufe der Perseiden 1953,"
Bull. astr. Insts. Czech., **8**, 135-142 (1957).

282.   F. E. Volz and R. M. Goody, "The intensity of the twilight and upper
atmospheric dust," J. atmos. Sci., **19**, 385-406 (1962).

283.   F. Volz and R. M. Goody, "Twilight intensity at 20° elevation. Results of observations," Blue Hill Met. Obs., Harvard Univ., Contract
AF 19(604)-4546, Sci. Rep. No. 2a and 2b, AFCRL-62-261(a, b)(1961).

284.   F. Volz and R. M. Goody, "Twilight intensity at 20° elevation. Analysis and discussion of observations," Blue Hill Met. Obs., Harvard Univ.,
Contract AF 19(604)-4546, Final Rep., AFCRL-62-856 (1962).

285.   M. S. Malkevich, Yu. B. Samsonov, and L. I. Koprova, "Water vapor
in the stratosphere," Usp. fiz. Nauk, **80**, 93-124 (1963) [Soviet Phys.
Usp., **6**, 390-410 (1963)].

286.   C. E. Junge, C. W. Chagnon, and J. E. Manson, "Stratospheric aerosols," J. Met., **18**, 81-108 (1961); C. E. Junge, "Vertical profiles of
condensation nuclei in the stratosphere," J. Met., **18**, 501-509(1961);
C. W. Chagnon and C. E. Junge, "The vertical distribution of sub-
micron particles in the stratosphere," J. Met., **18**, 746-752 (1961);
C. E. Junge and J. E. Manson, "Stratospheric aerosol studies," J.
geophys. Res., **66**, 2163-82 (1961).

287.   See, for example: F. Link, "Sur le rôle des poussières météoriques
dans l'atmosphère terrestre," Mém. Soc. r. Sci. Liège, (4) **15**, 35
(1955).

288.   N. N. Kalitin, "Cosmic dust according to actinometric measurements," Dokl. Akad. Nauk SSSR, **45**, 375-378 (1944).

289.   I. Zacharov, "Influence des Perséides sur la transparence atmosphérique," Bull. astr. Insts. Czech., **3**, 82-85 (1952).

290.   D. Deirmendjian, "The optical thickness of the molecular atmosphere," Arch. Met. Geophys. Biokhim.,B6, 452-461 (1955).

291.   L. G. Elagina, "Optical device for measuring the turbulent pulsations of humidity," Izv. Akad. Nauk SSSR, Ser. geofiz. (1962) 1100-1102 [Bull. Adad. Sci. USSR, Geophys. Ser (1962), 704-708].

292.   F. Link and L. Neužil, "Tables mondiales des réfractions et des masses d'air," Bull. astr. Insts. Czech., 9, 28-33 (1958).

293.   F. Link, "Sur quelques problèmes relatifs à la dioptrique des rayons crépusculaires," Mém. Soc. r. Sci. Liège, (4) 12, 151-154 (1952).

294.   F. Link, "Densité de la haute atmosphère calculée d'après les phénomènes crépusculaires," C. r. Acad. Sci., Paris, 200, 78-80 (1935).

295.   F. Link, "Quelques théorèmes relative aux effets crépusculaires dans les couches luminescents élevées," Mém. Soc. r. Sci. Liège, (4) 12, 155-159 (1952).

296.   F. Link, "L'influence de la saison et du climat sur les réfractions astronomiques et la répartition des masses d'air," C. r. Acad. Sci., Paris, 204, 1080-82 (1937); idem, Čas. Pěst. Mat Fys., 67, 62-66 (1937).

297.   See, for example: P. Gruner, "Dämmerungserscheinungen," Handb. Geophys., 8, 432-526 (1942).

298.   N. B. Divari, "Brightness distribution of the twilight sky in the solar vertical" [in Russian], Scattering and Polarization of Light in the Earth's Atmosphere (All-Union conf., Alma-Ata, Oct. 1961)—Trudy astrofiz. Inst., Alma-Ata, 3, 188-210 (1962).

299.   G. V. Rozenberg, Usp. fiz. Nauk, 79, 441-522 (1963) [Soviet Phys. Usp., 6, 198-249 (1963)].

300.   N. B. Divari, "Photometric analysis of twilight from sky-brightness measurements along the solar vertical," Geomagnetizm Aéronomiya, 2, 720-731 (1962) [Geomagnetism Aeronomy, 2, 600-609 (1963)].

301.   N. B. Divari and L. S. Trofimova, "Analysis of the twilight glow from the brightness observed at two points of the solar vertical," Geomagnetizm Aéronomiya, 3, 657-665 (1963) [Geomagnetism Aeronomy, 3, 529-536 (1964)].

302.   N. B. Divari, "Brightness of the twilight sky as a function of lunar phase," Geomagnetizm Aéronomiya, 3, 1136-37 (1963) [Geomagnetism Aeronomy, 3, 915-916 (1964)].

303.   N. B. Divari, Geomagnetizm Aéronomiya, 4, 886 (1964).

304.   N. B. Divari and L. I. Plotnikova, "Computed brightness of the twilight sky," Astr. Zh., 42, 1090-1103 (1965) [Soviet Astr., 9, No. 5 (1966)].

305.   G. Dietze, "Messungen der Himmelslichtpolarisation während der
        Dämmerung also Indikator für hochatmosphärische Trübungen kos-
        mischen Ursprungs," Z. Met., 15, 36-43 (1961).

306.   L. N. Bondarenko, "Spectral polarimetric measurements of the twi-
        light sky polarization at the zenith," Astr. Zh., 41, 383-386 (1964)
        [Soviet Astr., 8, 299-302 (1964)].

307.   F. F. Yudalevich, "On calculating the intensity distribution and
        degree of polarization in the twilight sky" [in Russian], Scattering
        and Polarization of Light in the Earth's Atmosphere (All-Union
        conf., Alma-Ata, Oct. 1961) — Trudy astrofiz. Inst., Alma-Ata, 3,
        237-240 (1962).

308.   A. Kh. Darchiya, "Spectrophotometry of the twilight sky" [in Rus-
        sian], ibid., 241-244.

309.   V. M. Morozov, "Primary scattering at twilight," Izv. Akad. Nauk
        SSSR, Ser. geofiz., (1964) 787-793 [Bull. Acad. Sci. USSR, Geo-
        phys. Ser., (1964) 476-480].

310.   C. L. Mateer, "On the information content of Umkehr observa-
        tions," J. atmos. Sci., 22, 370-381 (1965).

311.   R. Goody, "Note on the refraction of a twilight ray," J. atmos. Sci.,
        20, 502-505 (1963).

312.   D. R. Barber, "Optical evidence of the lunar atmospheric tide," J.
        atmos. terr. Phys., 24, 1065-71 (1962).

313.   K. Bullrich, "Scattered radiation in the atmosphere and the natural
        aerosol," Adv. Geophys., 10, 99-260 (1964).

314.   V. E. Pavlov, "Atmospheric indicatrix of scattering in the region
        of small and large scattering angles," Astr. Zh., 41, 122-127
        (1964) [Soviet Astr., 8, 91-95 (1964)].

315.   V. E. Pavlov, "Atmospheric scattering law in the optical and ultra-
        violet," Astr. Zh., 41, 546-549 (1964) [Soviet Astr., 8, 435-437
        (1964)].

316.   V. E. Pavlov, "An empirical formula for the atmospheric scattering
        law which takes into account the circumsolar aureole," Astr. Zh.,
        42, 433-436 (1965) [Soviet Astr., 9, 340-342 (1965)].

317.   I. N. Minin and V. V. Sobolev, "Theory of radiation scattering in
        planetary atmospheres," Astr. Zh., 40, 496-503 (1963) [Soviet Astr.,
        7, 379-383 (1963)].

318.   G. Guilino and H. K. Paetzold, "Twilight sodium observations at
        Lindau/Harz, Germany," J. atmos. terr. Phys., 27, 451-456 (1965).

319.   Z. V. Kariagina and V. E. Mozhaeva, "Sodium emission at twi-
        light," Vest. Akad. Nauk kazakh. SSR, 21, 75-77 (1965).

320. M. Gasden, "On the twilight sodium emission. I. Observations from a southern hemisphere station," Annls. Géophys., **20**, 261-272 (1964).

321. W. Swider, Jr., "N_2^+ ions at twilight," Planet. Space Sci., **13**, 529-540 (1965).

322. E. R. Johnson, "Twilight resonance radiation of AlO in the upper atmosphere," J. geophys. Res., **70**, 1275-77 (1965).

323. H. M. Sullivan and D. M. Hunten, "Lithium, sodium, and potassium in the twilight airglow," Can. J. Phys., **42**, 937-956 (1964).

324. Draft reports, Commission 21 (Luminescence du Ciel), Trans. internatn. astr. Un., **12A** (in press).

325. K. D. Cole, "The predawn enhancement of 6300 A airglow," Annls. Géophys., **21**, 156-158 (1965).

326. A. B. Meinel and M. P. Meinel, "Height of the glow stratum from the eruption of Agung on Bali," Nature, Lond., **201**, 657-658 (1964).

327. J. B. Edson, "The twilight zone of Venus," Adv. Astr. Astrophys., **2**, 1-42 (1963).

AUTHOR INDEX

345

SUBJECT INDEX